T0254447

Springer Praxis Books

More information about this series at http://www.springer.com/series/4097

Roberto Manzocco

Transhumanism - Engineering the Human Condition

History, Philosophy and Current Status

Published in association with

Praxis Publishing
Chichester, UK

Roberto Manzocco
Gorizia, Italy

SPRINGER-PRAXIS BOOKS IN POPULAR SCIENCE

ISSN 2626-6113   ISSN 2626-6121   (electronic)
Popular Science
ISBN 978-3-030-04956-0   ISBN 978-3-030-04958-4   (eBook)
Springer Praxis Books
https://doi.org/10.1007/978-3-030-04958-4

Library of Congress Control Number: 2019930046

This Springer imprint is published by the registered company Springer Nature Switzerland AG
The registered company address is: Gewerbestrasse 11, 6330 Cham, Switzerland

Bill Gates: "So is there a God in this religion?"
Ray Kurzweil: "Not yet, but there will be."

You know what they say the modern version of
Pascal's Wager is? Sucking up to as many
Transhumanists as possible, just in case one of
them turns into God.
                            —Greg Egan, *Crystal Nights*

You have evolved from worm to man, but much
within you is still worm.
    —Friedrich Nietzsche, *Thus Spoke Zarathustra*

Introduction: Gilgamesh Versus the Dragon-Tyrant

One Man Against Mortality

People enjoy stories, apparently. Just look at human history, with all of its myths, poems, and legends. Alternatively, for the laziest among us, just skim through the list of movies and TV shows offered by Netflix or the comic books published by Marvel and DC Comics. So, stories are really part of our deep nature, and some of them can reveal something fundamental about us.

And stories, or at least some of them, will definitely help us to understand the Transhumanist narrative, as it represents the perfect fit for the legends and works of art that we are going to mention in this introduction.

My favorite story? The *Epic of Gilgamesh*, an ancient poem coming straight from Mesopotamia and considered by many specialists to be the earliest example of epic literature to have survived the ravages of time. But let's skip the philological details and jump into the core of the story. A story full of drama blessed by a cinematic style, so much so that someone should try to adapt it into a movie or, even better, a TV series. Anyway, the main character is Gilgamesh, King of the Mesopotamian city of Uruk – the timing is unclear, but we know that different stories, dating back as far as 2100 BC, were merged together into a single poem and that the earliest available version of this poem was composed more or less during the eighteenth century BC.

To make sure that the readers understand the real nature of Gilgamesh, which is that of a demigod, the poem stresses that our King is one-third human and two-thirds divine. In fact, he is the offspring of a goddess and a king, but this, of course, doesn't explain the weird proportion; maybe someone here was not that good at math or just wanted to underline that, in Gilgamesh, the divine part was predominant. Full of energy and fury, our hero is oppressing his own people and forcing them into endless battles; to answer the lamentations of all of these young warriors' wives and girlfriends, the Gods create a wild man, Enkidu, maybe the symbol of the original animal nature of humanity; Enkidu is tamed, and thus civilized, through sexual initiation by a prostitute. And so, Gilgamesh finally finds someone able to stand up to him; the two fight, the King wins, and, in the end, Gilgamesh and Enkidu become best friends. Of course, this is just the beginning, as Gilgamesh and Enkidu will together face many more challenges and journeys. Like the one to the Cedar Forest, a legendary place, where the two kill its monstrous guardian, Humbaba – also known as "The Terrible" – and cut the sacred trees, what could possibly go wrong? Later on, the two troublemakers kill the Bull of Heaven, sent to punish them by the Goddess Ishtar – the main reason for this is actually the fact that Ishtar, a quite vindictive goddess, had her sexual advances dismissed by Gilgamesh. The Gods decide to show their disapproval for the actions of the two, and, as a form of retaliation, they sentence Enkidu to death.

And this is when things start to become interesting. With the death of his best friend, Gilgamesh becomes aware for the first time of his own mortality. In spite of being two-thirds a god, he will have to surrender to the grim reaper. There is nothing he can do about that. Or maybe there is: there might be a man – Gilgamesh is told – that somehow managed to escape death. Just a rumor, but more than enough for Gilgamesh to undertake a long, lonely, and dangerous journey through the world, with the purpose of finding the secret to a never-ending life. Do you see the pattern here? Since the very beginning of our written history, we can find narratives about human beings trying to defeat death. Not such a naïve, childish dream, apparently, but a quintessential part of our cultural DNA. But let's go on with our story.

Gilgamesh faces many dangers and has many fascinating adventures – he meets the Scorpion Men, creatures featured in the Babylonian version of the Epic; he crosses the tunnel that the Sun traverses during the night below the Earth, which was flat, according to the Sumerians; and so on. And in the end, Gilgamesh finally meets the first and only immortal human being. His name? Utnapishtim, a kind of Sumerian Noah, a man who survived the Great Flood the Gods sent to punish humanity. Utnapishtim reveals his secret to Gilgamesh: the Gods, repenting for the excessive harshness of the punishment they released upon humanity, decided all together to make the Sumerian-Babylonian Noah immortal, as a form of compensation. This is far from good news for Gilgamesh: in fact, this is the bottom line; it would take a new general meeting of the Gods – and so another Great Flood – to grant immortal life to Gilgamesh. As a consolation prize, Utnapishtim offers Gilgamesh a magical plant that will keep him young throughout his mortal life. But luck is not on our hero's side: while Gilgamesh is drinking water from a river, a snake steals the magical plant, and the only thing left for Gilgamesh is a sad return home.[1]

---

[1] Stephen Mitchell, *Gilgamesh: A New English Version*, Atria Books, New York 2006

Hubris, Anyone?

For our purposes, other mythical and legendary figures are worth mentioning. One of them is a character belonging to Greek mythology and also very dear to many Transhumanists, the Titan Prometheus, a true cultural hero, credited with different things – according to the different sources of the myth. So, Prometheus is either the creator of humanity, or the one who stole the fire from the Gods to donate it to mankind, or the one who – against the Gods' will – taught humanity all of the arts and the knowledge that they have.

One way or the other, in the Western tradition, he became the symbol of both humanity striving for knowledge, progress, and civilization and the risk of surpassing the limits set by the laws of nature, paying a price in terms of hubris and the related unintended, usually bad, consequences. During the Romantic era, Prometheus became the embodiment of the figure of solitary genius, whose attempts to create something great and to ameliorate the human condition would inevitably end up in tragedy. The classic history – as told for the first time by Hesiod – represents him punished by Zeus, king of the Gods, bound to a rock, with an eagle eating his liver every day, only to have it regrow every night, ready to be eaten again the following day, forever. Prometheus's myth appeared for the first time in the eighth century BC, with his story told by the poet Hesiod in his *Theogony*. In this case, it has a happy ending: after few years of punishment, another iconic Greek hero, Heracles, kills Zeus's eagle and frees Prometheus.

Hesiod's epic is just one of the many poems and works of art devoted to Prometheus, and we have at least to mention the fifth-century BC tragedy *Prometheus Bound*, in which Aeschylus widens the field of transgression perpetrated by the Titan against Zeus. Besides stealing the fire from the Gods, Prometheus stands accused of such charges as teaching human beings arts like medicine, agriculture, mathematics, and so forth. So, it is not a coincidence that, in 1818, Mary Shelley chose, for her novel, *Frankenstein*, the subtitle *The Modern Prometheus*.

Frankenstein is very important to us for many reasons. Unlike other novels of that age, the story told by Mary Shelley (1797–1851) – a young and ambitious scientist, Victor Frankenstein, decides to tinker with life itself and creates a monstrous but intelligent creature that in the end rebels against his own creator and all humankind – is not based on magic or some other supernatural device. It is human science that decides to challenge the laws of nature, in order to bring a human – if "composite" – being back to life. Yes, the idea for the novel came to Mary Shelley after a journey near Frankenstein Castle, where, a couple of centuries earlier, some obscure alchemist had attempted his own experiments. And yes, Shelley conceived her monster during a competition with authors like Polidori, Byron, and her future husband Percy Shelley, to see who was capable of writing the best horror story. But the scientific background that she tried to give to her work qualifies it as the first modern science fiction novel – according to the sci-fi writer Brian Aldiss and others.

Besides, to her contemporaries, Mary Shelley's idea of reanimating dead bodies using electricity would not have looked so extravagant, as shown by Sharon Rushton – Chair in Romanticism at Lancaster University – in her essay *The Science of Life and Death in Mary Shelley's Frankenstein*.[2] In her paper, Rushton surveys the scientific background of Shelley's novel and shows us how one of that age's main obsessions was the difficulty of distinguishing clearly between the states of life and death and the connected fear of being buried alive. To study the phenomenon, in 1774, two London physicians, Thomas Cogan and William Hawes, created the Royal Humane Society, whose initial name was the Society for the Recovery of Persons Apparently Drowned. Every year, the society would organize a procession of those saved and reanimated by the two doctors' resuscitation methodology – among them, Mary Shelley's mother. The successes of the Royal Humane Society spread the idea that distinguishing between life and death was impossible and that, rotting corpses excluded, everybody was at risk of being buried alive – a fact that even inspired a market for "life-preserving coffins," fully equipped with devices for easily opening them from the inside and holes for breathing.

Diderot and d'Alembert's *Encyclopédie* distinguished between "absolute death" – characterized by a state of putrefaction – and "incomplete death," comparable to states of coma, suspended animation, sleeping, and fainting. And, of course, in the same years, someone even tried to animate truly dead animals and people. During the second half of the eighteenth century, the Italian scientist Luigi Galvani managed to make a dead frog's legs twitch using electricity – a phenomenon that came to be known as "galvanism."[3] Just as it would happen in a horror novel – and, in some aspects, in Shelley's novel itself –, Galvani's nephew, Giovanni Aldini, tried to reanimate a human body, more specifically, the corpse belonging to a hanged criminal – a procedure allowed by the 1752 "Murder Act," which established that a murderer's corpse could be dissected for research purposes. And, incredible as it may sound, Aldini apparently succeeded – partially – as his corpse contorted its face, opened one eye, raised one hand, and moved its legs.[4] During those years, the nature of life itself was open to debate, with a member of the Royal College of Surgeons, John Abernethy, defending the idea that "life" was a separate material substance, basically a vital principle "super-added"

---

[2] https://www.bl.uk/romantics-and-victorians/articles/the-science-of-life-and-death-in-mary-shelleys-frankenstein

[3] Luigi Galvani, *De viribus electricitatis in motu musculari commentarius*, 1792, https://archive.org/details/AloysiiGalvaniD00Galv

[4] Giovanni Aldini, *An account of the late improvements in galvanism*, 1803, http://public-domainreview.org/collections/an-account-of-the-late-improvements-in-galvanism-1803/

to the organized body – an idea in line with all of the fantasies about reanimating dead bodies through electricity.

As we are talking about people filled with hubris, can we really avoid mentioning Faust? Of course not. He is the protagonist of a popular German legend, apparently based on a semi-historical character, Johann Georg Faust – who was born more or less in 1480 and died more or less in 1540, even though some historians believe that this character actually represents the fusion of two different historical figures. At any rate, according to the classic legend, Faust is a charlatan – an astrologer, a magician, an alchemist, and a necromancer – very much dissatisfied with his life, to the point that he sells his soul to the Devil in exchange for superhuman knowledge and, of course, a considerable amount of physical pleasure. His story inspired many kinds of artwork, ballads, novels, movies, and so on. Even better, his name became an adjective, "Faustian," used to indicate any person who renounces their moral principles in order to achieve powers and knowledge normally forbidden to human beings. A fact that, besides the moral part – Transhumanists talk a lot about ethics –, looks very Transhumanist. With a side effect, though, normally, Faustian characters end up losing their soul – of course, it was in the contract – and burning in Hell forever and ever. Anyway, the adaptations of the legend of Faust that are worth mentioning are, of course, Christopher Marlowe's *The Tragical History of Doctor Faustus*, a play likely published around 1587; Johann Wolfgang von Goethe's *Faust*, a play in two parts published between 1808 and 1832; and Thomas Mann's *Doctor Faustus*, a novel published in 1947. Mann's novel represents a reframing of this old legend and its adaptation to modern times. The novel tells the story of a fictional musician, Adrian Leverkühn, who sells his soul and his sanity to the Devil in exchange for 24 years of superhuman creativity. The description – or actual lack of it – of what awaits every Faustian Transhumanist in Hell is particularly effective:

> That is the secret delight and security of hell, that it cannot be denounced, that it lies hidden from language, that it simply is, but cannot appear in a newspaper, be made public, be brought to critical notice by words-which is why the words 'subterranean,' 'cellar,' 'thick walls,' 'soundlessness,' 'oblivion,' 'hopelessness,' are but weak symbols. (…) that is what the newcomer first experiences and what he at first cannot grasp with his, so to speak, healthy senses and will not understand because reason, or whatever limitation of the understanding it may be, prevents him from doing so, in short, because it is unbelievable, so unbelievable that it turns a man chalk-white, unbelievable, although in the very greeting upon arrival it is revealed in a concise and most forcible form that 'here all things cease,' every mercy, every grace, every forbearance, every last trace of consideration for the beseeching, unbelieving objection: 'You cannot, you really cannot do that

with a soul'-but it is done, it happens, and without a word of accountability, in the sound-tight cellar, deep below God's hearing, and indeed for all eternity.[5]

Of course, this will not stop the Transhumanists, for the following fundamental reason: Transhumanists are not religious types – at least in most cases – and, to them, the idea of hubris is nothing to be ashamed of. On the contrary, floating in a sea of nihilism, Transhumanism sees in hubris the only hope of salvation, the only chance human beings have of avoiding something that they fear even more than an imaginary Hell: pure nothingness.

The Fable of the Dragon-Tyrant

What I would like to stress here is the complete overturn of the traditional hubristic narrative, so radical that, to the Transhumanist mind-set, characters like Gilgamesh, Prometheus, Victor Frankenstein, and Faust would unquestionably be seen as positive figures. So, let's press the forward button and take a look at a contemporary tale, written by one of the most prominent Transhumanist thinkers, Nick Bostrom (more about him later on): *The Fable of the Dragon-Tyrant*.[6] "Once upon a time – the tale begins – the planet was tyrannized by a giant dragon. The dragon stood taller than the largest cathedral, and it was covered with thick black scales. Its red eyes glowed with hate, and from its terrible jaws flowed an incessant stream of evil-smelling yellowish-green slime. It demanded from humankind a blood-curdling tribute: to satisfy its enormous appetite, ten thousand men and women had to be delivered every evening at the onset of dark to the foot of the mountain where the dragon-tyrant lived. Sometimes the dragon would devour these unfortunate souls upon arrival; sometimes again it would lock them up in the mountain where they would wither away for months or years before eventually being consumed."

A pretty horrifying story, if you ask me. In the imaginary world depicted by Bostrom, tens of thousands of human beings die every day, which is exactly what happens in our world. In fact, as you have probably already figured out, the Dragon-Tyrant is none other than the personification of the aging process and, of course, death. The rest of the story represents, step by step, the actual struggle of humanity – existential, religious, psychological, and technological – to fight mortality, to cope with it, to accept it, and, at least if you are a Transhumanist, to ultimately defeat it. And, of course, just like every fairy tale worth its name – and

---

[5] Thomas Mann, *Doctor Faustus*, Vintage Books, New York 1999, pp. 260–261

[6] Nick Bostrom, *The Fable of the Dragon-Tyrant*, «Journal of Medical Ethics», 2005, Vol. 31, No. 5, pages 273–277 https://nickbostrom.com/fable/dragon.html

unlike classic moral fables – this Fable has a happy ending, which is exactly what Transhumanists expect: a never-ending "Happily Ever After." In fact, after musings, reflections, meetings, and so on, the King and his people figure out a way to defeat the horrible monster that mortality is: they build – after many attempts – a superweapon and kill the Dragon-Tyrant.

The moral that we can extract from this Transhumanist Fable is quite straightforward: aging and death are bad. Period. We could not do anything about it, for the better part of human history, except accept it and rationalize it. But now things are changing, and technology is about to make the defeat of the Dragon-Tyrant possible, or at least thinkable. And this can be done only collectively, if we get rid of our religion-bound "hubris" mentality and decide that we have nothing to lose but our beliefs of a punishment for our "arrogance" in an afterlife that probably does not even exist. Even worse, stories like the ones mankind's main religions tell are not harmless; their acceptance-based mentality actually represents a concrete obstacle for the development of weapons capable of killing the Dragon-Tyrant. The Fable ends with the King musing and talking about the need for the whole society to reorganize.

Negentropy Versus Thermal Death

Surely, defeating individual death is not and cannot be the only goal for Transhumanists. After all, if you are planning to stay around for at least a few billion years, eventually you will have to face the issue of the death of the Sun and, after that, the thermal death of the universe, which right now seems the most likely scenario. This brings us to another Fable, courtesy of a well-known postmodern philosopher, Jean-François Lyotard. Trigger warning: in his most famous book, *La Condition postmoderne: Rapport sur le savoir*, Lyotard defines the concept of "postmodern" as "incredulity towards meta-narratives."[7] These are large-scale visions of the world, scientific theories, and philosophical systems about the world as a whole. For example, the idea that science will be able to answer every question and know everything, the belief in the possibility of absolute freedom, the concept of unstoppable progress, and so forth. So, Lyotard seems to disqualify the Transhumanist dreams as mere "meta-narratives" – and, in this case, let me stress the abysmal difference between "disqualify" and "debunk."

In some of his writings, Lyotard outlines a "Postmodern Fable" that he thinks constitutes the real fabric of the contemporary human techno-scientific endeavor, an implicit story that, maybe unconsciously, we, the Postmoderns, like to tell ourselves. Let's call it the "Fable of Negentropy." So, let's imagine a very distant

---

[7] Lyotard, Jean-François, *La Condition Postmoderne: Rapport sur le Savoir*, Les Editions de Minuit, Paris 1979, p. 7

future – says Lyotard – in which our Sun, reaching the end of its life cycle, will turn into a nova and destroy everything around it, including our beloved home, the planet Earth. Besides Lyotard's scientific error – technically, the Sun will turn not into a nova but rather a red giant, swallowing many planets in the process, and then into a small, superdense white dwarf – the concept is very clear: life is going to end in our system, no matter what. It will die; that's what counts. Are we still going to be around? We, or our post-human descendants? If this is the case, how are we going to get out of this? Anyway, according to Lyotard:

> The narrative of the end of the Earth is not in itself fictional, it's really rather realistic. (…) Something ought to escape the conflagration of the systems and its ashes. (…) The fable hesitates to name the thing that ought to survive: is it the Human and his/her Brain, or the Brain and its Human? And, finally, how are we to understand the *'ought* to escape'? Is it a need, an obligation, an eventuality? (…) There remains much to be done, human beings *must* change a lot to get there. The fable says that they can get there (eventuality), that they are urged on to do it (need), that doing it is in their interest (obligation). But the fable cannot say what human beings will have become then.[8]

And it seems that the end of the world as we know it, plus the possible end of the human mind itself in the far future, is going to constitute a huge philosophical problem:

> It's impossible to think an end, pure and simple, of anything at all, since the end's a limit and to think it you have to be on both sides of that limit. So what's finished or finite has to be perpetuated in our thought if it's to be thought of as finished. Now this is true of limits belonging to thought. But after the sun's death there won't be a thought to know that its death took place. That, in my view, is the sole serious question to face humanity today. (…) In 4.5 billion years there will arrive the demise of your phenomenology and your utopian politics, and there'll be no one there to toll the death knell or hear it. It will be too late to understand that your passionate, endless questioning always depended on a 'life of the mind' that will have been nothing else than a covert form of earthly life. A form of life that was spiritual because human, human because earthly - coming from the earth of the most living of living things. (…) With the disappearance of earth, thought will have stopped - leaving that disappearance absolutely unthought of.[9]
> (…)

---

[8] Jean-François Lyotard, *Postmodern Fables*, University of Minnesota Press, Minneapolis 1997, p. 84

[9] Ibid., *Can Thought go on without a Body?*, in: Ibid., *The Inhuman. Reflections on Time*, Stanford University Press, Redwood City 1991, p. 9

The sun, our earth and your thought will have been no more than a spasmodic state of energy, an instant of established order, a smile on the surface of matter in a remote corner of the cosmos. You, the unbelievers, you're really believers: you believe much too much in that smile, in the complicity of things and thought, in the purposefulness of all things! (...) Once we were considered able to converse with Nature. Matter asks no questions, expects no answers of us. It ignores us. It made us the way it made all bodies – by chance and according to its laws. Or else you try to anticipate the disaster and fend it off with means belonging to that category – means that are those of the laws of the transformation of energy. You decide to accept the challenge of the extremely likely annihilation of a solar order and an order of your own thought.[10]

Human thought works analogically, according to Lyotard, not just logically, and this way of thinking depends on a body and its correlation with a reality that is perceived as inexhaustible. Even more, human thinking is inextricably tied to two other human "endowments": the ability to feel *pain* – no pain, no gain, we could say, even in the field of thinking and, most importantly, philosophizing – and the ability to *desire*. Will our descendants be able to build machines bearing consciousness, able to think logically *and* analogically and to feel pain and desire?

And here is where the issue of complexity has to be brought up again. I'm granting to physics theory that technological scientific development is, on the surface of the earth, the present-day form of a process of negentropy or complexification that has been underway since the earth began its existence. I'm granting that human beings aren't and never have been the motor of this complexification, but an effect and carrier of this negentropy, its continuer. I'm granting that the disembodied intelligence that everything here conspires to create will make it possible to meet the challenge to that process of complexification posed by an entropic tidal wave which from that standpoint equates with the solar explosion to come. I agree that with the cosmic exile of this intelligence a locus of high complexity – a centre of negentropy – will have escaped its most probable outcome, a fate promised any isolated system by Carnot's second law – precisely because this intelligence won't have let itself be left isolated in its terrestrial-solar condition. In granting all this, I concede that it isn't any human desire to know or transform reality that propels this techno-science, but a cosmic circumstance. But note that the complexity of that intelligence exceeds that of the most sophisticated logical systems, since it's another type of thing entirely. As a material ensemble, the human body hinders the separability of this intelligence, hinders its exile and

---

[10] Ibid. pages 10–11

therefore survival. But at the same time the body, our phenomenological, mortal, perceiving body is the only available *analogon* for thinking a certain complexity of thought.[11]

To sum it up, in the natural world, entropy – that is, disorder – tends naturally to increase, but there are "pieces" of this world in which exactly the opposite happens, that is, order, or negentropy, increases. For example, the evolutionary process somehow represents a "negentropic wave" that creates growing tides of order and complexity. As far as we know, the final battle between entropy and negentropy will be won by the former, even though science fiction – which Lyotard is probably not very fond of – has offered us some possible, indeed imaginative, solutions.

Entropy is definitely a serious issue, if you are planning to stick around for many billions of years – maybe killing time with philosophy or art in the meantime. Moreover, being very proactive people, Transhumanists will probably disagree with this passive interpretation of the human species as just a stage or a consequence of this "center of negentropy." Quite likely, they would like their post-human descendants to "take charge" of this process, to become the conscious and willing embodiment of this negentropic wave against universal entropy, not just "going with the flow" – that is, finding ways to establish "pockets" or "islands" of negentropy in a "sea" of entropy – but establishing its direction and its final destination. One thing is sure: Transhumanism is not just about surviving biological death or universal entropy, which would be a reactive attitude; in fact, this movement has plans and projects that, among its most intellectually brave members, border and actually spill over into religion.

In other words, Transhumanism represents a movement that unashamedly embodies one of the most human of passions: the burning desire for life and knowledge against all odds and, of course, the related hubris, which – courtesy of the philosophically nihilistic era we live in – has become something not to be punished but to be rewarded.

Spoiler Alert

In spite of the fact that Transhumanism is quickly going mainstream, this movement still has many detractors, who see it as either dangerous or, more frequently, silly. So, if one wants to find critics of Transhumanist ideas, there are many out there, and one simply has to pick among a wide selection. My approach is different, though. I do not consider myself fully a Transhumanist, because, firstly, I do not like labels of any kind and, secondly, because there are quite a few Transhumanist

---

[11] Ibid. p. 22

ideas that I find questionable – not from a moral viewpoint but from a metaphysical one. But, yes, I am kind of sympathetic. In spite of all of the abysmal differences, I do like the Transhumanists, as they are a quite interesting strain of human being and also because, in this post-ideological era – I mean, you are not going to take the political ideologies available nowadays seriously, are you? – they found the ability to dream again. And dream big, I have to say, as I will explain in the following chapters.

So, here are the topics that I will cover in this book. In the first chapter, I will cover what I call, with some reservation, "precursors" of Transhumanism, while Chapter 2 will deal with the Transhumanist movement itself, its main ideas, its main representatives, its organizations, and much more. I will devote the third chapter to a specific Transhumanist topic, the attempt to live as long as possible, maybe forever. Chapter 4 will cover the Transhumanist "plan B," that is, cryonics – a good idea, if your plan A, immortality, fails. Chapter 5 will analyze another of the "pillars" of Transhumanism, nanotechnology, while Chapter 6 will cover the actual research into enhancing the human body through technology attempted by single individuals, enterprises, and organizations. Chapter 7 will bring us inside the human brain and the possibility of it interfacing with machines, as well as modifying it with different technologies and fundamentally changing the human biological experience – mind uploading will also be considered – in the eighth chapter, we will explore the concept of "Paradise Engineering." Chapter 9 will cover extensively one of the most beloved Transhumanist concepts, the technological singularity and its consequences; Chapter 10 will examine the controversial relationship between Transhumanism and religion and the desire of the former to ascend to a God-like state. We do not know if Gilgamesh could possibly defeat the Dragon-Tyrant, as he did with Humbaba – aka "The Terrible" – but one thing we know for sure: a Hell of a ride is waiting for us.

Contents

1

Stairways to the Sky

1.1 Climbing the Slope, Looking for a Purpose

Overcoming the limitations of our short lives has always being among the deepest, most heartfelt of human desires, no matter how arrogant this may sound. And, by the way, just look at the original Latin root of the verb "desire," a combination of "de," which indicates "lack of something," and "sidus," that is, "star." To desire literally means "to miss the stars," to feel a need for them. Transhumanists, like many others, want to fulfill this "human, too human" desire to reach for the stars, and in a quite literal sense. Reaching the stars, living among them, becoming like Gods, and, of course, living forever, without ever having to meet the Grim Reaper.

Among the forerunners of the Transhumanist movement, we can mention people as diverse as the European alchemists of the Middle Ages, with their obsessive research into the Philosopher's Stone and the related Elixir of Life, able to grant, or so they say, eternal youth; the Chinese Taoists, with all of their meditative, medical and gymnastic practices; or even the Ancient Egyptians, promoters of the mummification of the Pharaoh's body, a kind of precursor to contemporary cryonics.

We could go on and on for hours, but this would be getting lost in the meanders of history. As an alternative, I decided to select a few moments from the history of philosophy that Transhumanists themselves consider "foundational" to their own cultural movement. So, what do Transhumanists think about their own family tree? If we ask Nick Bostrom, scholar at the University of Oxford and one of the

---

The original version of this chapter was revised with the late corrections from the author. The correction to this chapter can be found at https://doi.org/10.1007/978-3-030-04958-4_11

© Springer Nature Switzerland AG 2019
R. Manzocco, *Transhumanism - Engineering the Human Condition*,
Springer Praxis Books, https://doi.org/10.1007/978-3-030-04958-4_1

most prominent contemporary Transhumanist thinkers, he would tell us that the starting point can be found in the *Oration on the Dignity of Man* (1486) by Pico della Mirandola, in which the author states that man is "a creature neither of heaven nor of earth, neither mortal nor immortal," and, as the free and proud shaper of his own being, he should "rise again to the superior orders whose life is divine."[1,2]

Within the Transhumanist Pantheon, Bostrom also includes Francis Bacon, as, in his *Novum Organum*, the English philosopher didn't just propose a scientific method based on empirical data, he also stated that science should have been used to subjugate Nature and improve men's lives, with the final goal of "making everything possible." Renaissance ideals, science and rationalism would then constitute fundamental ingredients of the Transhumanist mentality.

Among the spiritual ancestors of this movement, we then find the Marquis of Condorcet, who used to speculate on the possible indefinite, although not infinite, prolongation of life, through the improvement of the human race; and Benjamin Franklin as well, who dreamed of being preserved with some friends in a barrel of Madera, in order to see the future of the country that he founded.

Last, but not least, let's mention the French philosopher Denis Diderot (1713–1784), who, according to the Transhumanist George Dvorsky, believed that "humanity might eventually be able to redesign itself into a great variety of types whose future and final organic structure is impossible to predict."[3]

1.2   The Nietzschean Knot…

When we speak of a cultural, intellectual and political movement as articulated as Transhumanism, it is difficult to establish the starting point, or identify a date or character functioning as a divide between "before" and "after." Mostly because – and this is a classic issue in the field of the history of science – we would end up classifying those who share the idea we are analyzing but came before the chosen

---

[1] Linguistically speaking, Dante was the first to use a term similar to "Transhumanism." In the "Divina Commedia – Paradiso," he uses the verb "transumanar," that is, "to transcend the human condition." After reaching Heaven, the "father of the Italian language" meets Beatrice and, looking into her eyes, is "transhumanized," that is, he gets purified and transcends his human limitations. In the Twentieth Century, we find a similar term in Thomas Stearns Eliot's 1949 play "The Cocktail Party," in which the author speaks of the human efforts to reach enlightenment as a process in which the human is "transhumanized."

[2] Cf. N. Bostrom, *A history of transhumanist thought*, in: «Journal of Evolution and Technology», Vol 14, n.1, 2005 http://www.nickbostrom.com/papers/history.pdf

[3] G. Dvorsky, *Revisiting the proto-transhumanists: Diderot and Condorcet*, https://ieet.org/index.php/IEET2/more/dvorsky20101111

date as "precursors," a category based on a retrospective view – that is, biased by our privileged position of "people of the present day."

Sadly, we cannot excuse ourselves from this task: for our purposes, we need to track down a figure or character that functions as a "point of origin" or "father" of the Transhumanist movement.

And this operation is made more difficult by the fact that Transhumanists are not dead and forgotten, but rather alive and well, and they work hard to meticulously back up their theories – and, in doing so, try to enlist as many "mainstream" thinkers as possible, in order to acquire indispensable academic credentials. So, instead of picking one character, I am going to indulge in one of yours truly's favorite hobbies, creating, in this chapter, a *list* of characters and movements that can be considered "precursors" of Transhumanism. I will call them "Stairways to the Sky," because they all represent attempts to raise us from our mortal condition and reach for the stars – *per aspera ad astra*, as the Ancient Romans used to say. Anyway, first of all, we have to tackle an issue produced by this typical Transhumanist way of doing things retroactively, which is: does the oft-quoted, but seldom understood, Friedrich Wilhelm Nietzsche and his concept of the "overman" have anything to do with Transhumanism?

German philosopher Jürgen Habermas has defined Transhumanists as a group of eccentric intellectuals who refuse what they consider the illusion of equality and aim to put biotechnology at the service of their super-humanistic fantasies originated from Nietzsche.[4] Is this true? And, above all: is there any connection between Transhumanism and Nietzsche's thought?

Born in 1844, Friedrich Nietzsche is too famous a philosopher to go on about his thought in detail; thus, I will provide you just a few hints. With a philological background, Nietzsche is considered by most schools of philosophy – certainly at least by the "Continental" one, which today refers to France – an "epochal" thinker, a "breaking point" from the previous tradition, from Plato on. The synthetic and non-appealable statement of Nietzsche in *Gay Science*, "God is dead," constitutes the first diagnosis ever made of the condition in which the West dwells, that is, nihilism, which consists of the historical-metaphysical process by which "all higher values devalue themselves."

In other words, consolidated religions, and especially Christianity, fade, and with them the faith in the existence of an after-life in which evil is punished and good is rewarded. And not only that: even the idea of God and, more generally, of a higher metaphysical reality guaranteeing the validity our knowledge fades. In fact, the death of God is not just the verification of God's and the after-life's non-existence. It is the awareness that there's no objective parameter of knowledge – that is, there is no truth, and everything is interpretation –, no moral principle or

---

[4] J. Habermas, *Die Zukunft der menschlichen Natur. Auf dem Weg zu einer liberalen Eugenik?*, Suhrkamp, Frankfurt 2001, p. 43.

guideline able to give meaning to our lives and allow us to face the nothingness waiting for everybody.

Nietzsche contrasts all of this with his well-known and controversial doctrines of the Eternal Recurrence and the Overman. Man is something that must be overcome, and the Overman is a new anthropological type, able to embody an aristocratic ethos, to see beyond the shallows of nihilism and accept, even welcome with joy, life as it is, with all of its beauty and ugliness, so much so that he would wish to repeat it eternally. Of course, this is a simplified version; in Nietzsche, there are a number of controversial points and open loops, in part because of his way of writing aphoristically and inorganically. For instance, it is not clear what Nietzsche really means by Eternal Recurrence – is it an "ironical" device, aimed at finding a new post-nihilist system of values, or does the thinker really believe that time is circular in nature? But, of course, the most controversial concept is that of the Overman, because it was suspected of supporting racist and eugenicist interpretations (the use the National-Socialists made of it was itself enough to foster these suspicions), beyond the more classic/para-Existentialist one: the Overman as a person who can face the Nothingness, accepting it and ingraining his own freely created values into it.

So, should or should not Nietzsche be considered the forefather of Transhumanism, that is, the idea that human beings should take control of their own biologic evolution, freely designing it through technology, in order to reach a post-human stage? Yes, according to the German scholar Stefan Lorenz Sorgner. An expert in Nietzschean thought, the philosopher, who teaches at John Cabot University, Rome, created a provocation in 2009 in the online Transhumanist *Journal of Evolution and Technology*. In his paper *Nietzsche, the Overhuman, and Transhumanism*,[5] Sorgner supported this connection, stressing the similarities between the Transhumanist concept of genetic enhancement and the Nietzschean concept of education. Basically, while Bostrom rules out Nietzsche from belonging to the list of Transhumanism's ancestors, Sorgner tries to identify similarities between the two. According to the latter, both promote a dynamic vision of life and ethics, and the Nietzschean notion of Will to Power would for Transhumanist purposes work nicely. More specifically, the impulse towards self-improvement and "the feeling that power is growing" dear to Nietzsche would embody the "technological enhancement" of the human faculties desired by the Transhumanists. Sorgner tries to institute a parallelism between the educational process, which Nietzsche sees as the main tool for creating the Overman, and the few types of genetic enhancement sought by the Transhumanists, which Nietzsche couldn't have known, but that he possibly would have liked, or at least considered acceptable as a means of education.

---

[5] S. L. Sorgner, *Nietzsche, the Overhuman, and Transhumanism*, «Journal of Evolution and Technology», Vol. 20 n. 1 – March 2009 – pages 29–42, http://jetpress.org/v20/sorgner.htm

Contrastingly, Bostrom connects Transhumanism and the Utilitarian and Pragmatist thought belonging to the Anglo-American philosophical tradition. What worries Bostrom the most is stressing the democratic nature of Transhumanism and distancing it from any tradition of thought – Nietzsche and eugenics *in primis* – that had ties to the twentieth century's tragedies. To Bostrom, the main point is to promote a liberal interpretation of eugenics, that is, to see Transhumanism as a set of proposals that every individual can choose among or refuse to choose.

To Sorgner, Transhumanism is not very well developed from an ethical viewpoint, and Nietzsche might provide Transhumanists a chance for reflection and better self-knowledge. On the other side, Transhumanists might help to concretize the figure of the Overman, which Sorgner doesn't mean in metaphorical or ironical terms, but as an actual enhanced human.

The Transhumanist philosopher Max More comes to the rescue of Sorgner; according to the former, between Nietzscheanism and Transhumanism, there are no mere parallelisms: the first has directly influenced the second. The proof? Max More himself, whose musings have been inspired directly by the reading of Nietzsche.[6] According to More, Nietzschean thought contains different conflicting concepts, and some of them, like that of Eternal Recurrence, are not compatible with Transhumanism. Other conceptions are, though, and the latter, like those of the Overman and Will to Power, are the ones that inspired More.

More tells us that studying Nietzsche's thought is what pushed him to write, in 1990, his essay *Transhumanism: Towards a Futurist Philosophy* and to elaborate his "Extropic Principles," which we'll mention again later. So, maybe not every Transhumanist has been inspired by Nietzsche, but some were – the correctness of their interpretation is, as they say, a horse of a different color. Likewise, other authors, like Bostrom, borrowed from completely different traditions, like Enlightenment Rationalism.[7]

The lively debate on the relationship between Nietzsche and Transhumanism has enriched the pages of the *Journal of Evolution and Technology* with many other interventions. So, for example, Bill Hibbard, researcher at the University of Wisconsin, has tried to read the Nietzschean conception of Eternal Recurrence in physical terms, showing that, if time really has a circular structure, if every event is really destined to repeat itself, this would very strongly connect the Transhumanists' scientific worldview and Nietzsche's thought.[8]

---

[6] M. More, *The Overhuman in the Transhuman*, «Journal of Evolution and Technology» – Vol. 21 n. 1, January 2010, http://jetpress.org/more.htm

[7] N. Bostrom, Op. cit.

[8] B. Hibbard, *Nietzsche's Overhuman is an Ideal Whereas Posthumans Will be Real*, «Journal of Evolution and Technology» – Vol. 2, n. 1, January 2010 – pages 9–12, http://jetpress.org/v21/hibbard.htm

As Sorgner belongs to the Nietzschean community, could the latter avoid joining the debate? Of course not. In fact, in 2011, the journal of Nietzschean studies *The Agonist* published a few analyses, by as many Nietzsche scholars, regarding the relationship between Nietzscheanism and Transhumanism.

With some distinctions, the Nietzscheans' responses to Sorgner seem to be relatively negative, starting with the one by Keith Ansell Pearson,[9] scholar at the University of Warwick, who has stressed the distance between Nietzsche and Transhumanism in a book as well.[10] Also, Babette Babich, from Fordham University, New York, disagrees with Sorgner; to her, the Nietzschean Overman does joyfully accept every single aspect of existence, including the most cruel, banal and sad, which is something quite different from the Transhumanist ambition of redesigning human life, starting with its main characteristic, which they dislike the most, that is, mortality.[11] In other words, to Babich, Transhumanists' dreams are "human, too human," and their vision is nothing more than a form of that renunciation of the real world so criticized by Nietzsche. Paul S. Loeb, University of Puget Sound, expresses some sympathy toward Transhumanism; to him, the advent of a new post-human species would require the incorporation into Transhumanism of Nietzschean concepts such as the Overman and Eternal Recurrence – as the core of Transhumanism is precisely the desire to somehow control time, just like the Nietzschean Overman embraces Eternal Recurrence with the purpose of *willing* the past.[12] For Loeb, the Nietzschean conception of Eternal Recurrence should be taken very seriously, as it would describe, for Nietzsche, the world as it is; actually, it would be the only possible way for the Overman to *assume real control over time*, as, if time really flows circularly, then our will, directed toward the future, would end up including the events of the past inside its sphere of influence – and so those past events would not be necessary, unavoidable or superimposed on our will, but would rather be simply a consequence of the latter.

At the end of the debate, Sorgner reaffirms the purpose of his work, which is to stress the few structural similarities between Nietzscheanism and Transhumanism.[13] The bottom line? Nietzsche is not exactly an *ante litteram* Transhumanist, but you can make him one if you wish; you can get inspired by him in your Transhumanist

---

[9] K. Ansell Pearson, *The Future is Superhuman: Nietzsche's Gift*, in: «The Agonist», Vol. IV, n. 2, Fall 2011.

[10] K. Ansell Pearson, *Viroid Life: Perspectives on Nietzsche and the Transhuman Condition*, Routledge, New York 1997.

[11] B. Babich, *Nietzsche's Post-Human Imperative: On the "All-too-Human" Dream of Transhumanism*, in: «The Agonist», Vol. IV, n. 2, Fall 2011.

[12] P. S. Loeb, *Nietzsche's Transhumanism*, in: «The Agonist», Vol. IV, n. 2, Fall 2011.

[13] S. L. Sorgner, *Zarathustra 2.0 and Beyond. Further Remarks on the Complex Relationship between Nietzsche and Transhumanism*, in: «The Agonist», Vol. IV, n. 2, Fall 2011.

speculations and musings, and you can even merge Transhumanism and Nietzscheanism, thus creating a kind of "philosophical cyborg."

Our story is not over, though. In 2017, Cambridge Scholars published an anthology with all of the papers of the first "round," plus other contributions that would make our debate even more lively: *Nietzsche and Transhumanism: Precursor or Enemy?*, edited by Yunus Tuncel.

As pointed out by one of the contributors, the Australian philosopher Russell Blackford, Transhumanism is a broad intellectual movement with no body of codified beliefs and no agreed-upon agenda for change; it is a cluster of philosophies, based on a few assumptions (human beings are in a state of transition, change is desirable and it will happen through technological means, and so forth).

Transhumanism is a grassroots movement, an aggregate of loosely tied ideas, concerning the possibility of enhancing human capabilities through technological means, the radical extension of human life, youth, and health, and, of course, the opportunity and desirability of self-directed human evolution, that is, the opportunity for our species to take human evolution in our own hands. Accordingly, Transhumanism is compatible with any ideology, religion, or philosophy that is willing to accept, or at least not oppose, these goals. This is why we can find blends of Transhumanism with Liberalism, Anarchism, Socialism, Communism, Fascism, Atheism, Christianity, Mormonism, and so forth. Similarly, we can blend Transhumanism with any philosophical view of reality, for example, with materialistic reductionism, naïve realism, post-humanism, and, of course, with the thought of Friedrich Nietzsche – as attempted by Ted Chu in his 2014 book *Human Purpose and Transhuman Potential* (Origin Press).

It is difficult to summarize the dense philosophical content of the anthology; so, permit me to mention a few interesting suggestions that the reader can find and benefit from.

Ashley Woodward compares and confronts the concept of education in Nietzsche, which he identifies with the "Technologies of Self" mentioned by Foucault, such as reading, writing, meditation, dietary regimes, physical practices, etc., with the technologies that Transhumanists are very fond of, the "GRIN" technologies of genetics, robotics, information technology and nanotechnology. Woodward hints at a future in which these two expressions of the human spirit might interact and interlace.

Paul S. Loeb – whom we already mentioned – gives us an interesting take on the topic of the Overman/post-human and its relationship with time. So, let's go back to his perspective, with some more details. Far from being a prison, Eternal Recurrence represents – when taken as a real feature of the world, and not as an ironic device – a powerful ontological tool, a way for the Overman to will himself backwards in circular time, an Eternal Recurrence-enabled mnemonic control of the past. The Overman is thus able to defeat the contingency that informs our lives, gain complete control over time, autonomy, self-affirmation, and self-knowledge. After all, if you are able to will yourself backwards and turn your past, including any minimal detail, into a personal choice, you can know absolutely everything about yourself, your life, your relationship with your social and cultural context. This entails absolute self-knowledge and absolute autonomy (and freedom from any form of contingency, any type of external causation). This is quite an evolutionary jump, for the Nietzschean Overman!

In *Nietzsche's Overhuman is an Ideal Whereas Posthumans will be Real*,[14] Bill Hibbard analyzes a famous Nietzsche quote:

> Here man has been overcome at every moment; the concept of the 'overman' has here become the greatest reality – whatever was so far considered great in man lies beneath him at an infinite distance. (Nietzsche 1888, 305.)

Hibbard takes this quote very seriously, stressing the importance of this "infinite distance." The Overhuman is real, or he/she/it(?) will be, and is defined by Hibbard as an individual that "has no need for improvement, having achieved satisfaction with life." The Overhuman is an ideal that post-humans of the near and far future will struggle to achieve. There is an assumption here, quickly recognized by Hibbard: that humans and different post-humans will agree on what improvement means and on this ideal of the Overhuman, which, in fact, might not be the case. Anyway, the sentient being that we call the Overhuman has achieved such a state of satisfaction with life that he/she would gladly repeat it forever and ever. An infinite degree of satisfaction, as infinite as the distance between what is considered great in humanity and the Overhuman her/himself. So, taking the statement by Nietzsche very seriously and literarily, Hibbard introduces this element of infinity into the human struggle for satisfaction. Which appears also to be a synonym for evolution: infinite is the distance between the man and the Overman, and this distance is occupied by an infinite series of post-human existences, which we can call s_1, s_2, s_3, and so forth. What Hibbard is trying to do here is to treat this infinite sequence of improving post-humans mathematically; the point is to show that, in a finite universe, improvement is finite, and there is an achievable state beyond which any improvement is physically impossible. Not ideally, just physically. But, the universe being what it is, the post-human that achieved that ultimate state would be a *de facto* Overman, and would rest satisfied with his own life. There is another assumption here, which is that there is only one universe, and not an infinite series of universes, or a higher order multiverse, or a ladder of more and more complex universes, leading to an ultimate, infinitely complex reality – if there is such a thing. Anyway, if we abandon this assumption, we reintroduce the idea of infinity into the human struggle for improvement in all of its full force.

---

[14]Y. Tuncel (ed.), *Nietzsche and Transhumanism: Precursor or Enemy?*, Cambridge Scholars Publishing, Cambridge 2017, p. 37.

And we need to deal with it. If this idea of "infinite distance" has to be taken seriously, then we have to deal with the idea that either the state of the Overhuman is ultimately non-achievable – too vast is the distance between it and the human – or the Overhuman *is* achievable, as thought by Nietzsche himself, according to some interpretations. If the latter is the case, then the post-humans of the far future should try to imagine a way to bootstrap themselves toward/into the infinite complexity that lies ontologically ahead of us. Short of some form of mystical union, which I am not even going to attempt to outline, it being outside the purpose of this writing, I would say that I have no idea as to how to do that. That is, how to bootstrap humanity into an infinite complexity. Maybe the post-humans, in their superior wisdom, will know better.

We have to raise another question though: is this what Nietzsche was thinking? Quite likely, not. I am no Nietzsche scholar, but I think that is a safe bet. Should we care? No: unless you are a Nietzsche scholar – which is fine with me, of course –, what really matters is to develop *brand new ideas* about our future evolution, and this confrontation between Nietzsche and Transhumanism seems a good breeding ground.

If we want, we can metaphorically consider Transhumanism a kind of "stair" extended toward the Sky – a Tower of Babel 2.0, basically – and all of the "precursors" of this movement as the many attempts to climb Olympus. According to this view, Nietzsche could be considered, if we wish, a first stair. An ambiguous stair, though, and even a bit of a shaky one, a stair that we can shore up right away, with another cultural movement that opened the way to Transhumanists: Futurism.

1.3 …and the Futurist One

Comparing Transhumanism and Futurism is easier; if you have ever covered the latter, you'll at least vaguely remember the Futurists' enthusiasm toward speed, machines, technology and human ingenuity, able to rule over Nature. Founded by the Italian poet Filippo Tommaso Marinetti, this was a movement born against the worship of the past, so much so that it provocatively asked for the closing down of museums and universities, which the new movement accused of being mere keepers of the past.

Futurism was officially launched in 1909, with the publication of the *Futurist Manifesto* – the first of a long series –, in which Marinetti explained the principles underlying his view of art, from contempt for the past to the worship of technology and machines, to the search for a style representing a break with everything done so far. Many painters joined the movement – Umberto Boccioni, Carlo Carrà, Giacomo Balla, Gino Severini, Lucio Russolo –, but Futurism actually invaded every area of art, from architecture to music, and even fashion and cuisine.

The Futurists even attempted to enter the political arena, cultivating contradictory positions and alternating their attraction towards Fascism with one towards Communism. And so, sometimes Futurism is patriotic and war-prone and sometimes it is close to the working class and animated by internationalist feelings – a stance stemming from the fact that its influence reached even Russia. Initially favoring Fascism, Futurism went on to distance itself from it, because of its cult of the past and the attempts by Mussolini to build good relationships with the Church – hated by Marinetti and his colleagues.

It's worth noting that Futurism arose in conjunction with a period of strong technological development characterized mostly by power and speed, while Transhumanism became established during an age characterized by progress of an even more radical sort, that of biotechnologies, which, as we see every day, allow us to get into the control room of life. Now, the question is: besides some superficial analogies, do the Transhumanist ideology and the Futurist movement have something deeper in common? In other words, if Nietzsche represents the first, shaky philosophical "Stairway to the Sky" of the current world, does Futurism represent the second one?

The problem here lays in the fact that Futurism has always been considered "just" an artistic movement, and not a "total" and "proactive" worldview. Or is it?

According to Riccardo Campa – professor of sociology of science at the Jagellonian University of Krakow and a well-known Italian Transhumanist – this is exactly the case. In his interesting *Trattato di filosofia futurista*, Campa tries to identify the philosophy underlying Futurism, which, according to him, represents a consistent and complete form of Transhumanism *ante litteram*.[15] So, Futurism would represent, in particular, a real philosophy of technology, a movement that doesn't see the latter as "dehumanizing," but, on the contrary, as something that should be welcomed in an ecstatic way – and I am not exaggerating when I mention the notion of "ecstasy," as there has been a lot of talk among historians about the "technological sublime" of Futurism, that is, the fact that the power of technology would provoke in the soul of the Futurist that mixture of wonder and terror provoked by natural forces in the soul of the Romantics.

Superhuman and demiurgical tendencies can be detected as far back as the title of a 1915 text by Giacomo Balla and Fortunato Depero: *Ricostruzione futurista dell'universo*; but it is Marinetti himself who, in *Fondazione e manifesto del futurismo*, states the hope for "a violent assault against the unknown forces (of Nature), in order to force them to bend the knee in front of Man."[16] Anticipating the modern culture of advertisement, Futurists coined several effective slogans, from

---

[15] R. Campa, *Trattato di filosofia futurista*, Avanguardia 21 Edizioni, Rome 2012.

[16] F. T. Marinetti, *Fondazione e Manifesto del futurismo*, in Various authors, *I manifesti del futurismo*, Edizioni di «Lacerba», Florence 1914.

"challenge the stars" to "rebuild the Universe," from "climbing the Sky" to "create the mechanical man with interchangeable parts."

We cannot deny that there is a great degree of self-glorification in Futurism, which borders on a form of delusion of omnipotence – even though it is not clear how seriously the Futurists took themselves. The interpretative key offered by Campa is very clarifying, though: to him, Futurism is a philosophy of Becoming, just like that of Heraclitus. And, like all philosophies of Becoming, Futurism is also aware of the impermanence of things, of the fact that everything is dragged away and corroded by the time flow; in the case of Marinetti and colleagues, the only way to contrast this irresistible process of annihilation would be to welcome it, intensifying it in every possible way. Not through the adoption of a Dionysian lifestyle, though, but through the development of a technology able to confer upon us a demiurgic role.

And, if we look carefully, Futurism does have an explicit wish to create a post-human being. For example, in 1910, Marinetti writes, in *L'Uomo Moltiplicato ed il Regno della Macchina*, that the goal of Futurists is exactly the creation of a "non-human type" or the "identification of the Man with the Engine"; that countless human transformations are possible; that, as the future world will be characterized by speed, humans will have "unexpected organs, adapted to the needs of an environment made by continuous impacts." Last, but not least, the multiplied man "will not know the tragedy of aging." A very Transhumanist idea indeed.

Poet Paolo Buzzi talks about the "Impossible Children of the Future," Fedele Azari states that surgery and chemistry will produce a standardized kind of man-machine "resistant, un-consumable and almost eternal," and Futurists in general aim to create a "mechanical anthropoid" able to merge Dionysian instincts, speed and superb technological progress. Speed is actually the main symbol of the Futurist post-human, in the sense that technology, faster and faster, confers upon us a kind of *sui generis* immortality: so, for example, Azari stresses – in *Vita simultanea futurista* – that everyday life is consumed mostly by banal activities, from personal hygiene to personal beauty care, from feeding to moving from one place to another, from dressing to doing the chores; but the speed provided by technological progress will free us from these needs, making them temporally more compact, and freeing more time than we have now for intuition, art, sport and creative activities.

So, this is the Italian Stairway to the Sky, which follows the Nietzschean one. From the viewpoint of Transhumanist history, these are ambiguous characters and topics, whose role as Transhumanist precursors deserves more research. If you are looking for a group of thinkers openly connected and connectable to Transhumanism, we have to look at Russia, and especially at the Cosmists.

1.4 The "Project"

When we think about Russian history, normally, the first things that come to mind are the splendor of the Czars and the seventy-year long Communist dictatorship that spanned through the twentieth century, polarizing the second half. But if you take a second, less superficial look at that country, you would notice right away a strong and sometimes eccentric spirituality – at least from our Western viewpoint – that permeates the Russian culture.

A *sui generis* culture, even from a philosophical viewpoint, as a specific Russian philosophical tradition never existed, or, put better, in Russia, philosophy has always formed a unity with literature and theology – just think about Dostoevsky, Tolstoy, Solov'ev, Bulgakov, Florensky, Berdyaev, to name only those most familiar to the Western public. Characterized by a strong interest in ethics and eschatology, Russian thought has always been obsessed by the question "what should be done?" For Russians, individual and collective lives have always been something you "should do something about," something that has to be animated by a superior and super-individual goal. Obsessed by a linear conception of time, these thinkers have always waited for a metaphysically exceptional experience at the end of it all – the Kingdom of God on Earth, the perfect society promised by Marxism, and so on.

Russian spirituality is peculiar, as we have said; in particular, because, after the fall of the Byzantine Empire, Russia became the crucial center of Orthodox Christianity. As is known, the spiritual life of the Eastern Church included forms of mystical asceticism as well, and especially Hesychasm, a meditation practice aimed at providing the practitioner with a form of interior peace and communion with God. Just as with the mystics of other religions, magical powers were attributed to the Christian-Orthodox ones. For example, according to the tradition, the Venerable Sergius of Radonezh – who lived in the fourteenth century – could perform miracles, heal the sick, bring the dead back to life and, just like Saint Francis, reconcile with wolves and bears. Another Russian mystic, Seraphim of Sarov – who lived between the eighteenth and nineteenth centuries – could levitate, be in multiple places at the same time and emit a light so powerful that he could blind people – a condition that he then healed afterwards.

Russian mysticism was not simply imported from abroad; in reality, the Christian mystical tradition was inserted into a rich native shamanic tradition, of Slavic origin. This syncretic spirituality offered a good terrain for the growth of a further peculiar element of Russian thought, that is, Esoterism. We are not talking of the popular, low-level magic loved by the masses here, but of the huge interest shown by the dominant class toward the masonic tradition, occult sciences, the Rosicrucian Order, alchemy and the more general "higher" esoteric thought that came from Europe. All of this mixed with chronologically younger esoteric doctrines, like Helena Petrovna Blavatsky's Theosophy and Rudolf Steiner's Anthroposophy, which spread widely in Russia.[17] This is the cultural context in which the forefather of Russian Cosmism would work: Nikolai Fedorov.

Born near Tambov – in Southern Russia – in 1828, and dying in 1903, Fedorov was and wanted to be an obscure figure: despite publishing almost nothing, he elaborated an original and visionary thought, which we know relatively well today only thanks to the posthumous publication of many of his writings, a collection – edited by three of his disciples – entitled *Philosophy of the Common Task*. Stern, tall, skinny, with spirited eyes and a neglected beard, this thinker worked for 25 years as bookkeeper at the Rumiantsev Museum in Moscow, which included the main library of the city; in spite of his radical ideas, Fedorov regularly attended mass, prayed and followed the orthodox religious calendar. His lifestyle was more rigid than a monk's; he used to drink only tea, slept in a small rented room, on a wooden surface, with a book instead of a pillow, covering himself sometimes with a few newspapers; he despised money and used to get rid of it as soon as possible, donating it to the poor. Known directly or indirectly by some of the main Russian intellectuals and writers of that age – from Tolstoy, to Dostoevsky, to Solov'ev –, this Moscow bookkeeper lived an ascetic life, during which he developed a plan of action – christened the "Project" or "The Common Task" – quite ambitious to say the least: the physical resurrection of the dead through scientific means.

Besides this ambitious goal, Fedorov's thought represents a strange union of pre-Transhumanist ideas and Orthodox Christianity, a reactionary mentality mixed with revolutionary ideas. Worried – and rightly so, I must add – that his ideas could be too radical for his contemporaries, Fedorov raised the interest of characters like Dostoevsky, who said he completely agreed with him, and Solov'ev, who declared that he completely accepted the Project as a great leap forward for the human spirit toward Christ. For his part, Tolstoy said he was proud to have lived in the same age as Fedorov.[18] According to the primary Western Fedorov scholar, George M. Young, Fedorov was a thinker with a single great idea. He believed that all of the issues known to man were rooted in the problem of death, that no other solution to whatever social, economic, political or philosophical problem one could think of mattered until the problem of death was solved. But if we could find the solution to the problem of death, then all of the solutions to every other problem would follow.[19]

---

[17] To have an accurate portrait of the cultural origins of Cosmism, cf.: G. M. Young, *The Russian Cosmists. The esoteric futurism of Nikolai Fedorov and his followers*, Oxford University Press, Oxford 2012.

[18] For a good introduction to Fedorov and his relationship with his contemporaries, see: G. M. Young, *Nikolai F. Fedorov: An Introduction*, Nordland Publishing Company, Belmont 1979.

[19] Ibid. p. 13.

So, for Fedorov, the only true enemy of mankind is death, and with it Nature, that is, the first cause of our mortality; thus, we have to organize all of our possible resources in service of the widest enterprise ever conceived: the defeat of the grim reaper. All human beings, regardless of party, ideology, nation or religion, must unite as brothers and sisters in this fight against death. Crystal clear, I would say.

If the universal rule is death and disintegration, then the Common Task of every human being will be reintegration, which doesn't mean fusion, but rather the formation of a totality, as fusion – in which every unity loses its individuality and its peculiarity – is, for Fedorov, another kind of death. Basically, the thinker here attacks every philosophy that would like to reduce the individual to a super-individual entity, be it Society, History, the Hegelian Absolute Spirit, or whatever. The world as it is swings between disintegration and fusion, the two principles that rule it, and everything is disconnected in particles or amalgamated in collective entities with no life.

The purpose of man is then to turn upside down the natural flow of life toward these two opposite aspects of death and restore everywhere a totality that assures the integrity of unity. The model or "icon" of the Universe as it should be is, of course, the Trinity, perfect superior and contemporary synthesis of Three and One; death is undesirable, while a unity that is contemporarily in-divisible and in-fusible is desirable.

Federov's project of universal resurrection is, at the same time, political, religious, scientific, artistic and, obviously, philosophical. The political vision of Fedorov – who was a sincere slavophile patriot – consists of an enlightened and Russo-centric autocracy, in which the Czar, as a father-autocrat, must assume the task of unifying Europe and Asia. The goal of art is to offer representations of the idea of resurrection to humankind; it must cooperate with science and theology, and turn itself into an activity or rearrangement of the Cosmos.

Fedorov called his worldview "supramoralism," to be understood as a universal synthesis of the three objects of knowledge and action, that is, God, Man and Nature; Man is a tool of the divine reason and he is himself the reason for the Universe. Thought and action form a superior unity, while religion blends with science and art. Knowledge is neither subjective nor objective, but rather "projective"; the whole natural world is not an object, but a "project" that we must try to lead. "Projectivism" – which we might define as an epistemology for artists rather than critics, for engineers rather than theoreticians – wants to be a bridge between materialism and idealism; ideas exist in our mind as projects to concretize in the material world.

About the nature of the world, Fedorov seemed to consider it a mechanism so simple that it could come into existence without the need for a Creator; his is thus a materialist system that accepts God but doesn't consider it strictly necessary; Man is composed of atoms like everything else and is the only entity equipped with rationality.

Even Fedorov's vision of universal history is syncretistic in nature; in particular, it is a mixture of Russian legends from the Middle Ages, folklore regarding the Pamir plateau – which would have been humanity's cradle and where Adam's grave could be found –, and science fiction-like speculations, all of this arranged into a conceptual frame based on a modified version of Hegelian dialectics. Fedorov basically pays tribute to Hegel, and he is inspired by him in his attempt to develop an omni-comprehensive project, which includes it all, just like the Spirit subsumes everything. To him, Hegel is the last thinking philosopher, and, from Fedorov on, there must be only acting philosophers. The Russian thinker knows Nietzsche, and appreciates his musings on the Will to Power, but criticizes the lack of an objective; to him, Nietzschean thought is to be considered a glorification of adolescence.

Even if he is not openly a mystic, Fedorov insists on the idea that the human mind, which is finite, potentially has a range of action that is infinite, and an infinite ability to learn (we must add that, fragmentary and un-systematic as it is, Fedorov's thought presents several contradictions and obscure aspects).

Practically speaking, the Project is a single, very complex idea, composed of sub-projects that range from the development of small local museums to the unification of all peoples under a Russian autocrat. All of these projects are interconnected. And not just that: every job, even the humblest, must be seen as a contribution of the Project, which incorporates and so ennobles it. When everybody can see every single activity as a simple part of the Common Task, then and only then will the solution for every problem come to light. The Project is thus a kind of "universal deployment" – a form of universal enlistment, on a voluntary basis, that will involve all of the peoples of the Earth – a military aspect that is stressed by Fedorov himself. Soldiers will use their skills in the war against Nature, and, in order to prevent the military from using the impartial knowledge gathered by the wise to fight against each other, the latter will have to become a temporary task force with the purpose of developing practical means to regulate Nature; at the end of the day, armed forces will be converted into an "experimental force."

During this "War on Nature," the armies of the whole world will cooperate to free humanity from natural forces and all weapons will be turned into tools for the benefit of Man. As an example, Fedorov mentions the research of some American scientists who managed to make the clouds produce rain by shooting at them with cannons. To put it simply – both in a metaphorical and in a concrete sense – it is all about changing the orientation of the weapons, pointing them vertically. Verticality and horizontality are typical categories of Fedorovian thought: horizontal means death – corpses lie horizontally –, art and architecture should celebrate verticality, and we should all move vertically, launching vehicles into space. In the context of the Project, the space enterprise has a fundamental role. Part of

Federov's plan consists in tracking the "particles" of people who died a long time ago and were dispersed throughout the universe, recreating those human beings – of course, here, by "particles," he doesn't mean our contemporary subatomic particles, but rather refers to no-better-defined corpuscles that, according to the quasi-mystical and holistic conception implicit in his thought, would conserve some kind of relationship with the dead to which they belonged. Rescue teams will venture into the cosmos, both to look for these particles in every corner of the Universe and to land on other planets for the benefit of the dead – who will colonize them, and whose bodies will be adapted to conditions impossible for us.

Even if he doesn't know how the biologists of the future will actually synthesize human bodies, Fedorov thinks that the creative potential of Man is limitless; with enough will and effort, any solution can be found. People will learn to create and re-create life, and so sex will not be necessary anymore – our hyper-sexualized culture begs to disagree, but, in Fedorov's time, chastity still enjoyed some popularity. Even the bodies of human beings will be modified, in order to eliminate the need to feed on organic matter – which would mean eating the particles of the ancestors. Moreover, the recovery of these particles will be facilitated by very small animals – not visible to the naked eye – equipped with powerful microscopes, which will allow us to see and gather them. The Project includes the regulation of Nature in order to produce enough food and the transformation of Christianity into a project of Universal Sonship and an effort to resuscitate the fathers.

Federov's goal of creating a paradise on Earth consists of a series of very futuristic ideas, from space travel to genetic engineering, so it's no wonder that he decided not to publish them; they would have seemed crazy to everybody. Today, many of his proposals don't look that eccentric, starting with the substitution of natural organs with artificial ones, the search for substitutes for the food we eat, the vision of Nature as a system that we have to regulate and for which we are responsible, and the direct use of solar energy.

Maybe because of his job, Fedorov promoted the role of museums in the context of the Project; in fact, those institutions, and especially the local ones, will have a central role in the scientific resurrection of the dead. Museums don't just archive or transmit information, they re-create whole representations of several aspects of life. Today, museums gather inanimate objects; in the future, they will take on a more active role. In this case, Fedorov is thinking of museums-temples-schools-laboratories for the study and practice of the sacred and scientific art of resurrection. Meanwhile, local museums will serve as warehouses for every fragment of information about the people living around them. Basically, an articulated web of museums – destined to become "centers of resurrection" located in every city and every village – will allow us to preserve all of the information we need on the dead that we will bring back to life; to Fedorov, nothing printed should be thrown away, because, as he used to say – and this is also true for the book you are holding in your hands – "behind the book a man is hidden."[20]

In what looks like a cosmological version of psychoanalysis – that is, to enlighten the unconscious, turning it into conscious –, Fedorov promotes the extension of thought to every material force, until we reach what he calls "psychocracy." It is an ambiguous conception; for example, when he speaks of "populating other planets," Fedorov uses the term "animation" – almost suggesting that the matter will be equipped with a soul. Anyway, psychocracy means to "instill" the spirit in every material function; it also means a state of interior universal consensus through which everyone will want to freely join the Common Task, without external obligation. Fedorov talks of unifying matter and spirit, but one cannot be reduced to the other, and vice versa, and neither can we talk of division and fusion; anyway, the thinker reiterates the fact that he is not a mystic and that that union will not be achieved through unknowable, supra-rational means, but rather through human knowledge and effort. The final goal of his thought reminds us of the typical aim of some forms of mysticism, the *theosis*, that is, the ascension of man to a divine level – "meeting God face to face," says Fedorov. Such ascension will also include omniscience, that is, Man will acquire absolute knowledge, free from space and time.

Man will finally become able to redesign the Universe. Thanks to science, we will free the whole universe from the slavery of the force of gravity, and so rearrange all of the particles of matter into an order determined not blindly and naturally, but consciously and rationally; we will "unhook" the Earth from its orbit and move it as we like, moving and rearranging the stars, and so on. In describing the future paradise created by the Project, Fedorov indulges in poetic metaphors, and speaks of a life always new, in spite of its antiquity; a Spring without Fall; a morning without evening; youth without old age; resurrection without death. Darkness will remain, but only as a representation, as a grief that has been overcome, and it will raise the value of the bright day of resurrection.

One last note: several sources suggest – maybe rightly – that Fedorov is indebted to the esoteric and alchemic literature imported from the West; the Fedorovian project of universal resurrection has many points in common with the masonic ideal of rebuilding themselves and rebuilding the world, and with the alchemic ideal of transmutation of the Self through the transmutation of the matter.

If we have to pinpoint the legacy of Fedorov, we can find it in this idea: death doesn't have to be considered unavoidable, and immortality is not a divine gift, but a human project. Death is the enemy, it is present everywhere, and to defeat it, we have to restructure the Universe, and we have to do it all together. Not bad, as a

---

[20] G. M. Young, *Nikolai F. Fedorov: An Introduction*, Nordland Publishing Company, Belmont 1979, p. 31.

goal – and it makes you think that, in the Middle Ages, the alchemists would have just settled for eternal life.

1.5   Russian Visions…

Fedorov is just the first in a long series of visionary thinkers delivered by Mother Russia, the so-called "Cosmists," a heterogeneous group with fuzzy borders, still alive and kicking, and representing a curious mixture of futuristic scientific theories, esoteric doctrines and utopian fantasies. Today's Cosmism presents many points in common with Transhumanism, even though it has peculiar traits that prevent the complete assimilation by the latter. We have just mentioned some of them; others include the obsession with the centrality of the Russian homeland, a subterranean opposition to Western thinking – considered too individualistic and self-destructive – and the closeness to Orthodox spirituality and, in some cases, even the world of the Paranormal.

Anyway, if I had to define the essence of Cosmism, I would mention the definition given by the main contemporary promoter of this movement, Svetlana Semenova, and I would talk in terms of "active evolution."[21] Besides the Cosmists' fondness for the Occult, their central idea seems to be closer to Transhumanism than to New Age or Theosophy.

So much for the contemporary phase of Cosmism; about the beginnings, we can talk more or less of two phases, the Fedorovian and the actual Cosmist. The first one mostly regards the direct disciples of Fedorov; in fact, the thinker used to hold informal seminars at the Rumiantsev Museum; they were attended by, among others, Konstantin Eduardovic Tsiolkovskii (1857–1935), that is, the father of astronautics – primarily the Soviet branch –, a visionary scientist whose ideas brought about the first Sputnik in 1957.

While it has been suggested that even the initiative to preserve the body of Lenin was influenced by Fedorov's thought, it is well known that, during the Soviet era, several thinkers reprised the work of this thinker but concealed their interest – any distancing from the Marxist orthodoxy could have been dangerous. One of the "centers of Fedorovism" was the "Commission for the study of the productive natural forces of Russia," established by the Academy of Sciences in 1915. Managed by Vladimir Vernadskii, it proposed attempts to use the Sun and electromagnetism as sources of energy. Even among the members of the Party bureaucracy, Fedorovian followers could be found – even though nobody dared to name him, and the ideas promoted excluded the religious aspects of his thought.

---

[21] Cf: G. M. Young, Op. Cit., Oxford University Press, Oxford 2012, p. 8.

Let's start our review of the Fedorovians with the father of astronautics, which promoted a form of cosmic messianism as well, according to which it is our destiny to populate the stars.

The work of the "eccentric of Kaluga" – as he was known in the village southwest of Moscow where he lived – contains, in a nutshell, the entire Soviet space program, detailed step by step. Tsiolkovskii is basically a theoretician of astronautics; inspired by Jules Verne, during his intellectual life, he developed methodologies for performing aerodynamic tests on rigid aerial structures; solved the problem of rockets' flight in a uniform gravitational field; calculated the quantity of fuel necessary to overcome the Earth's gravitational attraction; invented gyroscopic stabilization for space rockets; and discovered how to cool the combustion chamber using compounds contained in the fuel itself. Basically, as a scientists what he did was to translate space travel into mathematical equations, to actively promote and popularize the idea of space travel and, in the end, to inspire in the new generations enthusiasm for rocket science. In 1903, Tsiolkovskii published his most important work, *The Exploration of Cosmic Space by Means of Reaction Devices*.

According to him, in order to survive, mankind must expand into the Solar system, that is, it must expand into the Cosmos in order to avoid the dangers of its trespassing on Earth; everybody knows his famous quote "Earth is the cradle of humanity, but one cannot live in a cradle forever" – even though the original quote went more or less like this: "A planet is the cradle of mind, but one cannot live in a cradle forever." Anyway, according to Tsiolkovskii, mankind is destined to colonize the whole galaxy.

Less known are his occult theories, which, in today's Russia, has made him popular among New Agers; they are about gnostic and theosophical speculations, and they especially promote Panpsychism – the view that everything in the Universe has a soul. In his work *The Will of the Universe. The Unknown Intelligence*, published in 1928, Tsiolkovskii states that the physical constituents of the universe and space possess mental properties, and that the cosmos itself possesses a soul with which we can communicate – the rays of cosmic energy are similar to the Pleroma, that is, the totality of divine powers Gnosticism talks about. The thinker imagines incorporeal beings with a higher-than-human intelligence that would inhabit faraway realms in space. Moreover, there is no individual eternal life, but rather the destiny of individuals is to merge with the Cosmos.

Among Federov's direct followers, let us mention Alexander Gorsky (1886–1943), who believed in the future androgyny of human bodies and in the idea that erotic impulses will be transformed into a new, regulated version, no longer a force for physical creation, but purely spiritual and cultural. Let's also remember Vasily Nikolaevich Chekrygin (1897–1922), an artist who, under the influence of this new Cosmist aesthetics, wanted to create a set of frescos for the Sistine Chapel of Cosmism, that is, the Cathedral of the Museum of Resurrection, with the purpose of illustrating the real resurrection of the dead and their ascent into the Cosmos. It's an artwork that he would never create; Chekrygin left more than one thousand and five hundred sketches, mostly about this project.

One of the most interesting Fedorovians is certainly Valerian Murav'ev, son of Count Nikolai Murav'ev, minister of foreign affairs of Nikolai II. Marxist, esoterist, philosopher, diplomat and poet, the young Murav'ev spent time with the Moscow Fedorovians, and promoted a total alchemical transformation of the individual and the Cosmos. Murav'ev wrote and published, at his own expense, a book, *The Control over Time*, in which he proposed his own version of the Common Task, which aimed to redesign the human being and defeat death through the control of time – such a conquest would have to be pursued through scientific means and such power would be used to resurrect the dead. Murav'ev started with the theories of Einstein about the relativity of time, but the way he developed them is not very clear, referring to a conception of time as conditional to the notions of change and movement; so, in a limited and restricted context, it could be possible to reverse its course – for example, by dividing and putting back together the elements that compose a certain object, and so on. The ability of reversing the course of time would depend on our ability to manage the multiplicity, that is, many things at the same time.

A supporter of Pythagoras, Murav'ev states that things are essentially numbers, and that even human beings are multiple sets of numbers – highly complex sets, which can be summed up in highly complex formulas, but still quantifiable and reproducible. We are multiplicities, and, like any multiplicity, our components can be recomposed and rearranged. As a consequence, control over time means control over ourselves.

Murav'ev imagines the past and the future as two lines departing from the individual in opposite directions; the past is the line of the given, the future is the line of the project, and it is up to us to find a way to decide which one of the two lines must be considered the past and which one the future – just like in mathematics, it is possible to change the sign of a numeric expression, reverting it. Murav'ev discriminates between external time – the time of the calendar and of the clock –, which is the time of necessity and cannot be inverted, and internal time, which can be inverted; essentially, we human beings can change, repeat and reverse the time of the psyche. The final goal is to extend this kind of control over the cosmos as much as possible, basically extending human consciousness.

Let's mention, *en passant*, his contemporary, the Cosmist poet Kraiskii – "Extreme", the *nom de plume* of Alexei Kuzmin –, who talks about rearranging the stars and building the Palace of World Freedom over the channels of Mars.

Among the most prominent Cosmists, the most scholarly one – that is, the one most devoted to the theoretical aspects and the consistence of his ideas – is, without doubt, Vladimir Ivanovich Vernadsky (1863–1945). As a geochemist, he is known for his idea of the "noosphere" – even though this term was coined by the French paleontologist Teilhard de Chardin, after attending the Russian scientist's lectures in France. The starting point is, for Vernadsky, the geosphere, that is, the terrestrial surface covered by inanimate matter, on which life is grafted and which forms a unity that the scientist christened the "biosphere" – the title of his homonymous 1926 book. From the biosphere, the noosphere evolves, which consists of a further "stratum" of life penetrated and ruled by the human mind. It is the third stage of Earth's development, and, as the birth of life fundamentally transformed the geosphere, the birth of cognition fundamentally transformed the biosphere. The noosphere, this emerging stratum of thinking matter, must be seen, in a certain sense, as a new geological phenomenon, in which mankind assumes, for the first time, the role of a great natural force. According to Vernadsky, no organism is really "free," as everybody is inextricably and constantly connected – first and foremost because their need for food and air to breathe – to the material and energetic environment that surrounds us; these observations somehow anticipate the concept of "Gaia," developed many years later by James Lovelock.

A Vernadskyian concept that is very interesting to us is that of autotrophic existence, which the scientist takes from Fedorov. According to Vernadsky, as our resources are going to run out, mankind cannot keep living as we do right now; we must transform ourselves, passing from a heterotrophic to an autotrophic state, that is, living as plants and bacteria do, subsisting on air and sunlight – and minerals, I would add –, rather than other living matter.

A botanist and naturalist, the Cosmist thinker Vasily Feofilovich Kuprevich (1897–1969) was a supporter of unlimited longevity and scientific immortalism. According to him, death is not necessary, and is actually contrary to human nature. To the optimist Kuprevich, very soon, maybe even by the twenty-first century, science will discover the means to prolong life indefinitely. In today's Russia, Kuprevich's musings are trendy again, as Russian immortalists want to shore up their theories and elaborate plans of action aimed at adapting society and the economy to the drastic changes generated by the advent of collective physical immortality.

Laid down this way, Cosmists' theories look like the prerogative of a few, isolated visionaries, without connection to the real world; it is surprising then to find out that, relative to the evolution of man, such a vision was shared by Comrade Trotsky.[22]

In particular, Leon Trotsky believed in a self-directed, self-aware evolution, inspired in this by Marx himself; in fact, contrary to many contemporary Western

---

[22] Cf the interesting article by R. Campa, *L'utopia di Trotsky: un socialismo dal volto postumano*, in: «Divenire. Rassegna di studi interdisciplinari sulla tecnica e il postumano», n. 1/2008, Sestante Edizioni, Bergamo 2008, pages 55–74.

communists – tendentiously ecologists and distrustful of modern technology, when not explicitly primitivist –, for the German philosopher, the achievements of the industrial society were not to be completely erased, but to be used in order to build a better society. Trotsky goes beyond the Marxian positions and speaks explicitly of the creation of a superman. Man is a deeply disharmonic creature, both on the physical and the psychical planes; he looks for material well-being, but he is afraid of dying and seeks solace in a belief in an after-life. The improvements proposed by Trotsky are thus physical, psychical and even aesthetic; the superman will be a more beautiful version of us; the man of the future's organs will move with more precision, functionality and sobriety, and so with beauty. Not just that: the Trotskian superman will take over control of the unconscious processes of his organism, like respiration, blood circulation, digestion, and so on, subjecting everything to the control of will and rationality. Social progress will be tied to biological evolution, and it will have to deal with the human fear of death. About this, Trotsky states – without further clarifying – that the uniform development of our tissues will allow us to reduce the fear of death to a normal reaction of the organism against a danger; to him, the anatomic and physiologic disharmony of our body and the imbalance between the development and the wear and tear of our internal organs confer to the vital instinct that hysteric, morbid and anguished form typical of the fear of death – which puts our intellect to sleep and stimulates our phantasies about a life after death. According to Trotsky, Man will become owner of his own sentiments, he will raise his instincts to the level of awareness, he will make them crystal clear, he will make the will completely conscious, and, in doing so, he will raise himself to the level of a superman.

To Trotsky, death is inescapable, and the only way to solve the issue is to modify ourselves in order to erase such a fear. Specifically, the superman will have to generate specific "biomechanical modifications" in his organism, overcoming both anguish and religious sentiment.

Accepting our mortality doesn't prevent the Soviet politician from proclaiming that there is nothing that cannot be penetrated by conscious thought, and that we will dominate everything and rebuild everything.

This is about Russia; before covering – in the next chapter – actual Transhumanism, we still have three "Stairways to the Sky" left: Stanisław Lem, Teilhard de Chardin and the Anglo-Saxon stairway.

1.6   ...and Polish Musings

Often neglected by historians of Transhumanism, Stanisław Lem (September 13, 1921 – March 27, 2006) was a Polish science fiction writer and, yes, also a philosopher and futurologist, who anticipated so many of the typical topics dear to

Transhumanists that he deserves a special place in their Pantheon. He wrote many novels – just check out Wikipedia to get the complete list,[23] but, more importantly, read them, they are all worthwhile. He is best known for his novel *Solaris* – a fascinating first encounter with an alien intelligence so different from us that any communication is impossible. The novel has been made into a feature film three times: in 1968 by Boris Nirenburg – the only version completely faithful to the original –, in 1972 by Andrei Tarkovsky and in 2002 by Steven Soderbergh. Anyway, for our purposes, what I would like to mention is his philosophical text *Summa Technologiae*,[24] published in 1964. The title is in Latin and it means "Sum of Technology"; it represents an allusion to two other truly classic works from the Middle Ages, the *Summa Theologiae* by Thomas Aquinas and the *Summa Theologiae* by Albertus Magnus. Lem's book deals with the remote and the not-so-remote future, and tries to envision our technological development, imagining technologies that, at least theoretically, nowadays definitely belong to the realm of possibility, like nanotechnology, virtual reality and artificial intelligence.

Let's take a look at its contents: in its eight chapters, the book – which also represents a challenge because of the vocabulary specifically coined by Lem – tries first to track a parallel between social, biological and technological evolution. After reviewing the contemporary – to the book – state of affairs regarding space exploration, and especially the SETI project, Lem introduces the field of "Intellectronics," which covers the topic of artificial intelligences equal or superior to that of humans. It is not the only new word coined by our futurologist. Let's also mention "Phantomology" or "Phantomatics," which corresponds more or less to our virtual reality, while "Cerebromatics" is basically neurosciences. Anyway, Lem imagines that, together, Phantomatics and Cerebromatics can offer full stimulation of our brain, thus providing an actual cyberspace as imagined by science fiction novels like William Gibson's *Neuromancer*. And let's not forget "Ariadnology," of course, a discipline that provides a guide into the labyrinth of consolidated knowledge – a kind of Google, more or less.

But Lem goes beyond that, and, in Chap. 5 – *Prolegomena to Omnipotence* – he asks whether technology will give us more and more powers, until we reach a point at which we'll be able to do anything. But it's not over yet: Lem introduces the concept of "cosmogonic engineering," asking whether we will be able to create artificial worlds populated by synthetic creatures and – why not? – whole universes, and maybe even an artificial after-life.

---

[23] https://en.wikipedia.org/wiki/Stanisław_Lem

[24] Stanislaw Lem, *Summa Technologiae*, University of Minnesota Press, Minneapolis-London, 2013,

1.7   Christian Superman

French Jesuit and paleontologist Pierre Teilhard de Chardin (1881–1955) is remembered for his posthumous works, in which he very innovatively mixed theology and evolutionary theory. From this viewpoint, his main work is *The Phenomenon of Man* (1955), to which we must add *The Future of Man* (1959) and *Human Energy* (1952).

In 1920, Teilhard de Chardin became professor of geology and paleontology at the Catholic Institute of Paris, but his attempt to reconcile the Christian doctrine of original sin and the theory of evolution antagonized the Ecclesiastic hierarchies, who removed him and transferred him to China – where he lived from 1926 to 1946, joining several scientific expeditions. He spent his last years in New York City.

His most famous conception is the "Omega Point"; according to this, universal evolution consists in the constant growth of the degree of complexity and consciousness of everything – a process that he calls "complexification." A slogan frequently used by this thinker is "everything that rises must converge," to indicate the fact that evolution tends towards universal unification at the end of time – the "Omega Point" christened by Teilhard de Chardin and identified by him with the Christian *Logos* – in other words, Jesus Christ. Therefore, evolution goes through several stages, from the geosphere to the biosphere to the noosphere, and it is literally attracted by the Omega Point, which simultaneously constitutes the cause and the effect – and which represents the maximum degree of psychic unification, complexity and conscience achievable by the evolutionary process.

The Church accused him of Pantheism, but, to Teilhard de Chardin, this idea represented a misunderstanding of his thought; the unifying center at the end of time must be intended as pre-existing and transcendent, not immanent. According to the Jesuit, the degrees of evolutionary complexity correspond to degrees of consciousness; as a consequence, we can say that everything possesses a minimal degree of consciousness. Another phenomenon stressed by the scientist is the so-called "involution," which consists in the consciousness directing itself toward the inside – in other words, it's the self-consciousness, or the ability to refer to oneself. Even this trait appears in degrees, and it has reached its maximum development in our species – for now. About human nature: Man is not the center of Creation, but rather the arrow that points toward the final unification of the Universe; he is thus the last and most complex of a series of strata of life. Evolution reached the noosphere, then, and technology represents a fundamental aspect of the latter, almost like an organism provided with limbs, a nervous system, sensory organs and memory. Machines have a role in the creation of a true collective consciousness – Teilhard de Chardin explicitly mentions TV and radio programs, and computers as well, which free our brains from the most boring tasks and allow for the acceleration of thought. All of the machines of this planet, taken together, form a single, vast organized mechanism, a single giant web that envelopes the Earth. In this moment, we are seeing a quick rise in the "psychic temperature" of the Earth, caused by the activity of a continuously accelerating economic-technological web. Human intelligence is destined to grow too; super-intelligent humans will emerge, and we will see the birth of what Teilhard de Chardin calls "super-life." The technosphere will turn itself into the noosphere, a pan-terrestrial sentient organism. Mankind is getting close to a critical point at which it will enter a super-human stage – the Jesuit uses the term "Trans-Human."

Eric Steinhart, philosopher at William Paterson University, has tried to enlist the thought of Teilhard de Chardin for the goals of Transhumanism. If the latter were to grow widely, it would likely encounter resistance from Christian organizations; at this point, the thought of the Jesuit could be used as a starting point for dialogue with the more open sectors of Christianity.[25]

1.8   The Anglo-American Dream

Maybe because of the achievements of Victorian science, maybe because the Industrial Revolution began in England, maybe because science fiction saw its greatest development in the US; in any case, it is in the Anglo-American cultural and linguistic context that Transhumanism was born and has largely developed.

Transhumanism didn't appear out of the blue, though: the door was opened by the sub-cultures of sci-fi and comics lovers, by the world of nerds and computer geeks and, many years before, by scientists and thinkers who, with great intellectual courage, dared to imagine our far future. We can find these intellectuals in the UK, acting during the first half of the twentieth century. Here you have them: John B. S. Haldane, Julian Huxley, John D. Bernal.

Haldane (1892–1964) was a well-known British biologist. Besides his scientific merits, we care mostly about one of his articles – actually the text of a lecture – published in 1924: *Daedalus, or Science and the Future*.[26] In this short essay, Haldane reads the Greek myth of Daedalus as a symbol of the revolutionary nature of science, especially regarding biology. In the field of physics or chemistry, the inventor is always a kind of Prometheus, be it fire – stolen from the gods by Prometheus himself – or flight – performed by Daedalus. In these fields, there is no invention that hasn't been seen as an insult to some deity. So, if the

---

[25] Eric Steinhart, *Teilhard de Chardin and Transhumanism*, «Journal of Evolution and Technology», Vol. 20 Issue 1 –December 2008 – pages 1–22. http://jetpress.org/v20/steinhart. htm

[26] J. B. S. Haldane, *Daedalus, or Science and the Future*, E. P. Dutton and Company, Inc., New York 1924.

inventions of physics and chemistry are seen as profanities, according to Haldane, the biological invention is perceived as perversion. In particular, the scientist analyzes in his text the possible future development of biology, contemplating – among other things – the possibility of one day developing ectogenesis – that is, the production of human beings in the lab, outside of the maternal womb. Another possibility considered is the control of our evolution through direct mutation – in today's terms, through genetic engineering. Haldane's musings then consider the development of no-better-specified methods for the control of passions and for the stimulation of imagination way superior to the ones available today, and the connected possibility of developing new addictions, deeper than the ones we suffer today because of alcohol and drugs. Haldane is not a supporter of immortality: death will continue to exist, but the abolition of diseases will make it a sleep-like physiological phenomenon. Basically, everybody would have the same life expectancy and a generation that lived together would die together. The purpose of this chronological upward "leveling" is clear: according to the scientist, the human longing for an after-life is due to two things, that is, the sensation of having lived an incomplete life and the grief for the premature loss of a friend. Basically, living together and dying together would fix this. This article also influenced the work of his friend Aldous Huxley, who, in *Brave New World* (1932), described a dystopic society based on a few of Haldane's ideas, including the process of ectogenesis; this novel would eventually be used as a "boogeyman" by Transhumanism's critics.

And now, it's the turn of Aldous's brother, Julian Huxley (1887–1975), a famous evolutionary biologist, who plays a fundamental role for us: he has the merit of being the first to use the actual term "Transhumanism." In a 1927 text, *Religion without revelation*, Huxley says that, because of the development of the scientific worldview:

> It is as if man had been suddenly appointed managing director of the biggest business of all, the business of evolution —appointed without being asked if he wanted it, and without proper warning and preparation. What is more, he can't refuse the job. Whether he wants to or not, whether he is conscious of what he is doing or not, he is in point of fact determining the future direction of evolution on this earth. That is his inescapable destiny, and the sooner he realizes it and starts believing in it, the better for all concerned.

The first thing to do, then, is to explore human nature, discover its possibilities and its limits. But this exploration has just begun, as – Huxley says – "A vast New World of uncharted possibilities awaits its Columbus." And so:

> The human species can, if it wishes, transcend itself —not just sporadically, an individual here in one way, an individual there in another way, but in its entirety, as humanity. We need a name for this new belief. Perhaps

transhumanism will serve: man remaining man, but transcending himself, by realizing new possibilities of and for his human nature.

"I believe in transhumanism": once there are enough people who can truly say that, the human species will be on the threshold of a new kind of existence, as different from ours as ours is from that of Pekin man. It will at last be consciously fulfilling its real destiny.[27]

Here you have it, the secular faith promoted by Huxley and by his spiritually akin contemporaries; and among them, we have one left on our list: Bernal.

John D. Bernal (1901–1971) was a pioneer of the use of X-ray crystallography in molecular biology; a Marxist and Soviet sympathizer, although a British citizen, he offered several contributions to the military field, particularly in regards to D-Day. For our purposes, the Bernal text that we are most interested in is *The World, The Flesh and the Devil. An Enquiry into the Future of the Three Enemies of Rational Soul*,[28] published in 1929 and defined by Arthur C. Clarke as "the most remarkable attempt to predict the future of scientific possibility ever made, and certainly the most stimulating."

And predictive it is, this book by Bernal: in it, the author proposes, among other things, the use of space sails, the creation of a space habitat able to host people permanently – named the "Bernal Sphere" – and the idea that technological progress is accelerating – an idea many Transhumanists are very fond of, as we are going to see.

His idea of colonizing the cosmos is farsighted; it is not a vague idea, but an articulated project based on the construction of space stations, which would allow us, among other things, to transfer industrial production into space, restoring the original habitat of our planet. Furthermore, space colonies would allow people to freely associate in communities of their liking. Bernal is decidedly a visionary and, from a speculative viewpoint, has it all: according to him, Man will colonize the Universe, but he will not be satisfied with a parasitic role toward the stars; rather, he will invade and re-organize them for his own purposes.

In Bernal, we can find several sparks of what will later be called "astronomic engineering," a set of speculations about the possibility – for a technology incredibly more advanced than ours – of modifying or moving stars, planets or even bigger "pieces" of matter. The future man of Bernal will rearrange the matter of the cosmos, optimizing it, and will then be able to prolong the life of the Universe by a factor of several millions of millions.

---

[27] J. Huxley, *Transhumanism*, in: *Religion without revelation*, E. Benn, Londra 1927. Revised edition in: *New Bottles for New Wine*, Chatto & Windus, Londra 1957, pp. 13–17. https://web.archive.org/web/20110522082157/http://www.transhumanism.org/index.php/WTA/more/huxley

[28] http://www.santafe.edu/~shalizi/Bernal/

The man will modify itself as well; in this case, mankind will begin to actively interfere with its own shape, and in a very unnatural way. In particular, many of the technological instruments that we use nowadays might be transferred inside of us, through surgery; furthermore, our body might be altered through chemistry. For Bernal, our limbs are appendages that use up the better part of our energy and nourishment, forcing wear and tear on the internal organs in providing the later to the former; so, it's not difficult to envision which kind of modifications Bernal is considering.

Bernal's proposals for our senses would make Superman envious: the man of tomorrow will have a small sensory organ able to track radio waves; his eyes will see infrared, ultraviolet and X-rays; his ears will hear beyond our limits; he will have detectors for high and low temperatures, organs able to perceive electric potential and many kinds of chemical organ.

Human evolution will represent an adaptation to the devices that we will develop in the future as well: in fact, we will have machines for which two hands will not be sufficient to handle; moreover, there will be machines that will be operable with just our will power. Our nerves devoted to pain-perception could be calibrated in order to perceive malfunctioning and problems inside of our devices.

Bernal's proposal for a post-human society is quite interesting. The scientist thinks that, in the future, human beings will be produced by ectogenesis and will have a "larval" existence, non-specialized, which will last between 60 and 120 years. So, during the first stage of their life, human beings will live like the people of the present – a provision inserted by Bernal for the purpose of tranquilizing traditionalists and supporters of a more "natural" lifestyle. During the larval stage, human beings will be able to devote themselves to arts, pleasures and, if they want, to reproduction in the classic way. In the following stage, the "chrysalis," humans will undergo the implantation of new organs and new senses, and a period of reeducation will follow, after which they will re-emerge as new effective and integrated organisms.

It is difficult, according to Bernal, to identify a final evolutive stage for our species, both because it will remain fluid and improvable, and because not everybody will have to transform themselves in the same way; many individual variations will be possible, to a point that – I would like to add – it won't make sense to talk about a "new species," or unidirectional evolution, but rather of possible individual evolutions, potentially as many as the number of single individuals.

We are going to be biomechanical cyborgs, the scientist guesses, and we will have organs that we don't have right now, and that will allow us to manipulate and repair the other organs; we will have also tele-moving organs useful for manipulating things at a distance and that we would be able to exchange with other people – just like the Micronauts, the small Japanese toy-robots with magnetic limbs very popular among the kids during the 1980s. Those organs will be separable from the body more and more, reaching areas that the organic body cannot reach or survive in, like the interior of the Earth or the inside of the stars, whose movements might be precisely directed using those devices.

Therefore, the beings of the future might extend themselves throughout a very wide area of space and time, keeping their unity through a no-better-specified "ethereal communication"; and so, the stage of life won't be the dense and hot atmosphere of the planets, but the cold emptiness of outer space. Furthermore, our far future descendants might keep themselves busy with pure research and speculation, rather than satisfying our classic physiologic and psychologic needs. In other words, one day, we might reach a point at which we will live to think rather than – as happens today – thinking to live.

Bernal doesn't fail to note the psychological turmoil that this transformation will produce in those – and they might be many, maybe the majority – who will decide not to transform; the normal human beings will see those robotized humans as strange creatures, monstrous and inhuman. There is no other solution, though, according to the scientist: to him, the normal man is an evolutionary dead end, while the mechanical man is the future. Anyway, this evolution won't be enacted until we manage to overcome the disgust toward the mechanization of the body; maybe we will witness the most radical of all divisions inside humanity; on one side, we will find the "humanizers" – those who will want to preserve the human aspect as unchanged as possible – and, on the other, the "mechanizers" – favorable to our transformation into cyborgs. To Bernal, it is possible that scientists and a few other people might decide to continue their technorganic evolution in space, and that the "classic" mankind will remain on Earth, looking at the inhabitants of space with reverence and curiosity. Or the Earth might become a "human zoo," managed by the space inhabitants in such an intelligent and cunning way that the terrestrials won't even be aware of it.

The progressive "cyborgization" of humanity will go beyond the body, and will include the mind and the brain: we will be able to connect brains, even more than two, and such a connection might become a permanent condition. The "multiple" individuals will be *de facto* immortal, in the sense that the old members of the "community" will be progressively replaced by the new ones – just as happens to the clonal plants, those vegetal organisms that form continuously re-generating "collectives." The level of connection will be, for us, unimaginable: individual brains will feel a part of a Whole in a way far more profound than the members of a cult can feel. Difficult to understand, such a condition might look like a state of ecstasy.

As unattractive as it may seem – not everybody would like to lose their own individuality, as a matter of fact, it could be comparable to a form of personal death –, such a condition might have an advantage – whoever joined such a collective mind would be able to truly, directly communicate their own sensations and feelings, without them being blocked or deformed by the wall of language. Moreover, memories will be shared. Collective minds might be structured hierarchically, with an ability to comprehend that will transcend that of us mortals. *En passant*, I confess that I don't really understand Bernal's fascination with collective minds; joining his speculative game, we can imagine extending and enhancing the individual minds so much that they could reach levels comparable to those of the collective minds. But it's just my opinion.

No matter whether we decide to evolve into this collective intelligence or not, for the scientist, even our emotions, or, better yet, our "emotional tonalities" – which travel with us during our everyday life –, will be under our conscious control; a certain emotional tonality will be brought about in order to fulfil a certain purpose. Right now, it would be dangerous for us to have this ability, as the majority of us would choose to live in a state of perpetual apathetic bliss; but maybe the psychology of the mechanical man will be different, and he will be able to manage this ability.

In the post-human world of Bernal, even the sense of time could be altered: we will be able to slow down or accelerate events, having whole geological eras pass in one instant or being able to discriminate between events of super-short duration. Furthermore, the temporal faculties of the mechanical man might be very different from ours; while, physiologically speaking, our apprehension of time – which includes the short-term memory and the anticipation of the next instant – extends along one second, more or less, the mechanical man's might include years or centuries of past and future. The post-human instants might last for entire historical periods.

And, in the end, human beings might end up completely "etherealized," turning first into masses of atoms communicating through radiations, and then turning completely into light; Bernal stresses that this might be an end or a beginning, in any case, impossible for us to imagine.

The speculations of the visionary scientist go beyond that; for Bernal, it is possible that, in the far future, even our metaphysical relationship with time will change, and that our post-human descendants – human beings 3.0? 4.0? – will learn to move through time in a pluri-directional way, just as we do in space.

Let's end with some considerations of Bernal that are quite interesting and current. According to the scientist, every person, even the less religious ones, when thinking about the future, tends to welcome in her/his mind – maybe semiconsciously – the idea of a kind of *deus ex machina*; that is, the expectation of some transcendental superhuman event that will bring the universe to perfection or destruction. People have an unconsciously eschatological mentality, we might say. Now, Bernal thinks, for the first time, we are beginning to see the future as something relatively clear, and we are starting to interpret it as a consequence of our actions. How will we manage this change? Will we distance ourselves from our crypto-religious mentality or will we stay with it? Here, the idea is that human nature might include a "constitutional religiosity." This is a topic that has recently raised a lot of interest in the academic world; let's think, for example, of so-called "neurotheology"[29] or of the experiments by Persinger on religious feelings.[30]

Basically, our religious feeling, our sense of inferiority and dependence toward a higher, omnipotent power, might be printed in our brain; religiosity might have, for us, an adaptive value, for our survival. While contemporary scientific thought is inclined toward an evolutionary origin of this phenomenon, Bernal is less clear – that is, he doesn't specify whether, in his opinion, religious sentiments are biological or cultural. Anyway, his question still stands: will the post-human beings try to get rid of their religious sentiments – through genetic manipulations – reaching a kind of "metaphysical maturity"? And, just to reprise Trotsky, will we develop "genetically Godless" super-men? Genetic atheism, then, with a bonus: the lack of faith in a superior deity would not be perceived as a dearth.

These fanta-theological speculations end our analysis of the Anglo-Saxon precursors of Transhumanism. All of this, and much more, will be grafted onto a very fertile terrain, especially apt for the production of innovative and, let's face it, sometimes clearly crazy ideas: The United States of America.

Historian of technology Thomas Hugues defines the first half of twentieth century America as "a nation of machine makers and systems builders."[31] To him, "the remarkably prolific inventors of the late Nineteenth Century, such as Edison, persuaded us that we were involved in a second creation of the world. The systems builders, like Ford, led us to believe that we could rationally organize the second creation to serve our ends."[32]

So, America has been, since the beginning, a country deeply fascinated by technology, a fact confirmed by the technical progress that it has achieved, by its scientific conquests and even by its popular culture, saturated with visions of science fiction – actually, it is in the US that sci-fi has flourished the most. And let's not forget that, maybe because they are physically and symbolically separated from the pluri-millennial past of their European ancestors, Americans have preferred to turn their attention toward the future; not to speak of the fact that, the US being a nation built by immigrants for immigrants, it has always attracted people wishing to build something, to innovate, to find an opportunity – the famed "American dream," which, between ups and downs, still exists. Don't be surprised if such a cultural context ended up generating Transhumanism.

---

[29] Cfr. A. B. Newberg, *Principles of Neurotheology*, Ashgate Publishing, Farnham 2010.

[30] Cfr. M. Persinger, *Neuropsychological Bases of God Beliefs*, Praeger, Westport 1987.

[31] Thomas Hugues, *American Genesis: A Century of Invention and Technological Enthusiasm, 1870–1970*, Penguin, New York 1989, p. 1.

[32] Ibid. p. 3.

2

A New Tower of Babel

2.1 The Most Radical of the Revolts

And here we are, at the real Transhumanism; this strange hybrid movement that expressly wants to retrace the footsteps of the builders of the famous biblical tower, with the awareness that, this time, there won't be anyone to confuse the languages. Considered by some an ideology or a philosophy, by others a kind of faith, by others still a mixture of theories waiting for scientific validation or denial, this movement – precisely because of its interdisciplinarity and the way in which it mixes political agendas and theoretical speculations – is still difficult to define. Wikipedia considers Transhumanism "an international intellectual movement that aims to transform the human condition by developing and making widely available sophisticated technologies to greatly enhance human intellect and physiology."[1] Not bad, as a program, but then, neither were the forerunners kidding.

As for us, we like to consider Transhumanism as a coherent system of rational para-scientific fantasies that act as a secular answer to the eschatological aspirations of traditional religions. It is therefore not a set of pseudo-scientific theories – that is, they have nothing to do with the world of parapsychology and the paranormal in general – but para-scientific: they use and absorb into themselves the enormous amount of knowledge accumulated by contemporary science, although the latter does not accept them or refuse them, but rather leaves them in the waiting room. Essentially, many ideas of Transhumanism – such as

The original version of this chapter was revised with the late corrections from the author. The correction to this chapter can be found at https://doi.org/10.1007/978-3-030-04958-4_11

---

[1] Cf. http://en.wikipedia.org/wiki/Transhumanism

© Springer Nature Switzerland AG 2019
R. Manzocco, *Transhumanism - Engineering the Human Condition*,
Springer Praxis Books, https://doi.org/10.1007/978-3-030-04958-4_2

cryonics or radical longevism – are neither considered anti-scientific nor scientific; they are in the antechamber of science, and could be fully accepted in the near future – or not.

Whatever the case, the aforementioned process of transformation should lead, according to the Transhumanists, to a post-human evolutionary stage. It is difficult, however, to establish whether our possible post-human descendants are to be regarded as mere human beings, a new species, a variegated set of new species very different from each other, or, finally, something no longer biological, which eludes the classic taxonomy taught in biology courses. But what do Transhumanists mean by "post-human"? This term indicates a possible future being whose physical and mental abilities exceed ours at a level that cannot be classified as "human" anymore. A possible post-human being should therefore possess an intelligence superior to that of any human genius past or present, as well as being far more resistant than us to diseases and aging. Alongside these qualities – which are merely enhanced versions of what we already see among human beings – a post-human being should also have direct control over their own desires and moods; the ability to avoid tiredness, boredom, unpleasant emotions and sensations; to adjust their sexual inclinations to their liking; to accentuate their hedonistic and aesthetic experiences; to experience brand-new states of consciousness inaccessible to the limited brains of *Homo sapiens*. In short, from our point of view, it is very difficult to imagine what it means to be post-human; such beings could harbor thoughts inconceivable to us.

It is also possible that the post-humans decide to abandon their biological bodies, or even the non-organic artificial ones, to live as electronic entities in very powerful computer networks – in a virtual world far more realistic than Second Life. Such synthetic minds may have cognitive architectures quite different from ours – to the point of seeming alien – and be equipped with sensory modalities qualitatively different from human ones. Post-humans could shape themselves and their world in fantastic and unimaginable ways, so much so that any attempt to visualize this future reality could be doomed to fail.

This, therefore, with regard to the definition of "post-human". And for that of "trans-human"? This term indicates anyone who is in a phase of transition from the human to the post-human stage; we humans of the present, with our gymnastic, dietary and medical practices that force the limits imposed by nature, with our cosmetic surgery, the operations for the transitioning of gender, our vitamin and mineral supplements, with our glasses and hearing aids – not to mention artificial limbs and organs.

Beware, though: Transhumanism does not just want to promote certain technologies; in fact, it believes that, in some respects, these technologies will sooner or later be reached, and that many of them are already on their way – an almost "messianic" faith, not shared by all of the Transhumanists, to be honest – and it therefore proposes thinking about them in depth, so as to evaluate the ethical aspects and suggest strategies for preventing damage or global catastrophes.

If you meet a Transhumanist, be careful not to mention eugenics; if there is one thing that Transhumanists do with vehemence, it is to distance themselves from this movement. Far from the desire to create – in a more or less coercive way – a "superior race," Hitler-style, perhaps with selection, cross-breeding, sterilization and elimination of subjects judged "non-suitable," the Transhumanists desire the development of enhancing technologies that will allow normal human beings – regardless of their physical condition or ethnic group – to improve physically and mentally, and to live longer and happier lives. In short, the discourse is not around a non-existent "improvement of the race," but around the freedom for anyone to choose which technologies to use or not. Among the possibilities, we can list, for example, becoming more beautiful – what Nick Bostrom calls a "positional advantage," as it makes sense only if inserted within the context of inter-individual comparison – or to increase one's intellect – which, for the philosopher, is instead an absolute advantage.

But if there is one thing that makes the Transhumanists furious, it is the idea that death must be accepted as "natural." Who cares, they answer; naturalness has nothing to do with whether it is desirable or not. How to disagree with them? On the other hand – reiterate our thinkers – contemporary life-expectancy, much longer than that of the Paleolithic era, was obtained in ways that have very little to do with what is natural, and nothing prevents us from imagining that further extension is possible and desirable. The inevitability of death has led human beings to elaborate a series of rationalizations, which – though useful to the men of the past to make their limited life acceptable – today represent an obstacle on the road to earthly immortality. To the point that the Transhumanists have thought to attribute to all philosophies and conceptions that somehow rationalize and accept death the label – vaguely derogatory – of "deathism." Not that the Transhumanists want to force everyone to live forever; rather, the idea would be to abolish involuntary death, allowing everyone to choose if and when to take their last breath. Technophiles to the highest degree, the Transhumanist harbingers use a term – also a rather negative one – to indicate the despisers of technology and scientific progress, especially in the biological field: "neo-Luddite."[2]

---

[2] Ned Ludd is a semi-historical character – in the sense that his existence is not certain – who lived in the Eighteenth Century and came from the village of Anstey, near Leicester. According to the story, in 1768, this worker, in a fit of rage, broke a pair of mechanical looms in the factory where he worked. The twenties of the Nineteenth Century saw, in England, the birth of a protest movement among workers inspired precisely by the figure of Ned Ludd, so-called Luddism, whose main strategy was to sabotage industrial production and to destroy the machinery. With the term "neo-Luddism" or "bio-Luddism," Transhumanists refer to all of those thinkers who, in their opinion, would embody an anti-scientific and anti-technological vision of things. In reality, neo-Luddism is a composite movement, or rather a term that acts as an umbrella for

As regards ethics, Transhumanism seems compatible with a large number of different ethical systems; there are, however, some principles that seem to be shared by all Transhumanists, and that concern the individual freedom to choose whether or not to use any reproductive or enhancing technology as they become available; the desire to prolong life, or at least to have the possibility to choose if, when and how to die; the rejection of speciesism – the need to consider all sentient creatures on the same level, be they human beings, post-humans, cognitively enhanced animals or artificial intelligences. Not just human rights, therefore: Transhumanists also want to ensure the legal protection of entities that do not yet exist, but that maybe will.

It is certainly possible that the advent of Transhumanist technologies will generate further social inequalities, particularly between the countries of the First and Third Worlds. But this would not represent a novelty, but simply the repetition of a cycle already seen. One must not be pessimistic, however: as current technologies – for example, mobile phones or antibiotics – have reached or are reaching the Third World, so it could also happen for those of enhancement. Not only that: Transhumanists aim precisely to address the problem of disparity, and to propose possible solutions.

At this point, someone usually raises a finger and makes a criticism: dear Transhumanists, instead of looking – very childishly – for immortality or enhancement, would it not be better to devote your efforts to solving the problem of hunger in the world, illiteracy, the poor hygiene conditions of many peoples, and so on? Answer: you can do both. First of all, many of the things promoted by Transhumanists are *already* under development, so to think about them would certainly help in facing the future. Moreover, the study of enhancing technologies would be very useful in the fields of education, health, and more. Thirdly, the desire for immortality is as old as man, shared throughout history by respected scholars and thinkers – not to mention the billions of believers of this or that religion, who *already* believe that they are immortal. If this is infantilism, then we are in the presence of a psychological trait common to all of humanity – so, at least, the Transhumanists would respond.

And what about the ecological problems of our planet, from the perspective of Transhumanists? According to them, the current industrial civilization is certainly ecologically unsustainable, and the only solution is a "leap forward," towards new technologies – both those under development and those they expressly advocate,

---

figures and positions very different from each other, from ecologists to conservatives, from the religious to the anti-globalists. Of particular interest – for the criticisms brought to Transhumanism from a conservative perspective – is the quarterly magazine *The New Atlantis*, edited by three American think tanks, the Center for the Study of Technology and Society, the Ethics and Public Policy Center and the Witherspoon Institute. Cf. http://www.thenewatlantis.com/

such as nanotechnologies. This will not only maintain, and even strengthen, our economic growth, but also fully protect the environment.

These are the general ideas of Transhumanism. But now, we want to know another thing: beyond the discourse concerning the precursors, how was this movement born and developed?

2.2   How It All Started

We have seen the difficulty inherent in establishing a precise starting point for a given historical-cultural phenomenon, especially if the latter is alive and active; but we can identify, with sufficient certainty, the cultural background in which Transhumanism has developed. In this case, Transhumanists come out of the American circles of science fiction fans, from the world of nerds and, in general, of computer experts, from lovers of all that is technological – the so-called techno-geeks – and from the fans of astronautics who sprung up like mushrooms after the successes of the US space program; not only that, but an important role was played by the alternative and psychedelic subculture of California in the 1960s, with the so-called Human Potential Movement[3] – let's not forget, for example, that Timothy Leary, the well-known American researcher of altered states of consciousness, was, for a while, a supporter of cryonics.

And indeed, fans of science fiction literature aside, it is around the world of cryonics that the ideas of Transhumanism began to coagulate. Although we will see that things are a little more complex, this "fringe" movement was officially born in 1964, when Robert Ettinger published *The Prospect of Immortality*, in which he promoted the decidedly unorthodox practice of freezing clinically dead people to guarantee them a possible future resuscitation. In 1972, Ettinger published another book, *Man into Superman*, in which the author proposes a number of improvements to the standard human being, in what is, in effect, a Transhumanist proposal. Around that same time, a "mainstream" writer, Alan Harrington, published an essay dedicated to the little-known universe of cryonics, *The Immortalist*,

---

[3] The Movement of the Human Potential was born in America in the 1960s, starting from the idea that human beings have a great untapped potential within themselves. In some respects, this movement represents the most "serious" side of the New Age; the core of this current can be seen in so-called "Transpersonal Psychology," a movement on the borders of the academically accepted, which examines aspects of the human soul as mystical experiences, incorporating aspects of Eastern thought and heterodox forms of psychotherapy. One of the sources of inspiration for the Human Potential Movement was undoubtedly Aldous Huxley, especially in his "lysergic" and spiritualistic phase – that of the books *The Doors of Perception* and *Heaven and Hell*. On the practical side, this movement gravitated around the Esalen Institute, founded in California in 1962 by Dick Price and Michael Murphy.

whose ideals the author summarizes in the slogan "Death is an imposition to the human race, and one no longer acceptable."[4]

Apart from Ettinger, Transhumanism has another founding father, Fereidoun M. Esfandiary, a scholar of future studies who taught at the New School for Social Research in New York during the 1960s and who originated a current of optimistic futurologists known as Up-Wingers. The 1970s and 1980s saw the birth and growth of a real Futurist subculture, which gave rise to different organizations – definitely "on the edge" of the dominant culture, one might say, but then, Christians have also celebrated the mass in catacombs – made up of enthusiasts without specific scientific qualifications and academic credentials. These groups tended to be rather isolated from one another and, although ideologically similar, they did not have a common organic ideology. The first self-proclaimed Transhumanist activists began to meet formally at the University of California, Los Angeles, during the 1980s – this university would soon become the main center of Transhumanist thinking. Local militants embraced the futuristic ideology of Esfandiary, which was proposed as a "Third Way," alternative to the left and right – no right and left, but up, that is, upwards. In that context, the artist Natasha Vita-More presented, in 1980, her own experimental film, "Breaking Away," which revolved around the idea of human beings who overcome their biological and gravitational limits, heading into space. In Los Angeles, Esfandiary and Vita-More began to hold regular gatherings of Transhumanists, which included the students of the former and the public of the latter.

1986 was another epoch-making year for Transhumanists: in fact, Eric Drexler published *Engines of Creation*, a book in which the possibility of building so-called "nano-machines" – computerized robots as small as viruses or proteins and capable of manipulating matter at the atomic and molecular level – was assumed for the first time. If this were possible, then the Drexlerian nano-machines would allow Transhumanists to realize all of their dreams, from physical immortality to the reconstruction and resuscitation of bodies stored in liquid nitrogen.

In 1988, the roboticist and Transhumanist scholar Hans Moravec published *Mind Children*, in which he discussed the forthcoming development of intelligent machines – another important piece of Transhumanism. Also, in the same year, the English/Californian philosopher Max More published the first issue of *Extropy Magazine* – "extropy" represents a concept contrary to that of "entropy," indicating that Transhumanists pursue a growth of order rather than chaos – which would be followed in 1992 by the foundation of the Extropy Institute. The magazine and institute would act as a point of reference for the libertarian current of Transhumanism, that is, Extropianism.

---

[4]Cf. A. Harrington, *The Immortalist*, Random House, New York 1969.

The '90s also witnessed the explosion of the Internet, and tools such as forums and mailing lists allowed Transhumanists to forge closer and closer contact with each other, finally taking full consciousness of themselves as a movement.

In 1998, two British philosophers, Nick Bostrom and David Pearce, founded the WTA (World Transhumanist Association), the first world organization aimed at spreading Transhumanist ideas to every corner of the globe and across the political spectrum – with the awareness of the political peculiarity of Transhumanism, that is, an ideology that cannot be easily framed according to the traditional categories of "right," "center" and "left," but which lends itself to be freely mixed with any currently existing political ideology. The importance of the WTA is not so much in its wide diffusion, but rather in the fact that the founders focused its action on the academic world from the beginning, that is, trying to present Transhumanism as a "serious" discipline worthy of studying in academia. To this end, a technical journal of Transhumanist studies was also launched, the *Journal of Evolution and Technology*, which publishes articles submitted to the practice of *peer-review*.

Considering its mission "essentially completed," in 2006, the Extropy Institute closed; the same year, a political struggle inside of the WTA ended with the victory of the liberal-democratic left, whose ideals would characterize its activity from then on. In 2008, the WTA changed its name to Humanity+, also launching the official movement magazine, *h + Magazine*.

From the second half of the 2000s, Transhumanism began to take root more and more in Silicon Valley[5]; for example, in 2007, the WTA established its headquarters in Palo Alto, while several other organizations with similar ideas flourished in that area. Like it or not, what used to be a marginal ideology reserved for science fiction nerds is coming under the spotlight, thanks to the fact that the high-tech millionaires of that area are now entering the game. And where could Transhumanism take root, if not in California? Just think of its composite nature, which mixes technological optimism, the American Dream, science-fiction literature and almost sectarian religiosity, all things that fit perfectly into the local ecosystem. In fact, we cannot forget that the '90s witnessed the unstoppable development of the so-called California Ideology, a social phenomenon that, above all, interested the knowledge workers of the area and represented a mixture of hippie rebellion, lysergic fantasies, economic neo-liberalism and techno-utopianism. And, in recent years, the numbers of Alcor – the most important cryonics company active right now – have grown thanks to the arrival of new members from Silicon Valley. Not to mention the fact that the maximum Transhumanist

---

[5] D. Gelles, *Immortality 2.0: A Silicon Valley Insider Looks at California's Transhumanist Movement*, in: «The Futurist», January 2009.

theoretician of biological immortality, Aubrey De Gray, often holds meetings in the offices of Yahoo and Google. Even the central metaphor of so much of Transhumanism, that is, that which sees the body as a machine and the brain as a computer – and, as such, perpetually repairable and updatable, at least in principle – could not please the techno-optimistic Californian millionaires more. Peter Thiel stands out among the most generous supporters of Transhumanism. Co-founder and former CEO of PayPal, currently CEO of Clarium Capital – a two-billion-dollar hedge fund – and one of the first to believe in Facebook, Thiel is a very popular character throughout Silicon Valley. And he regularly takes part in the meetings of the Transhumanists.

And now, in Silicon Valley we find practically all of the main think tanks that gravitate around Transhumanist ideals, from the Foresight Institute to the Lifeboat Foundation, from the Methuselah Foundation to the Singularity Institute – we'll see them all soon. In the Bay Area, you can also find a flourishing Transhumanist-friendly social network – facilitated in this by the presence of numerous meeting places of all kinds –, which, thanks to daily lessons, conferences, exhibitions, and so on, is able always to attract new ambitious and technologized young nerds.

The recent spread of the Transhumanist movement, the "normalization" and the re-styling that is being undertaken by this or that member of the academic community and the attention that it is gaining in the eyes of the media and the world of high-tech business will establish, in a very short time, whether this ideology/philosophy is destined to affect the contemporary culture to a significant degree or if it should remain a bizarre and marginal phenomenon – but let's keep in mind that, if this latter circumstance occurs, we have no doubt that, in a few decades, someone else will come to collect its inheritance and ideals.

2.3   The One Hundred Flowers of Transhumanism

Maybe the currents inside Transhumanist thought will not be as numerous as the metaphorical philosophical-cultural flowers advocated by Mao Tse-tung, however, just like any self-respecting political and intellectual movement, Transhumanism also has a number of internal currents, not necessarily in contrast, but, more than anything else, focused on different objectives and aspects of the Transhumanist enterprise. Let's have a look, starting from the first of them, "immortalism."

Let us not be deceived: even if, in Transhumanist circles, there is much talk of enhancing human capacities, artificial intelligence, and the like, and although not all of the followers of this movement are in favor of true immortality, the number one enemy of Transhumanism is always him, the grim reaper. In any case, immortalism – or, rather, the most modest search for a radical extension of the duration of life – has assumed a personality of its own within Transhumanism, which, in recent years, has resulted in the birth of real political groups – little more than online discussion groups, in fact, but rather active – dedicated to this cause, along the lines of those political movements that coagulate around a single objective, or around a limited number of them, such as the Pirate Party. For example, we have an international organization, the International Longevity Alliance, founded by Ilia Stambler, an Israeli scholar, Transhumanist and member of the Institute for Ethics and Emerging Technology. The Longevity Alliance represents the culmination of a process that has led, over the past few years, to the emergence of various Longevity Parties, in Europe, the United States, Israel and Russia; have a look on Facebook and, if you like, sign up.

Another interesting current is "abolitionism," which has, as its starting point, the idea that science and technology should be used to abolish any kind of involuntary suffering. This elimination should then concern all sentient life, including animals. Obviously, the main technology that will allow us to achieve this goal is biotechnology, with which we will modify our bodies in order to eradicate suffering. The ultimate goal of abolitionism is so-called "Paradise Engineering," the construction of Paradise here and now, on Earth, and, in particular, in our mind. Followers of the utilitarian ethics of the philosopher Jeremy Bentham, but also animalists and often vegans, abolitionists see individual happiness as the ultimate goal of life, and assume that, far from having a spiritual nature, human emotions have a physical basis, and, as such, can be manipulated – and improved – by the science of the future. In fact, the speculations of the abolitionists are not so farfetched: we need only think of the antidepressants and other increasingly targeted psychotropic drugs regularly placed on the market, as well as TMS – transcranial magnetic stimulation, which uses a magnetic field applied to the head of the patient to alter the functioning of the brain, in order to treat many diseases of the mind – or to "deep brain stimulation," a technique based on a sort of "pacemaker" of the brain that is currently used to treat the symptoms of various neurodegenerative pathologies. These are still crude techniques, but, if properly financed, can certainly be refined more and more, revolutionizing the lives of healthy people. A final note: the term "abolitionism" was originally coined by the American scholar Lewis Mancini, in an article published in 1986 in the journal *Medical Hypotheses*, which proposed the idea above, subsequently quickly adopted – you had doubts? – by the Transhumanists.

Now, let's move on to a truly radical Transhumanist current, "Post-genderism." As you can guess from the name, the Post-genderists aim at the elimination of gender through the use of biotechnologies and advanced – not yet existent – reproductive technologies, such as the artificial womb and the like. Frankly, it is a step that I do not feel ready to take – despite the criticism that the male gender has suffered in recent decades, I am fond of my gender, and, for now, I prefer to leave it as it is. In any case, Post-genderism is a current that brings together the typical reflections of Transhumanism with those of feminist thought. According to some Post-genderists, the reproductive purposes of sex will become obsolete, while, for others, all post-human beings will be able to change sex at will and assume both the paternal role – with the fecundative functions – and the maternal – thus bringing a pregnancy to term. Among the Post-genderists, we have essentially those who promote androgyny – with the consequent mixture of the "best" traits of men and women –, those who promote the ability to freely and easily change sex and inclinations, those who promote a form of reproduction at will with or without technological devices and with or without partners,[6] and, finally, those who promote the possibility of having more than two sex types. Among the Transhumanists, the main promoter of Post-genderism is George Dvorsky.[7]

We have already encountered "Extropianism," and now we are going to deepen it a bit. It is a hyper-optimistic philosophy created between the late '80s and the early '90s by the British philosopher Max More. It revolves around a system of values that aims to overcome every limit, and, in particular, that of mortality. Extropian thought – or Extropic, if you prefer – has a pragmatic attitude towards natural and social reality and a proactive one towards progress and human evolution. It also espouses a neo-liberal and anti-statist conception of society. The term "extropy" was adopted in the Transhumanist context by Tom Morrow – real name, Tom Bell – and by Max More, who chose to use it in a metaphorical sense, to indicate the degree of intelligence, functional order, vitality and energy of a living or organized system, as well as its capacity and its desire for growth and further improvement. More summed up his vision of things in his *Principles of Extropy*,[8] which are summarized by: perpetual progress, self-transformation, practical optimism, intelligent technology, open society, self-direction, rational thinking.

Estropianism has been followed by a similar derivative current, "Extropism," which reprises the same themes, associating them with techno-Gaianism and singularitarianism. It should be noted that, in the future society dreamed of by the Extropists, among other things, we will finally be able to free ourselves completely – thanks to robots and artificial intelligence – from the slavery of work, which will become irrelevant for survival. The movement in question was launched in 2010 by Breki Tomasson and Hank Hyena, through a specific Extropist Manifesto.[9] "Singularitarianism" is an ideology that gravitates around the concept

---

[6] Essentially through parthenogenesis, that is, the phenomenon in which, under certain conditions, the females of some animal species are able to self-impregnate without the use of male sperm.

[7] G. Dvorsky e J. Hughes, *Postgenderism: Beyond the Gender Binary*, 2008. http://ieet.org/archive/IEET-03-PostGender.pdf

[8] http://www.maxmore.com/extprn3.htm

[9] http://www.knowledgerush.com/kr/encyclopedia/Extropism/

of Technological Singularity, which is a progressive acceleration of technological progress that will lead to the birth of an artificial intelligence superior to that of humans, with all that follows in terms of social and technological progress. For the proponents of this vision, the Singularity is not only possible or probable, but highly desirable, so that the institutes and organizations that refer to this idea operate in such a way as to facilitate – in their opinion, at least – the advent of all of this. We cannot deny that the mentality embodied by this current has many points in common with that which animates various groups of Christian fundamentalists awaiting the forthcoming apocalypse. In this regard, Robert M. Geraci – professor of religious studies at Manhattan College in New York – has recently published an interesting essay, *Apocalyptic AI*, in which he analyzes the idea of technological Singularity from this point of view.[10]

David Correia, journalist of the radical leftist magazine *Counterpunch*, underlines the way in which the Singularitarian movement is made up of academics, entrepreneurs and financiers linked to the army and the world of multinationals, and would represent – according to him – a "fig leaf" that hides the economic and strategic interests of these subjects and the intention of the latter to perpetuate social inequalities. DARPA – the Defense Advanced Research Programs Agency, the US federal agency in charge of military research – would fund dozens of projects inspired precisely by the high-tech dreams of the Singularitarians.[11] However, the term "Singularitarian" was originally coined in 1991 by an Extropic thinker, Mark Plus – real name, Mark Potts[12]; the origins of this idea are, however, more complex, and we will cover them in due course.

"Technogaianism" represents an embodiment of so-called "bright green environmentalism," a term coined by the American writer and futurist Alex Steffen to indicate an approach to environmental problems that are currently growing, based, above all, on the development of new technologies and on the improvement of design – in an eco-compatible sense – of the existing ones. The aim of the Technogaians is to use a Transhumanist type of technology to actively restore the ecosystem. In particular, as well as being a fan of alternative energy – hydrogen and the like – Technogaians would like to use bio and nanotechnologies to repair

---

[10] R. M. Geraci, *Apocalyptic AI: Visions of Heaven in Robotics, Artificial Intelligence, and Virtual Reality*, Oxford University Press, Oxford 2012.

[11] Cfr. D. Correia, *If Only Glenn Beck Were a Cyborg*, «Counterpunch», September 15, 2010, http://www.counterpunch.org/2010/09/15/if-only-glenn-beck-were-a-cyborg/

[12] The movement founded by Max More stands out for some of its very "American" aspects, that is, those typical of a culture fascinated by secret societies, slogans and associative symbologies. In particular, among them, the assumption of fictitious names – like Max More, Tom Morrow and Mark Plus – and a particular handshake that symbolizes their enthusiasm for the future. Cf. E. Regis, *Meet the Extropians*, «Wired», October 1994, http://www.wired.com/wired/archive/2.10/extropians.html

the environmental damage caused so far by Man. Contrary to radical environmentalists, who propose the abandonment of technology to a different extent, technogaianism promotes even greater use, without prejudice for the fact that the goal must be the protection of the ecosystem. One of the technologies most dear to the Technogaians is undoubtedly that of the biospheres, that is, all of those projects that aim to build closed ecosystems useful for scientific experimentation – this is the case, for example, of the "Biosphere 2" project, managed by the University of Arizona.[13] Transhumanists, it is well known, tend to think big, and that is why Technogaians are, above all, fans of so-called *terraforming*, a set of technological practices – currently all purely theoretical – that, in a more or less distant future, should allow us to change the atmosphere and all of the physical-chemical parameters of the surface of this or that planet of the Solar system, thus making them capable of hosting life – imported from the Earth, of course – and, in particular, human beings. In the past, many scholars have had fun creating – on paper, of course – one or more terraforming procedures, especially as regards Mars and Venus, and even Carl Sagan played this intellectual game for a while.[14] According to the Transhumanist thinker James Hughes, an example of Technogaianism is the Viridian Design Movement, founded in 1998 by cyberpunk science fiction writer Bruce Sterling,[15] a movement that mixes environmental design, techno-progressivism and the ideal of global citizenship.[16] Other members of the Viridian Design Movement include the aforementioned Alex Steffen and the Transhumanist Jamais Cascio. However, in 2008, Sterling officially closed the movement, as his ideas would have now become a consolidated part of ecological thinking. Also, Hughes counts, among the points of reference of the Technogaians, Walter Truett Anderson, an American scholar of ecology and author of *To Govern Evolution: Further Adventures of the Political Animal*, and Michael L. Rosenzweig, an ecologist at the University of Arizona and author of *Win-Win Ecology: The Earth's Species Can Survive In The Midst of Human Enterprise*, a fundamental text of the so-called "ecology of reconciliation" – an approach that aims to promote biodiversity within human-controlled ecosystems. We mentioned James Hughes; the latter is the promoter of so-called "Democratic Transhumanism," a term that refers to all

---

[13] Cfr. http://en.wikipedia.org/wiki/Biosphere_2

[14] Cf. C. Sagan, *The Cosmic Connection: An Extraterrestrial Perspective*, Cambridge University Press, Cambridge 2000.

[15] Among other things, Sterling is the author of a novel, *Schismatrix*, in which he imagines a Solar system of the future divided between two species, the "mechanists" – human beings transformed into cyborgs – and the "shapers" – enhanced through genetic engineering. Here, he introduces the idea that, in the future, human beings will be able to divide themselves into more species of post-humans, very different from each other. See B. Sterling, *Schismatrix*, Ace Books, New York 1986.

[16] http://www.viridiandesign.org/manifesto.html

of those Transhumanist thinkers who, like him, adhere to a generically leftist, socialist or democratic vision – in the American meaning of the term "democratic." For the Democratic Transhumanists, the supreme ideal is that of individual and collective happiness, achievable through the assumption of the rational control of the natural and social forces that normally determine our lives. Among the ideas of Democratic Transhumanism, we recall the non-anthropocentric theory of personality – which attributes the same ontological status to any type of mind, including the artificial post-human ones – and support of the social state, in particular, of public health, understood as the best way to democratize the technological enhancements hoped for by Transhumanists. In more general terms, Democratic Transhumanism wants to restore dignity – at least in Hughes' intentions – to the utopian tendencies present in so much political thought and quarantined at the end of the Twentieth Century after the bloody and authoritarian outcomes of this or that ideology. In this regard, Hughes makes an interesting observation, that is, the fact that, historically, it has happened on more than one occasion that works of fantasy have inspired many political movements: and so, Edward Bellamy's utopian novel *Looking Backward* inspired many Socialist groups active in the United States of the late Nineteenth Century, while *Atlas Shrugged*, Ayn Rand's fanta-philosophical novel, served as an inspiration for the Anglo-Saxon brand of Libertarianism. Likewise, utopian Transhumanism can be inspired by the most complex science fiction works, using them to elaborate scenarios and evaluate options. At the moment, Democratic Transhumanism is the dominant current within the WTA; among the main adherents of it, we include Jamais Cascio, George Dvorsky, Mark Alan Walker, Martine Rothblatt, Ramez Naam, Riccardo Campa and Giulio Prisco. And although he has declared that he is neither right nor left, in reality – from the point of view of his progressive ideas – we could also include Esfandiary on this list.

"Libertarian Transhumanism," as you can imagine, finally brings the Transhumanist ideals and those of the Libertarians together – and this reminds us very much of the Extropians, who, at the beginning of their history, used a notion of "spontaneous order" in an anarcho-capitalist sauce. The Libertarian Transhumanists base their vision of things, a bit like their "non-Transhumanist" cousins, on the idea that the individual is the center of reality and on a consequent egoistic and rational brand of ethics, which wants to reduce the role of the State in social life to a minimum. Media theorists Richard Barbrook and Andy Cameron – of the University of Westminster – associate Libertarian Transhumanists with what they have named *California Ideology*, which we have already encountered above, and the positions of which would be well represented – although this is just the opinion of Barbrook and Cameron – by *Wired*.[17] Another criticism of Libertarian

---

[17] Cfr. http://www.imaginaryfutures.net/2007/04/17/the-californian-ideology-2

Transhumanists – who, in many cases, are computer science experts and have a strong interest in mind-uploading – is that they would be motivated by a form of "body disgust," a term coined by the critic and journalist Mark Dery precisely to indicate their alleged desire to escape from their "meat doll" to a virtual world.[18] Among the main Libertarian Transhumanists, I'll mention Ronald Bayley, who works for the Libertarian magazine *Reason*,[19] and Glenn Reynolds, a law professor at the University of Tennessee and owner of the well-known blog *Instapundit*.[20]

We have said before that Transhumanism can be mixed with any kind of ideology that accepts its principles; this applies, above all, to progressive ideologies, but not exclusively. This is the case, for example, of the Italian "Superhumanists," who refer to the Nouvelle droite by Alain de Benoist, to the thought of Giorgio Locchi and to the "Archeofuturism" of Guillaume Faye – essentially to the radical right.[21] Or think of the conservatives, whose point of view has been defended by the same Nick Bostrom, that is, a progressive; for him, it is possible to imagine Transhumanist modifications or enhancements aimed at promoting traditional values, such as couple relationships and the family. There was also an attempted political proposal in this sense, baptized "Conservatism Plus."[22] And, on the list, there are also the Anarchists, who, through Anarcho-Transhumanism, want to prepare for "a social and technological insurrection."[23] Finally, it seems that the Transhumanists have also managed to "place" one of their men in the Italian parliament – until March 14, 2013, the date of the end of his mandate; we are speaking of Giuseppe Vatinno, elected with the Italia dei Valori Party.

Essentially, beyond all of the theoretical debates and political currents, Transhumanists have created a number of associations, institutes and think tanks, which we will now look at.

2.4 Journey Through the Transhumanist Galaxy

It is very articulate, the social universe of Transhumanism, to the point of making Al-Qaeda feel envy; here, we will only list the most important groups and institutions, or at least the most curious ones, leaving you the pleasure of losing yourself in the complicated network of sites, blogs, Facebook pages, forums and virtual associations that the Transhumanists have managed to put up over the years.

---

[18] Cfr. M. Dery, *Escape Velocity: Cyberculture at the End of the Century*, Grove Press, New York 1997.

[19] http://reason.com/

[20] http://pjmedia.com/instapundit/

[21] http://ieet.org/index.php/IEET/more/hughes20091004

[22] http://conservatismplus.ning.com/

[23] http://anarchotranshumanism.com/

Humanity+

The most famous and widespread Transhumanist organization is the World Transhumanist Association – now Humanity+,[24] founded as a non-profit organization in 1998 by Nick Bostrom and David Pearce, with the aim of promoting the presence of Transhumanism in politics and academia. Unlike the extreme optimism of other Transhumanists, the driving force of Humanity+ is the desire to understand and manage the social forces that could oppose the development and diffusion of the enhancing technologies that they want to develop. Together with the organization, in 1998, an academic journal, the *Journal of Transhumanism*, was launched, later renamed, in 2004, the *Journal of Evolution and Technology*.[25] At the same time, *H+ Magazine* was launched, a quarterly magazine of Transhumanist ideas and news.[26] Then, there are the annual conferences, named TransVision and held in different cities around the world: in Holland in 1998, in Stockholm in 1999, in London in 2000, and so on. Among the best-known members, we have Natasha Vita-More, Ben Goertzel, Nick Bostrom, David Pearce, George Dvorsky, Giulio Prisco, James Hughes, Aubrey de Grey, Max More and Michael Anissimov. Humanity+ consists of dozens of local groups spread across all five continents; obviously, the national groups include several local sections. From a numerical point of view, the association currently includes about six thousand members, distributed throughout one hundred countries – including nations such as Egypt and Afghanistan. Finally, there is also a youth and student association, the Transhumanist Student Network.[27] If you want to subscribe, remember that the membership fee amounts to four dollars and ninety-nine cents a month.

The Extropy Institute

After contributing to the foundation of the first European cryonics organization – Mizar Limited, later renamed Alcor UK – Max More moved, in 1987, to the University of Southern California, Los Angeles, where, the following year, he published *Extropy: The Journal of Transhumanist Thought*, the journal around which scholars and fans of longevism, genetic engineering, nanotechnology, robotics and Transhumanist ideas in general are concentrated. Starting from *Extropy*, More and Tom Morrow then founded the Extropy Institute, a non-profit organization founded with the aim of acting as a center of information and aggregation for Transhumanism sympathizers. From a political and philosophical point of view, the institute's goal is to elaborate a set of principles and values that favor

---

[24] http://humanityplus.org/

[25] http://jetpress.org/

[26] http://hplusmagazine.com/

[27] http://www.transhumanism.org/campus/

techno-scientific progress towards the post-human. In 1991, the Extropy Institute launched a mailing list, and, in 1992, began organizing the first conferences on Transhumanism. Later, in various parts of the world, local groups sprang up that were affiliated with the institute, which organized their conferences, parties, discussions and more. The Extropy Institute is able to make extensive use of the birth and global dissemination of the internet. Subsequently, the institute intertwined its activities with those of similar Transhumanist associations and, as already mentioned, in 2006, the board decided to dissolve the organization, as it considered the objectives of the Institute to have substantially been reached.[28]

The Terasem Movement

Judging from the official philosophy, the communicative style and the aesthetics of the site, the Terasem Movement recalls some para-religious New Age organization.[29] In reality, the organization in question aims to educate the public on the need to extend human life through nanotechnology and "personal cyber-consciousness" – and this is the strong point of the movement, as we will explain shortly. The Terasem Movement was launched in 2002 – based in Melbourne Beach, Florida – and was joined in 2004 by a parallel organization, the Terasem Movement Foundation.[30] The founder of both is Martine Rothblatt – born Martin Rothblatt – an American lawyer and entrepreneur. In reality, the mystical and religious aspects are not lacking, in Terasem, which, among other things, defines itself as a "trans-religion" and speaks explicitly of a "faith." Faith that consists, in reality, not so much in adoring and praying to a God, but rather, to put it bluntly, in creating one. The idea would be to create self-replicating machines to spread through the cosmos, so that they accumulate knowledge at an exponential rate and use it to convert randomly distributed and organized matter and energy into intelligent and homogeneously distributed matter and energy; this would lead to the creation of a cosmic network capable of acting as a force able to control the physical universe. This true "collective consciousness" should progressively approach omnipotence, omniscience and omnipresence, practically creating our traditional benevolent God. An ambition that rivals that of Fedorov, in short.

For the moment, however, Terasem focuses on preserving human consciousness, and this is where the CyBeRev project comes into play. Everything is based on the concept of "beme," created by Rothblatt along the lines of the gene and the meme. It indicates a fundamental unity of being. Unlike memes, which are culturally transmissible and changeable, bemes are highly individual elements, such as aspects of personality, traits and habitual gestures, feelings, memories, attitudes,

---

[28] http://www.extropy.org/

[29] http://www.terasemcentral.org/

[30] http://www.terasemmovementfoundation.com/

values and beliefs. The project in question then aims to test the comparability of the real consciousness of a single human being with a digital representation of the same person created by special software that uses a database containing the psychological profile of the original person – specially developed by professional psychologists. The final objective is to preserve the information about a certain person with a degree of reliability that allows for recovery or replication in the future. Project participants can store information about themselves in many different forms – photos, videos, texts, audio recordings, lists – and can also undergo intensive personality testing.

Institute for Ethics and Emerging Technologies

In 2004, the Institute for Ethics and Emerging Technologies was launched by Nick Bostrom and James Hughes, with the aim of creating a techno-progressive think tank that would contribute to an understanding of the impact that emerging technologies will have on individuals and society. To this, a more ambitious purpose was added, that is, to influence the development of public policies favouring the democratic distribution of the benefits of the aforementioned technologies. The Institute for Ethics and Emerging Technologies is affiliated with Humanity+, and, as such, it handles the publication of the *Journal of Evolution and Technology*. The fronts on which the institute operates are: the extension of the concept of "human rights," the identification of threats to the future of our civilization, the management – some, including your author, might say the dismantling – of the objections to longevism, the fight against ageism and ableism – that is, discrimination based on age or against disabled people – and, finally, the development of positive, negative and neutral scenarios in relation to post-humanity and non-human intelligences that we will perhaps create.

Future of Humanity Institute

Connected to the Philosophy Department of the University of Oxford and the Oxford Martin School – an institute dedicated to forecasting and futuristic technologies founded in 2005 – the Future of Humanity Institute is directed by Nick Bostrom and aims at the interdisciplinary study of certain issues of fundamental importance to humanity, such as the effects of future technologies on the human condition or the possible occurrence of future global catastrophes.[31]

---

[31] http://www.fhi.ox.ac.uk/

Methuselah Foundation

Founded in 2000 by Aubrey De Grey and David Gobel, the Methuselah Foundation is a non-profit organization that aims to develop methods for extending the duration of human life by significant margins.[32] Located in Springfield, Virginia, the organization's main activity is the management of the MPrize – also known as the Methuselah Mouse Prize – a prize given to those who make a substantial contribution to the fight against the aging process. In particular, the prize includes two categories: the first is open to those who are able to substantially extend the total life expectancy of a classic laboratory rat, the second concerns those who manage to genetically manipulate a middle-aged mouse, making it manifest signs of youth. This is obviously a competition that is still open, as it is always expected that some new research group will be able to overcome the previous records. Andrzej Bartke, a researcher at Southern Illinois University in Carbondale, currently holds the title – genetically stopping the growth hormone receptors, which also play a role in the aging process – he managed to cause a guinea pig to reach 1819 days of life – almost five years. The Methuselah Foundation also participates in other projects, one of which involves Organovo, Inc.,[33] a biotech company in San Diego, specializing in regenerative medicine; together with it, the De Grey Foundation hopes to be able to create – starting from the DNA of the patients – new organs through three-dimensional printing. Among the generous lenders of the Methuselah Foundation, we again mention Peter Thiel, who, in 2006, donated three and a half million dollars.

SENS Research Foundation

In 2009, De Grey created another foundation, the SENS Research Foundation, located in Mountain View, California.[34] The new organization has taken over a large part of the research activity of the Methuselah Foundation, and aims to promote the SENS project through the world of mainstream research. As we shall see, the latter represents the most organic proposal of the Transhumanists to defeat aging and – although they say it in a low voice, so as not to unnerve the general public – death. The SENS Research Foundation therefore has a dual purpose, involving both education and research. The latter, besides being carried out at the institute, is also carried out in collaboration with various American and non-American universities, such as Yale University, Harvard University and Cambridge University. The research activity of the foundation is divided into seven different initiatives, one for each of the types of biological damage of which, according to

---

[32] www.methuselahfoundation.org

[33] http://www.organovo.com/

[34] http://www.sens.org/

De Gray, the aging process consists; for each of these, the foundation always maintains at least one or two active projects. A curiosity: among the financiers of the SENS Research Foundation is Justin Bonomo, a well-known professional poker player, who decided to donate five percent of his winnings to the organization.

Foresight Institute

An important piece of the history of Transhumanism, the Foresight Institute is an organization that promotes nanotechnologies, or rather the creation and use of molecular assemblers, the nano-machines capable of manipulating matter at the atomic and molecular level theorized by K. Eric Drexler.[35] And, indeed, it was the latter who founded this institute in 1986, establishing its headquarters in Palo Alto and beginning the usual activism of Transhumanist associations, from conferences to publications to mailing lists. Among the prizes awarded by the Foresight Institute, there are the Feynman Prize, which includes several theoretical and experimental categories, plus the Feynman Grand Prize – which amounts to 250 thousand dollars for those who create two molecular machines capable of accurate nanometric positioning and computation. At the same time as the foundation, two twin organizations were also created, the Institute for Molecular Manufacturing and the Center for Constitutional Issues in Technology.

Center for Responsible Nanotechnology

Located in Menlo Park, California, the Center for Responsible Nanotechnology[36] is a think tank founded in 2002 by Mike Treder, a biologist, and Chris Phoenix, a Californian nanotechnologist, with the aim of analyzing the social implications and, above all, the risks associated with nanotechnology – with the idea that the nano-machines theorized by Drexler are destined to arrive much sooner than we believe and that it is essential that we are not being caught unprepared. Phoenix and Jamais Cascio are currently managing the think tank.

Machine Intelligence Research Institute

The idea that techno-scientific progress is accelerating, and that this will bring our world to a Technological Singularity, has also produced some dedicated think tanks. The Machine Intelligence Research Institute certainly stands out among them.[37] Founded in 2000 in Berkeley, California, the organization aims to develop

---

[35] http://www.foresight.org/

[36] http://crnano.org/

[37] http://intelligence.org/

an artificial intelligence that does not present risks to humans – for example, one that cannot lead to a "Terminator scenario." The institute is based on the model of friendly Artificial Intelligence elaborated by co-founder Eliezer Yudkowsky. Initially known as the Singularity Institute for Artificial Intelligence, the think tank has, as its executive director, Luke Muehlhauser, and, among its members, Nick Bostrom, Aubrey De Grey, Peter Thiel and Ray Kurzweil – in turn, executive director from 2007 to 2010. The idea on which Yudkowsky and his colleagues work is that of the so-called Seed AI, an artificial intelligence capable of progressively improving its design, up to a super-human intelligence that, in the Institute's intentions, should already be calibrated to be constitutionally friendly. Every year, the institute holds a conference, the Singularity Summit,[38] an event held for the first time in 2006, at Stanford University, which was also among the sponsors.

Singularity University

In spite of the name, Singularity University[39] is not an accredited academic body, but a charity – at least, it is registered as such – that aims to provide courses that complete the traditional American academic path – in the Transhumanist sense, of course. Founded in 2009 by Peter Diamandis – creator of the X-Prize Foundation – and Ray Kurzweil, it is located in Moffett Field, California, inside of the NASA Research Park facilities – practically in the Silicon Valley. The offering of courses – often introductory – is quite rich, from future studies to management, from bio- to nanotechnology, from robotics to medicine. The teaching staff includes experts from various sectors and many of the most popular Transhumanists. Academic programs range from a ten-week postgraduate course – from June to August, for a modest sum of twenty-five thousand dollars – to an intensive program for managers, along with many other initiatives. Added to this are the Singularity University Labs, a start-up connected to Singularity University that wants to act as an incubator and a meeting point for local start-ups, offering them entrepreneurial training courses and meeting places that stimulate the advent and use of futuristic technologies. The list of lenders to the institute is not bad at all: from Google to Nokia, from Autodesk to LinkedIn, from the X-Prize Foundation – those who want to promote private journeys in space – to Genetech.

Acceleration Studies Foundation

Created by the American futurist John Smart and located in Mountain View, California, the Acceleration Studies Foundation is a non-profit organization that, starting from the premise that the technological evolution of our society is

---

[38] http://singularitysummit.com/
[39] http://singularityu.org/

accelerating, proposes to map this process, developing strategies to favor the most promising and positive technologies and to slow down or, in any case, control the potentially dangerous ones. Among the donors is the ubiquitous Peter Thiel.

Alcor and Her Sisters

And then there are the cryonics organizations, which we will only give a brief mention to here, because we will deal with them more extensively in another chapter. If you like the idea of living forever, but you do not think you'll make it, then you can opt for a "cryonics suspension contract," in which you essentially choose to get frozen at the moment of your clinical death, in the hope that, in a more or less distant future, post-human science can bring you back to life. So, in Scottsdale, Arizona, we have the Alcor Life Extension Foundation,[40] the largest of such organizations, while, in Michigan, there's the Cryonics Institute,[41] founded by none other than Robert Ettinger, the father of the entire cryonics movement. The American Cryonics Society is based in Cupertino, California[42]; associated with the Cryonics Institute, it is primarily intended for educational and research purposes. Also in California, Trans Time Inc.[43] is based in San Leandro, while, near Moscow, you have the only cryonics suspension facility in the world outside of the United States, KrioRus.[44]

Besides the actual cryonics organizations, there are other collateral associations, such as the Immortalist Society,[45] also created by Ettinger, also located in Michigan and also serving both educational and research purposes. Another collateral organization, for information purposes – the collection and distribution of information on cryonics and longevity, as well as the establishment of structural relationships with all Transhumanist organizations working on life extension – is the Immortality Institute,[46] founded in 2002 by Bruce Klein. The institute in question has also sponsored small-scale experiments – through fundraising among members – and promoted a multivitamin preparation, Vimmortal, developed through crowdsourcing – practically gathering all of the suggestions of the participants in the associative forum. Despite their rather explicit mission – "to defeat the blight of involuntary death" – in 2011, the association chose to change its name to LongeCity, to prevent the reference to immortality from producing skepticism in the scientific community and the general public. Finally, we have the interesting

---

[40] http://www.alcor.org/

[41] www.cryonics.org

[42] americancryonics.org

[43] http://www.transtime.com/

[44] www.kriorus.ru/en

[45] http://immortalistsociety.org/

[46] http://www.longecity.org/forum/page/index.html

Brain Preservation Foundation,[47] founded by Kenneth Hayworth, a connectomics expert – that is, the exploration and mapping of nerve cell connections – at the Howard Hughes Medical Institute, and by John Smart. In this case, the goal is to promote scientific research on the preservation of the entire brain in the long term, preserving its structure at the nanometric level. As in other cases, here, too, there is a prize, the Brain Preservation Technology Prize, dedicated to those who manage to develop an inexpensive, reliable and usable surgical procedure for the preservation of 99.9 percent of the structural connections of the synapses of our brain.

Ascender Alliance

One organization that has particularly intrigued us is the Ascender Alliance – which had a dedicated discussion group on Yahoo that has now disappeared. Founded by British futurist and Extropian Alan Pottinger, it is the first Transhumanist association dedicated to disabled people, with a mission to remove political, cultural, biological and psychological limits to self-realization and empowerment. However, the Ascender Alliance strongly opposes not only classical eugenics – also rejected by other Transhumanists – but also any permanent modification of the human genome. In essence, Pottinger wants respect for the specificity of people with disabilities, and wants any change or enhancement to be the result of a conscious choice by each person, not a decision by others imposed before conception itself – or, to put it as he does, the right to self-determination begins before conception. The only case in which pre-natal manipulation is acceptable is when there is a need to prevent physical and mental deficits that endanger the life of the unborn child. If the disability must be canceled, it must be in the manner chosen by those directly involved. These are thorny issues, as we can see. It is, however, interesting to note, as Dvorsky argues, that disabled people who are put in a position to compensate for their disability through technology can be seen as a prime example of post-humans; or, better yet, from the point of view of enhancement, they are at the forefront, as they have a more open mind towards the man-machine mix, and may therefore be among the greatest supporters of high-tech human enhancement in the future. In addition, the disabled, accustomed as they are to using articulated devices, may be more favorable than the average person to body changes that move significantly away from the usual commonly accepted aesthetic standards.[48]

---

[47] http://www.brainpreservation.org/

[48] G. Dvorsky, *And the Disabled Shall Inherit the Earth*, 15 September 2003, in: Sentient Developments   http://www.sentientdevelopments.com/2003/09/and-disabled-shall-inherit-earth.html

Transtopia

Have you ever dreamed of retiring to an island of your own, where you can live in peace? No? Well, Transhumanists did it. Utopian by nature, Transhumanism could not ignore traditional utopian thought and all of the thinkers who, historically, have tried to devise a perfect society, as well as all of those people who have actually tried to realize these communities in various parts of the world. The Transtopia Island Project[49] essentially involves the grouping of a certain number of volunteers and the collective purchase of an island in the Bahamas or somewhere else, in order to create am international Transhumanist community – both for fun and to make some money, and, finally, to guarantee a place of survival in case the rest of the world falls into chaos. If things go well, then you can think about expanding the community, creating a floating island – in practice, an artificial platform – that leads to the formation of a "micronation," a term that indicates very small territorial entities – sometimes simple platforms abandoned by oil companies – proclaimed, at a certain point, independent and normally not recognized by the international community.[50] As an alternative to this kind of micronation, the Transtopia Island Project has thought of the so-called "Freedom Flotilla"; the plan, in this case, involves the purchase of several ships and their registration under the banner of some off-shore haven that guarantees freedom of action. In practice, it would aim to achieve a sort of "permanent mobile base," We have not been able to gather much data on the Transhumanists who adhere to this project; it seemed to us that this is a relatively marginal movement compared to mainstream Transhumanism, one that has chosen the name of "Prometheism"[51] for itself. As a matter of fact, this group is close to the survivalists, or "preppers," a very diverse international movement, present mainly in Great Britain and the United States – focused on surviving all sorts of catastrophe, those both natural and of human origin. Developed primarily during the 1960s – in the face of the nuclear threat – this movement includes several magazines, courses, techniques, books, and reference authors – including the science fiction writer Jerry Pournelle. There are many adherents to what is a true philosophy of life: from the most extreme libertarians – adversaries of every form of state and lovers of isolation – to groups of the American right, to Christian fundamentalists waiting for the end of the world, to conspiracy theorists. Manuals are particularly numerous, including books – often rather politically incorrect – that explain how to falsify one's own documents, how to escape this or that surveillance technology, how to

---

[49] http://www.transtopia.net/

[50] There are, or there have been, many micronations, some of which have issued their own currency. A list of micronations can be found on Wikipedia or, if you prefer, at the following address: www.dmoz.org/Society/Issues/Micronations/

[51] www.prometheism.net

prepare for a nuclear disaster, and much more.[52] Anyway, the idea of a utopian island community must please even the most orthodox Transhumanists, since it is a mainstream Transhumanist like Ben Goertzel who promotes it; the latter has recently proposed to gather fifteen thousand Transhumanists, organize a substantial collection of funds and use them to repay the debts of Nauru, a small independent island in the South Pacific. In return – at least in Goertzel's intentions – the inhabitants of the island will agree to host the Transhumanists in question, who will be able to rearrange the small nation and transform it into Transtopia.[53]

Biocurious & Co

When they were born, computers were huge devices that filled entire rooms and could only be managed by people with specific skills. Then, technological progress not only shrank them, but also made them easier to use, so much so that, today, almost everyone knows how to handle a PC, at least rudimentarily. And now, it seems that the same is happening with biotechnology: the price of laboratory instruments has collapsed – especially in the case of second-hand ones, which can also be found on eBay or on specific websites – and genetic manipulation technologies – at least the most basic ones – have been simplified and standardized to such an extent that some have decided to turn them into a hobby. And so, several private citizens decided to set up a biotech laboratory in their garage, more or less equipped, to perform this or that operation – such as the creation of fluorescent bacteria, to be used for "living" works of art so as to feel like a part of the biotech revolution in progress. Thus, a movement was born, that of "garage biology," also called DiyBio, Do-It-Yourself Biology: people without particular academic qualifications – sometimes university students of biotechnologies – who, alone or in small groups, modify simple forms of life, perhaps cultivating the dream of sooner or later starting a business. The "official" year of birth of the movement, if we want, was 2008, when two biology garage practitioners in Boston, Mackenzie Cowell and Jason Bobe, created DiyBio, a network that brings together amateurs, artists and entrepreneurs; the organization then spread globally, leading to, among other things, the birth of several other American offshoots, such as Biocurious[54] in San Francisco and Genspace or Biotech Without Borders in New York. Garage biology maintains clear links with the world and the language of hackers – just think of the term that DiyBio practitioners use to refer to

---

[52] If you want to know more about these topics, you can take a look at the Survival Preparedness Index. Cf. http://www.armageddononline.org/disaster-prep-help.html

[53] B. Goertzel, *Let's Turn Nauru Into Transtopia*, October 13, 2010. http://multiverseaccording-toben.blogspot.it/2010/10/turning-nauru-into-transtopia.html

[54] http://biocurious.org/

themselves, that is, "biopunk," along the lines of cyberpunk. Some of these garage biotechnologists aim one day to create a biotech version of file-sharing programs like E-mule or Torrent, but centered on genome manipulation – while others work on creating open-source lab tools. Of course, the dangers are not lacking – even if they are not high, since dangerous viruses are certainly not accessible to hobbyists – and, for this reason, Biocurious in San Francisco have asked for and obtained the creation of common and controlled areas in which the amateurs of Do-it-yourself biology can take care of simple manipulations in complete safety, keeping potential bioterrorists away. At the moment, amateurs can test the DNA taken from their own saliva or modify simple microorganisms – do not therefore expect the cure for cancer or the like from these amateur laboratories; in the future, however, they may be able to do more useful things, such as withdrawing and storing their stem cells for therapy. And what do the Transhumanists have to do with the bio-hackers? The fact that the two movements often tend to overlap – that is, the members of one are often in contact or identify with those of the other – and that do-it-yourself biology represents the Transhumanist desire not to leave one's biological destiny in the hands of others, but rather to try, in some way, to take control of it.

Order of Cosmic Engineers

This is certainly the most visionary – some would call it "crazy" – Transhumanist organization. The Order of Cosmic Engineers[55] is a group of activists that aims to "transform the universe into a magical realm," or, better yet, to "engineer magic in a universe currently devoid of God or gods." Atheists, in short, but eager to create a God in a more or less distant future. The underlying idea is actually to defend the radicalism of Transhumanism against any attempt to make it "fly low," to make it politically correct, moderate, focused only on the problems of the present. In short, rather than a group that *seriously* seeks to create God and spread magic throughout the universe, these Cosmic Engineers seem rather intent on nurturing and maintaining the visionary and imaginative aspect of Transhumanism, keeping it alive and vital. And their conceptual starting point is the all too often quoted "third law" by Arthur C. Clarke: "any sufficiently advanced technology is indistinguishable from magic."[56] Their vision of the human condition is apparently pessimistic – indeed, it's downright nihilistic: for the Engineers, human beings find themselves abandoned in a small watery corner of a cold, cruel, indifferent cosmos – when it isn't being openly hostile. Contrastingly, their objectives – those

---

[55] http://cosmeng.org/. Currently (2019), the site does not seem to be active anymore.

[56] See A.C. Clarke, *Hazards of Prophecy: The Failure of Imagination*, in: *Profiles of the Future: An Inquiry into the Limits of the Possible*, Harper & Row, New York 1973, pp. 14, 21, 36.

declared, at least – are extremely ambitious: the infusion of intelligence into inanimate matter, throughout the universe, which will be optimized for computation; the answer to all final questions about the origin, nature, purpose and fate of reality; the creation of new universes with controlled physical parameters. In spite of the fact that the language adopted by our Cosmic Engineers recalls that of an initiatory group along the lines of the Rosicrucian Order or Freemasonry – for example, the members of the board have assumed the title of "architects," as if to recall the "Great Architect" of the "Masonic" universe – they reiterate that they are neither a religion nor a belief, a cult or a sect. On the contrary, they define a middle ground between a Transhumanist association, a spiritual movement, a group advocating for space enterprise, a literary salon and a think tank. And yet they certainly resemble a "non-religion," offering the same advantages of religion, without the negative and obscurantist aspects.

2.5 Transhumanism, That Is, Never Let Anything Remain Untried

What's the purpose of obtaining eternal life if an asteroid hits you in the head, or you get erased by any other event that could make us end up like the dinosaurs or, even worse, destroy our fragile planet? A reflection of this kind could not help but come to mind to Transhumanists as well, people who do not like bad surprises and tend to be more proactive than average, evaluating every possibility. And here is their answer: the Lifeboat Foundation. We decided to dedicate a separate section to this organization, because it seems representative of a typical aspect of the Transhumanist mentality, which is worth underlining: the habit of *thinking about any possibility*.

The Lifeboat Foundation[57] is a think tank that, although not expressly Transhumanist, actually includes among its ranks almost all of the "top" people of the movement – as well as a large number of researchers more or less known in every field of science and technology. Among the Transhumanists, we have Michael Anissimov, José Luis Cordeiro, Aubrey de Grey, Robert A. Freitas Jr., George P. Dvorsky, Terry Grossman, J. Storrs Hall, Ray Kurzweil, David Pearce, Michael Perry, Giulio Prisco, Martine Rothblatt, Eliezer S. Yudkowsky, Natasha Vita-More, and Riccardo Campa. The list of "mainstream" scientists, philosophers and scholars includes, among others, Cristiano Castelfranchi, Patricia S. Churchland, Robert Cialdini, Daniel Dennett, Stanislav Grof, Nobel Prize winner Daniel Kahneman, Louise N. Leakey, Michael Shermer, Peter Singer and Stephen Wolfram. The list is also rich in science fiction writers, with names like Catherine Asaro, Greg Bear, Gregory Benford, David Brin, David Gerrold, James

---

[57] http://lifeboat.com/ex/main

Gunn, Ian MacDonald, Jerry Pournelle, Robert J. Sawyer, Allen M. Steele, and Fred Alan Wolf.

This organization was founded – after the events of September 11, 2001 – by Eric Klien. A long-time cryonics supporter and member of Alcor, Klien has long worked as a stockbroker, accumulating a certain amount of money, which allowed him to withdraw, in part, from business. A supporter of a libertarian vision of the world and politics, Klien launched the Atlantis Project in the 1990s, with the aim of creating an independent floating city in the Caribbean Sea, which was to be called Oceania; after having initially arousing strong media interest, the Atlantis Project ended up in oblivion, and Klien decided to abandon it.[58]

And so we come to the Lifeboat Foundation, a non-profit organization that aims to encourage scientific research and reflection on the "existential risks" theorized by Bostrom, so as to develop adequate protocols to prevent any kind of catastrophe. The headquarters are located in Minden, Nevada, but the Lifeboat Foundation has branch offices virtually everywhere.

The organization consists of a huge number of research programs, which cover *any* type of risk, from the best known – nuclear war, destruction of the ecosystem, killer asteroids, and so on – to those specifically feared by Transhumanists – like a Technological Singularity hostile to Man – to those more classically science fiction-like – alien invasions, black holes, and so forth. In this case, the programs are divided into two groups, the "shields," which aim to avoid or stop various catastrophes, and the "preservers," which instead aim to ensure the survival of life and our civilization in the event that a catastrophe cannot be prevented. Let's take a look.

AIShield is the program specifically dedicated to preventing the advent of a hostile artificial super-intelligence; we do not know if this is possible, but in the case that it was – and since Murphy's Law almost enjoys the status of a physical law – better to be prepared, working to prevent a scenario like *The Terminator* or *The Matrix*. We cannot exclude the possibility that a fully automated society managed by artificial intelligences will end up bringing our species to a level of laziness never seen before – and this too is a subject discussed by the members of this program.

AsteroidShield, as you can imagine, aims to prevent a celestial body like the one that destroyed the dinosaurs from striking us again. It is not a risk to be underestimated: if, in fact, an asteroid, even a relatively small one, directed towards the Earth were to escape our detection systems, it would not just be Houston that would have a problem. So much so that the Lifeboat Foundation is not the only organization to worry about this risk. BioShield aims to protect humanity from biological weapons and pandemics that regularly affect the planet; since the Net is

---

[58] http://oceania.org/

assuming a fundamental importance for any type of activity, InternetShield wants instead to develop procedures to defend it from possible attacks or from a possible collapse.

The LifeShield Bunkers program aims, as you can guess, to develop shelters that are bomb-proof – indeed, everything-proof. Very detailed, this program aims to develop possible bunkers in the style of Biosphere 2, able to permanently host families with children. It does not end there, however: LifeShield bunkers also provide for the development of measures and procedures capable of transforming public buildings of various kinds, above all hospitals, as well as offices and houses, into structures capable of surviving possible disasters, with the addition of other proposals, such as the use of gene therapy to enhance our resistance to radioactivity.

ScientificFreedomShield recognizes the importance that scientists with radical and controversial ideas have had for the history of science and technology, and it proposes to promote this source of innovation, protecting it from the blocs that a bureaucratic mentality – typical of much research today – could impose. NeuroethicsShield serves to prevent abuse – above all of a political nature – in the fields of neurotechnology and neuropharmacology, i.e., areas that could prove more and more intrusive and dangerous for freedom and self-determination in the near future.

NanoShield represents the response of the Transhumanists to one of the threats deriving from the futuristic technologies that they themselves propose, that of nano-machines; devices that could one day get out of hand, literally starting to devour and reshape the entire planet. The solutions proposed range from the development of a nanotech immune system to be inserted into ourselves, in order to make us invulnerable to this threat, to the creation of nanotechnological filters able to sift and clean the atmosphere.

While NuclearShield proposes to develop new systems of protection and evacuation to protect our cities from nuclear explosions as much as possible, ClimateShield aims not only to monitor climate change, but also to develop that frontier area of research known as "climatic engineering," that is, a set of procedures literally apt to control the climate. Which means learning to produce rain on command, dissolving cyclones, diverting hurricanes, managing the global hydrosphere, controlling the seasons, using orbital screens to control sunlight, and using swarms of nano-machines to clean the air of the particulates. In short, at the moment, it is a decidedly sci-fi goal, but it is not to be dismissed that, if climatic engineering progresses in a constant way, sooner or later, Man will arrive at a direct and conscious management of the earth's climate.

But there are even more visionary projects. For example, AlienShield, which wants to prevent the annihilation of humanity due to an alien race. The destruction could be involuntary: in fact, if an alien species manages to reach us, this implies that it has a technology more advanced than ours, and history teaches us that when a less advanced human civilization meets a more advanced one, the former ends up losing its cultural autonomy and being assimilated. In addition to developing a strategy to prevent the destruction of war and cultural erasure, the program in question aims to develop an adequate protocol to manage a possible "first contact."

As the name implies, the ParticleAcceleratorShield aims to prevent possible catastrophes related to the use of near-future particle accelerators, including the unlikely formation of artificial mini-black holes and other amenities related to high-energy physics. AntimatterShield aims to prevent the annihilation of our planet through antimatter weapons; likewise, BlackHoleShield wants to monitor the skies for any black holes that threaten us, and develop procedures – at the present time unthinkable – to prevent this threat. Then, we have GammaRayShield. Gamma ray bursts are very powerful bursts whose duration ranges from a few milliseconds to several tens of minutes; it is the most powerful energy emission in the universe. Finally, we have SunShield, which aspires to develop a permanent monitoring system for the Sun and devise possible solutions for when, in our distant future – in five billion years or so – it will turn into a red giant, probably swallowing a few planets in the process. Even here, the Lifeboat Foundation members are not devoid of ideas – from moving Earth to transferring humanity elsewhere – and the fact that the threat in question is so remote demonstrates how far-sighted the Transhumanists are. On the other hand, it is certainly possible that some of them are planning to still be around in that remote future, so – from this point of view – to worry far in advance certainly makes sense.

And then, as we said, we have the preservers. SecurityPreserver wants to prevent terrorist, biotechnological and nanotechnological attacks using the most modern forms of high-tech surveillance. During the course of its long history, our planet has witnessed five great mass extinctions, events in which the disappearance of a very high percentage of living *species* occurs. The causes are various and, according to some scholars, the Earth is heading towards a sixth great extinction, in which, this time, Man would play a central role.[59] For this purpose, the Lifeboat Foundation has launched the BioPreserver program.

CommPreserver aims to develop new communication systems capable of surviving catastrophic events that would destroy traditional ones – as would be the case with nuclear explosions. EnergyPreserver obviously aims to study the possible solutions to our energy needs of the future. InfoPreserver, a particularly meritorious initiative, wants to preserve – along the lines of other non-profit

---

[59] R. Leakey; R. Lewin, *The Sixth Extinction: Patterns of Life and the Future of Humankind*, Anchor, New York 1996.

organizations such as the Alliance to Rescue Civilization[60] and the Long Now Foundation[61] – all of the information created by our civilization, such as art, culture, and scientific knowledge. LifePreserver wants to promote the study of one of the primary topics of Transhumanism, life extension. PersonalityPreserver, on the other hand, is a program that examines all of the possible ways to preserve people – or even individual personalities – including the favorite techniques of Transhumanists, i.e., cryonics and mind-uploading. SeedPreserver aims to preserve life and bio-diversity, promoting initiatives similar to that planned by Norway and five other countries bordering the Arctic – that is, the construction of a "Noah's Ark" located in a disused mine in the Svalbard islands with the aim of preserving millions of different seeds from all over the world, thus sheltering them from possible catastrophic events. Finally, the Space Habitats program aims to encourage human expansion into space, both through the colonization of other planets and the construction of autonomous spatial habitats. In this regard, the Lifeboat Foundation is working on Ark I, a project for the development of a self-sufficient space habitat able to guarantee the survival of humanity in the event that the Earth becomes uninhabitable.

2.6   The *Who's Who* of Transhumanism

After this long theoretical and organizational introduction, it's time for a nice round of presentations, taking into account the fact that, among the Transhumanists, there is a bit of everything: philosophers, especially the analytical variant, economists, entrepreneurs, futurologists, artists, journalists, science fiction writers, such as Greg Bear and Gregory Benford, theorists of artificial intelligence, computer science scholars, and even actors – like William Shatner, who you will certainly know from his role as Captain James T. Kirk of *Star Trek*. Transhumanism is now a global movement, and includes numerous representatives. Put another way: Transhumanists number many, many more than I thought at the beginning of this work; some we have already met, and we will meet others. In this section, we will indicate – according to our humble, but unquestionable judgment – the most important. Let's meet them, then, these theorists of the future of human evolution, and try to understand what kind of people they are.

---

[60] This organization, founded by the journalist William E. Burrows – also a member of the Lifeboat Foundation – and the biochemist Robert Shapiro, proposes the creation of a "backup" system of our civilization on the Moon. See http://arc-space.wetpaint.com/

[61] The Long Now Foundation, created in 1966, aims to promote a mentality based on the very long term. See http://longnow.org/

FM-2030

Let's start with a dean of the movement, a Transhumanist even in the name he has chosen for himself: FM-2030. Born in Brussels in 1930, and dying in New York in 2000, Fereidoun M. Esfandiary distinguished himself during the '60s/'70s for his unconventional positions, for which he coined the term *up-winger*. The use of the prefix *up* would represent, for Esfandiary, the truly radical nature of his thought, aimed precisely at the promotion of human technological development – utilized in, among other things, a specific book, *Up-Wingers: A Futurist Manifesto*.[62] And the choice, in the mid-'70s, to legally change his name to FM-2030, indicates Esfandiary's aspiration to celebrate his 100th birthday – in a year in which, according to him, thanks to scientific progress, everyone would enjoy the possibility of living forever – and be freed from the tribal conventions linked to the traditional attribution of proper names. In 1970, Esfandiary published *Optimism One. The emerging radicalism*,[63] in which he argues, among other things, that, after the present Space Age, a Temporal Age will occur, in which death will be transformed from an imperative into a solvable problem – probably within thirty to forty years, in his too optimistic estimation – thus guaranteeing us freedom from the pressure of time. To those who challenge him on the non-naturalness of immortality, Esfandiary replies that, if dying is natural, then to Hell with Nature. The defeat of death represents, for him, the next evolutionary step, through which we will no longer be at the mercy of time. 1989 saw the publication of *Are You a Transhuman? Monitoring and Stimulating Your Personal Rate of Growth in a Rapidly Changing World*, a book in which Esfandiary deals with the nascent anthropological type of Transhumans, who are the first manifestation of a new evolutionary stage. Among the signs of Transhumanity, FM-2030 includes the use of prosthetics, cosmetic surgery and reproduction *in vitro*, the intense use of telecommunications, the absence of religious beliefs, and so on.[64] Unfortunately for him, his dream of becoming immortal was cut short by a pancreatic tumor, and his body is now cryo-preserved at Alcor's facilities. FM-2030 used to say: "I am a twenty-first century person accidentally thrown into the twentieth. I have a deep nostalgia for the future." Finally, we would like to underline the fact that, far from being a visionary lacking ties with the official culture, Esfandiary taught at various universities, including New York's prestigious New School for Social Research and UCLA.

---

[62] F.M. Esfandiary, *Up-Wingers: A Futurist Manifesto*, John Day Company, New York 1973.

[63] F.M. Esfandiary, *Optimism One. The Emerging Radicalism*, Norton & Company, New York 1970.

[64] F.M. Esfandiary, *Are You a Transhuman? Monitoring and Stimulating Your Personal Rate of Growth at a Rapidly Changing World*, Warner Books, New York 1989.

Robert Ettinger

The father of cryonics, Ettinger was born in 1918, died in 2011, and, in accordance with the vision he promoted, immediately underwent cryonic suspension procedures. Nourishing himself with strong doses of science fiction as a boy, Ettinger believed early on that science was in the process of defeating aging and death. While coming of age, the young *ante litteram* nerd, began to realize that this venture would take too long; fortunately, science fiction literature came to his rescue, in the form of a story he had read a few years earlier, *The Jameson Satellite*, written by Neil R. Jones and published in July 1931 in *Amazing Stories*, the first science-fiction magazine – founded by Hugo Gernsback. The short novel tells the story of a certain Professor Jameson, whose corpse – launched into space and stored for millions of years – is recovered and brought back to life by a powerful alien race. In 1947, while he was hospitalized for war wounds, Ettinger became aware of cryogenics – the study of the production of very low temperatures and the effects of these on organic and inorganic materials – and, in particular, of the research carried out in that field by the French biologist Jean Rostand. In 1948, Ettinger published a short story in *Startling Stories*, called *The Penultimate Trump*, in which he proposes, for the first time, the paradigm of cryonics: the criteria for determining the death of a person are partially relative, the corpses of today could be the patients of tomorrow, and therefore it is important to use cryogenics to preserve their bodies until medical science has found a way to solve the problem. Now, the die had been cast; however, Ettinger would still need a few years to test his vision and get moving. In 1962, he privately published the first version of *The Prospect of Immortality*,[65] the book that would launch the cryonics movement; the well-known science and science fiction writer Isaac Asimov read the book and told Doubleday – a large publishing house, which commissioned him to read it – that Ettinger's theories were scientifically meaningful. In 1964, *The Prospect of Immortality* was republished, enjoying great success and being translated into many languages. Ettinger – whose pursuit of an academic career had led to his appointment as professor of physics and mathematics at Wayne State University in Detroit – became a celebrity, participating in various talk show and being interviewed by a number of magazines and newspapers, not all of them American. To be honest, Ettinger actually shares the title of father of cryonics with Evan Cooper,[66] who, in 1962, published, at his own expense, *Immortality: Scientifically,*

---

[65] Free download at the following address: http://www.cryonics.org/book1.html

[66] Cooper was a simple activist, and he was responsible for the formation of the first, true cryonics organization, the Life Extension Society; in 1969, however, he abandoned cryonics activism and, in 1983 – in a poetic way, in fact – disappeared at sea.

Physically, Now, a text devoid of technical and scientific rigor.[67] However, 1966 saw the birth of the Cryonics Societies of California and Michigan – with Ettinger being elected president of the latter. During the '70s, the father of cryonics transformed the company he presided over into two organizations, the Cryonics Institute and the Immortalist Society. The first patient to be cryo-suspended by Ettinger – in 1977 – was his mother Rhea.

Max More

And here we arrive at Max T. O'Connor. Born in Bristol in 1964, he graduated with degrees in philosophy and economics from the University of Oxford, obtained his PhD from the University of Southern California, and can be considered one of the leading contemporary Transhumanists. It was he who re-introduced, in a modernized form, the term "Transhumanism," in the 1990 article *Transhumanism: Toward a Futurist Philosophy*.[68] In More's view – vaguely inspired by that of Nietzsche – the end of religion has plunged us into a desperate nihilism, and Transhumanism represents an alternative to both of these visions of reality. The final aim is, obviously, to abolish the greatest of all evils: death. It certainly does not stop the progress of intelligent beings taken collectively, but it does obliterate the individual, depriving them of a meaningful existence. Individual death makes life meaningless, as it disconnects us from everything we value, whatever it is. Seen as an eternal recurrence of entities and events, not even the act of becoming has much meaning; a person can have it only if he/she has direction, that is, if he/she is directed towards the creation of a growing order – what More calls "extropy." Hence, the slogan typical of the Extropians: "forward, up and outward." Humanity must be seen only as a temporary stage in the evolution of life, and the time has come for us to take this process into our hands and accelerate it. To do this, however, we must free ourselves of a typically human trait, that is, our constitutive need for certainties, which naturally leads us to submit to religion or, in any case, to dogmas and ideologies; here, the solution proposed by More – very Transhumanist, no doubt about it – is to "re-engineer" our conscience, so that it can do without the desire for dogmatic certainty and withstand error and doubt, getting rid of blind faith once and for all. For the sole purpose of encouraging technological and scientific development in the direction he advocated, More elaborates, at a certain point, the "Proactionary Principle," as opposed to the classic Precautionary Principle. There is no unequivocal formulation of the latter, but there are some particularly influential ones, such as the one contained in the 1992

---

[67] See http://www.evidencebasedcryonics.org/ev-cooper-immortality-physically-scientifically-now/

[68] See. Http://www.maxmore.com/transhum.htm

Rio Declaration or the one from the 1999 Wingspread Declaration, which More uses: "When an activity raises threats of harm to human health or the environment, precautionary measures should be taken even if some cause and effect relationships are not fully established scientifically. In this context the proponent of an activity, rather than the public, should bear the burden of proof."[69] For More, if the Precautionary Principle had been applied in the past, technological and cultural progress would have remained completely blocked – think of practices such as water chlorination, the use of X-rays, or even just the development of mechanical means of transport. Precisely for this reason, the Proactionary Principle[70] includes not only anticipation *before* action, but also the philosophy of *learning by doing*. More also recommends risk assessment using a strictly rational method of analysis – thus excluding any concession to emotionality. The burden of proof goes to those who propose restrictive measures; it is also necessary to consider not only the possible damages, but also the losses due to the abandonment of a certain technology or a certain line of research. Priority must be given to the prevention of risks already well known, rather than only hypothetical risks; finally, the risks deriving from the technology must be treated in the same way as natural risks, avoiding diminishment of the latter or exaggeration of the former.

Natasha Vita-More

American, real name Nancie Clark, born in 1950, artist and designer, lecturer at the University of Advancing Technology of Temple, Arizona, and wife of Max More. Vita-More took an interest in matters relating to the future from an early age, but began to take this interest very seriously only after suffering an ectopic pregnancy.[71] Her academic qualifications include a master's degree in *future studies* from the University of Houston. A body-builder and nutrition expert, Vita-More represents an interesting cross between the worlds of art, Transhumanism and fitness – the latter being a subject about which many Transhumanists are passionate, in particular, as an anti-aging function. A member of many of the organizations related to the movement in question, she is currently Chairwoman of the Board of Directors of Humanity+. Apart from her artistic and academic works, Vita-More is very present in the media, never failing to use these occasions to promote the Transhumanist cause. It is with Natasha Vita-More that the notion of Transhumanist art officially begins. In 1982, she launched the *Transhumanist Arts*

---

[69] See http://www.sehn.org/wing.html

[70] See M. More, *The Proactionary Principle*, version 1.2, July 29, 2005, http://www.maxmore.com/proactionary.htm

[71] A pathological condition in which the embryo implant takes place in a different location from the uterus, generally with very low probability of success – and with possible risks, also considerable, for the mother.

Statement,[72] in which she proclaimed that Transhumanist artists want to extend life and defeat death, pursuing an infinite transformation and the exploration of the universe. The following year, she published the *Transhuman Manifesto*,[73] in which, among other things, she promotes a Transhumanist conception of values – centered on freedom and diversity – and so-called "morphological freedom," a "Transhuman right" concerning the freedom to modify one's own body at will. Among other things, her artistic production has included the development of *Primo Posthuman*, simultaneously a work of art and a theoretical model of how the post-human body could be redesigned.[74] According to Vita-More, the essence of human nature is the solution of problems through the development of ever new design methods; this idea is also at the center of her artistic perspective, which merges Conceptual Art[75] with the disciplines most loved by Transhumanists, namely, biotechnology, nanotechnology, robotics, neuroscience and information theory. The artist also recognizes other influences, such as Futurism and Dadaism. The artistic movement she launched has subsequently been divided into sub-genres, like the "Automorph" – the art of sculpting, in a conscious and complete way, one's own psychology and physiology – "Art as Being," in practice. The sub-genre "Exoterra" incorporates into the works of art – be they figurative, musical or otherwise – elements that refer to space, science fiction, and so on.[76] In 1995, 301 artists and scientists signed the *Statement*; also, during the '90s the association Transhumanist Arts & Culture – in fact, a fluid community of visual artists, performers, multi-media artists, directors, video-producers, scientists and techno-enthusiasts – became a hub for those who recognize themselves in the movement. On January 1, 1997, Vita-More launched a new Manifesto, the *Extropic Art Manifesto of Transhumanist Arts*, in which she declared: "I am the architect of my existence. My art reflects my vision and represents my values. It communicates the essence of my being. [...] As we enter the Twenty-first Century, the Transhumanist arts and the extropic art will pervade the universe around us."[77] In October of the same year, the Manifesto was launched into space on board the Cassini Huygens probe, en route to Saturn.

---

[72] http://www.transhumanist.biz/transhumanistartsmanifesto.htm

[73] http://www.transhumanist.biz/transhumanmanifesto.htm

[74] http://www.natasha.cc/primo.htm

[75] Conceptual Art is an artistic movement that has developed in the United States since the 1960s, starting from the premise that ideas and concepts embodied in artistic work take precedence over traditional aesthetic and material considerations. Conceptual Art is linked to the practice of "installations" – as the artistic works produced are called – which can often be built by anyone, following a *set* of written instructions.

[76] See http://www.transhumanist.biz/

[77] See http://www.transhumanist.biz/extropic.htm

Kim Eric Drexler

American engineer, class of 1955, Drexler is considered one of the greatest gurus of Transhumanism, and, in particular, of that part of the movement – absolutely majoritarian – that relies on nanotechnologies as a means to realize their own post-human dreams. During the '70s, while he was a student at the Massachusetts Institute of Technology, Drexler began to develop his ideas on molecular nano-technology. At that time – and influenced in this by the famous report by the Club of Rome, *The Limits to Growth* – he also became interested in the issue of resources available outside of the Earth, and came into contact with Gerard K. O'Neill, a physicist at the University of Princeton known for his theoretical work on the colonization of space. All of this led to Drexler's involvement in the L5 Society, an association for the promotion of space colonization, which, in 1980, managed, *inter alia*, to prevent the ratification by the US of the Treaty on the Moon, which, according to the association in question, provided restrictions that would prevent the creation of extraterrestrial colonies. But nanotechnologies would be Drexler's true vocation, and, in particular, the theoretical study of molecular nano-machines and all of their possible applications. This passion would lead him to publish the 1986 book *Engines of Creation: The Coming Era of Nanotechnology*,[78] with a preface by Marvin Minsky, illustrating the risks and opportunities related to nano-technologies, including the famous *gray goo,* which consists in the danger that out of control nano-machines might devour the planet and reduce it precisely to the state in question. In the same year, Drexler founded, together with his then-wife Christine Peterson, the Foresight Institute, whose *mission* is, very laconically, that of "preparing for nanotechnology." In 1991, he published, with Peterson and Gayle Pergamit, *Unbounding the Future*, in which he deepened the application scenarios of his molecular nano-machines.[79] In 1992, *Nanosystems: molecular machinery, manufacturing, and computation* came out, a very technical book taken – with some adaptations – from Drexler's doctoral thesis. 2007 saw publica-tion of *Engines of Creation 2.0: The Coming Age of Nanotechnology – Updated and Expanded*, a new version of the book that made him famous.[80] In 2013, he published *Radical Abundance: How a Revolution in Nanotechnology Will Change Civilization*, which, right from the title, bodes well for human destinies. In this regard, it is interesting to note that Drexler espouses one of the variants of the Fermi Paradox – in which it is wondered why, if the universe is swarming with intelligent life forms, no one has ever contacted us – and argues that there proba-bly aren't evolved civilizations in the neighborhood. As a result, the resources of

---

[78] http://e-drexler.com/p/06/00/EOC_Cover.html

[79] http://www.foresight.org/UTF/download/unbound.pdf

[80] http://www1.appstate.edu/dept/physics/nanotech/EnginesofCreation2_8803267.pdf

the vast universe are all for us, and, by combining space travel with nanotechnology, we can get around these limits to potentially infinite growth.

Hans Moravec

Austrian, born in 1948, Moravec works at the Robotics Institute of Carnegie Mellon University, Pittsburgh. Very well known for his work in the field of robotics and artificial intelligence, he is also one of the gurus of Transhumanism, in particular, for his scenarios related to the future of robots. In 1988, he published *Mind Children*,[81] in which he argues that robots will evolve into several new species from the '30s and '40s of the twenty-first century. In *Robot: Mere Machine to Transcendent Mind*,[82] from 1998, he further analyzes the possibility that a robotic intelligence will develop within a short time and that a rapidly expanding superintelligence will emerge from it.

James J. Hughes

A sociologist and bioethics scholar, James Hughes teaches health policy at Trinity College in Hartford, Connecticut. A former Buddhist monk, he was executive director of the World Transhumanist Association from 2004 to 2006, and now holds a similar role at the Institute for Ethics and Emerging Technologies, which he co-founded with Nick Bostrom. In 2004, he published *Citizen Cyborg: Why Democratic Societies Must Respond to the Redesigned Human of the Future*, in which he presents his vision of "Democratic Transhumanism."

David Pearce

A British utilitarian philosopher, Pearce has, as his main point of reflection, the abolition of every form of suffering from every sentient creature, human or animal. To achieve this goal, Pearce proposes the use of genetic engineering, pharmacology and nanotechnology to erase any unpleasant experience, within the framework of what he calls "Paradise Engineering" – a conception detailed in his online manifesto *The Hedonistic Imperative*.[83] An animalist and a vegan, the philosopher believes that our post-human descendants will have to take responsibility for the elimination of suffering, even among wild animals. In 1995, Pearce created

---

[81] H. Moravec, *Mind Children: The Future of Robot and Human Intelligence*, Harvard University Press, Cambridge 1988.

[82] H. Moravec, *Robot: Mere Machine to Transcendent Mind*, Oxford University Press, New York 1998.

[83] http://www.hedweb.com/hedethic/tabconhi.htm

a website, BLTC Research,[84] in which he collects and proposes materials related to biochemical and biotechnological methods for the abolition of suffering. In 2002, he founded the Abolitionist Society,[85] in order to promote his own Transhumanist view of suffering.

Gregory Stock

A biophysicist and biotech entrepreneur, Stock is a complex character, and, although he cannot be called a Transhumanist *tout court*, he still holds positions very close to those of the Transhumanists. In short, a strong sympathizer and, moreover, one well positioned at the academic and political levels. Founder and former director of the Medicine, Technology and Society Program at the University of California School of Medicine, Los Angeles, Stock has long taken heed of the ethical-political and evolutionary implications of avant-garde research sectors such as biotechnology and computer science. Former CEO of Signum Biosciences – a biotech company that works to develop therapies for Alzheimer's, Parkinson's and other diseases – he is a member of the California Advisory Committee on Stem Cells and Reproductive Cloning, as well as associate director of the Bioagenda Institute and the Center for Life Science Studies at the University of California at Berkeley. His is a highly respected curriculum, to which is added the fact that, in 1998, Stock organized a fundamental conference on life sciences at UCLA dedicated to the theme of *Engineering the Human Germline*, which was attended by the likes of James Watson. Another important conference organized by the scholar at UCLA in 1999, *Milestones on Aging*, has helped scientifically legitimize research into the significant extension of human longevity. This event was followed by another conference he organized in Berkeley, together with a well-known *mainstream* biogerontology scholar, Bruce Ames, and Aubrey De Grey – who, following this event, launched the Methuselah Foundation. A known adversary of the leading US anti-biotech intellectuals, such as Francis Fukuyama, Jeremy Rifkin and Leon Kass, Stock has always criticized any kind of restriction on biotechnological research and anti-aging studies. Among the works he has published, we particularly remember the 1993 book *Metaman: The Merging of Humans and Machines into a Global Superorganism* – which, as you can guess from the title, concerns the birth of a super-organism that includes humanity and its technology – and *Redesigning Humans: Our Inevitable Genetic Future*,[86] from 2002.

---

[84] http://www.bltc.com/

[85] www.abolitionist-society.com/abolitionism.htm

[86] G. Stock, *Redesigning Humans: Choosing our genes, changing our future*, Mariner Books, Boston 2003.

Nick Bostrom

Born in Sweden in 1973, and now a professor at the University of Oxford, he is one of the more philosophically sophisticated minds of Transhumanism and stands out for his efforts to give this movement philosophical and academic dignity. He is a well-known supporter of *human enhancement*,[87] and for his work on "existential risks,"[88] that is, all of those risks that, once realized, could cancel out the intelligent life on Earth or drastically and permanently reduce its potential.[89] Another philosophically interesting theme for which Bostrom is known is that related to the "problem of simulation," i.e., the possibility that we already live in a simulated reality[90] – a bit like what happens in the movie *The Matrix*, let's say.

Aubrey De Grey

Among the Transhumanists, De Grey is a true living legend. British, born in 1963, he is a biogerontologist, and he is also the promoter of the most articulate system ever developed to achieve – in the near future, his supporters hope – a life expectancy that is practically unlimited. A form of scientific immortality, essentially – but do not say that out loud, because bio-conservatives might be frightened; much better to talk about the "fight against the aging process for humanitarian reasons." Certainly, his appearance and the way that he presents himself – thin, tall, with a thick and very long beard that makes him look a bit like a shaman – perhaps do not help him to promote his cause; despite this, over the years, he has acquired a good number of followers, some of them quite influential and generous. A graduate in computer science from the University of Cambridge, De Grey worked, for a period, as a programmer, ultimately opening up his own software company. After meeting his future wife, geneticist Adelaide Carpenter, at a graduation party, he began to collaborate with the Department of Genetics at the University of Cambridge, as a manager of the fruit flies database studied at the university. Learning by himself, De Grey also began to study the biology of aging processes on his own; in 1999, these personal studies led him to write an essay, *The Mitochondrial Free Radical Theory of Aging*, in which De Grey claims that the removal of the damage suffered by mitochondrial DNA could significantly extend the lifespan – despite such damage being an important, but certainly not exclusive, cause of the aging process. Based on this book, the University of Cambridge

---

[87] N. Bostrom, J. Savulescu (ed.), *Human Enhancement*, Oxford University Press, Oxford 2009.

[88] http://www.existential-risk.org/

[89] See N. Bostrom; M. Cirkovic (ed.), *Global Catastrophic Risks*, Oxford University Press, Oxford 2011.

[90] N. Bostrom, *Are you living in a computer simulation?* http://www.simulation-argument.com/simulation.html

awarded the PhD to De Grey in 2000. We will have the opportunity to speak in depth about De Grey's theories later; for the moment, it is enough to remember the steps that led him to become – despite the fact that his true work, until 2006, was that of administrator of a database – an important and controversial theoretician of biogerontology, but, above all, the greatest contemporary promoter of the Transhumanist theory of scientific immortality. In 2000, he founded the Methuselah Foundation and, in 2009, the SENS Research Foundation. In 2005, he was the protagonist of a lively debate published in the MIT *Technology Review*, which we will present in detail. In 2007, he published, together with Michael Rae, *Ending Aging*, a book in which he offers a detailed account of his project. In his writings, De Grey also addresses the political and psychological objections to the idea of a very long or potentially infinite life; among other things, he also coined a term, "pro-aging trance," to indicate the psychological strategy that people commonly use to come to terms with the fact that we are all inevitably destined to grow old and die. For example, we tend to automatically interpret life extension as an extension of old age, rather than as a cure – a phenomenon that De Grey calls "the Tithonus error," a tribute to a character from Greek mythology, condemned to immortality and perpetual decrepitude. In the same way, we tend to defend old age, rationalizing it and considering it a natural and, indeed, desirable process. Finally, it is worth mentioning that, despite his controversial theories, De Grey is nevertheless included in the circuit of academic gerontology; for example, he is a member of the Gerontological Society of America and the American Aging Association.

Ray Kurzweil

And here, we are finally at Kurzweil, the real king of the Transhumanists and the supreme baby-boomer.[91] We will also speak extensively of this author, given the role that he played in the elaboration and diffusion of some of the main contemporary Transhumanist ideas, such as Longevism and Technological Singularity. It is hard to make a list of the many enterprises that Kurzweil has spearheaded. Born in New York in 1948, Kurzweil is the son of a visual composer and artist. Since he was very young, he has worked on computers, developing, among other things, a program that can analyze classical music works, break them down into patterns and produce other stylistically similar works. During his University years – at MIT in Boston – he worked as an inventor and entrepreneur, which then led him to open several companies and develop his most famous inventions: the first system able to recognize written texts with any type of character, named the Kurzweil Reading Machine, whose first purchaser was Stevie Wonder; Kurzweil K250, a machine

---

[91] The definition of "supreme baby-boomer" fits perfectly with Kurzweil, representing the desire – incarnated by much of his generation – to remain in play indefinitely.

capable of imitating various musical instruments; the first speech recognition program with a broad vocabulary, launched in 1987; the Kurzweil VoiceMed series, which allows doctors to compile their medical records verbally; Kurzweil Educational Systems, a line of computerized products designed to facilitate learning in people suffering from blindness, dyslexia and other disabilities; the K-NFB Reader, a pocket-sized tool with a digital camera capable of reading any text. His untiring productivity earned him, in 1999, the National Medal of Technology and Innovation – in practice, a sort of Nobel for technology – and, in 2002, introduction into the National Inventors Hall of Fame, under the aegis of the US Patent Office. Although noteworthy and certainly destined to be remembered, it is not Kurzweil's technological contributions that interest us, but his equally untiring efforts to elaborate and spread his peculiar vision of the future. A vision that includes the possibility of creating artificial minds similar to ours – in *The Age of Spiritual Machines*, 1999 – and the advent of an artificial intelligence superior to that of humans, which, in turn, will produce an "explosion of intelligence" and will allow us to overcome the limits of our humanity – in *The Singularity is Near*, 2005. We must not let ourselves be fooled by the technological wonders promised by Kurzweil, however: his real goal – not so hidden – is to be able to live forever, or at least a very long time, jumping just in time on the wagon of a technological revolution, which, according to him, is destined to explode in the coming decades. And, to maximize his *chances* of being able to see this epoch-making event, Kurzweil has developed a program that combines all of the medical knowledge currently available, plus some ideas from the world of alternative medicines, in order to optimize your health. This program materialized in two books, *Fantastic Voyage: Live Long Enough to Live Forever*, 2004, and *Transcend: Nine Steps to Living Well Forever*, 2009, both written with the doctor and alternative medicine practitioner Terry Grossman. Kurzweil's interest in medical issues arose after discovering – at the age of thirty-five – that he suffered from an early form of type 2 diabetes. Together with Grossman, the inventor then developed his radical health regime, which includes the daily ingestion of 250 supplement pills (subsequently reduced to 150), periodic injections of vitamins, and more, in the hope of minimizing the risks of getting sick of this or that pathology. In the case that his program fails – that is, he cannot "live long enough to live forever," reaching a time when biotechnology and nanotechnology will guarantee immortality to us – Kurzweil already has an ace up his sleeve, that is, a cryonics suspension contract stipulated with Alcor. In 2009, in an interview for the magazine *Rolling Stone*, Kurzweil expressed the desire – after the advent of Singularity – to bring his father, Frederic Kurzweil, back to life, using a sample of paternal DNA, as well as archival material he kept in a special store and his memories – among other things, it was his father's untimely death that pushed young Kurzweil down the path toward *life extension* and Transhumanism. Given his ideas, Kurzweil has joined various organizations and initiatives that promote Transhumanist thinking and, in particular, the Technological Singularity, such as the Singularity Institute for Artificial Intelligence and the Lifeboat Foundation. Not only that: in 2009, in collaboration with the NASA Ames Research Center and Google, Kurzweil created the aforementioned Singularity University. Kurzweil's popularity and contributions obviously guarantee him a place in the broader techno-scientific and strategic context, such as the Army Science Advisory Board, which deals, among other things, with the development of a rapid reaction system to possible biotechnological threats.

More People

We would like to acknowledge all of the other people who made important contributions to Transhumanism, but the list is so long. Let's at least address a few more. First, I would like to mention *Zoltan Istvan*, journalist, futurist and Transhumanist candidate in both of the last two US Presidential Elections – for which he campaigned through the country driving the "Immortality Bus," which was coffin-shaped – and the most recent Election for Governor of California. Istvan published *The Transhumanist Wager*, a philosophical novel that promotes the ideal of Transhumanism. *Anders Sandberg* is a super-prolific Swedish Transhumanist, whom we will mention multiple times during the course of this book. *Riccardo Campa* is an Italian professor of Sociology at the Jagellonian University of Cracow whose work is philosophically interesting because it breaks away from the typical analytical approach that characterizes the average Transhumanist. He is also author of several books, among which is *Still Think Robots Can't Do Your Job?: Essays on Automation and Technological Unemployment*. Lastly, we would like to mention a character whom we haven't heard anything about for a while, but who is still interesting: Romana Machado, who in the '90s, proposed her *Five Things You Can Do To Fight Entropy Now* – which are: Care for your mind; Care for your body; Care for your financial security; Empower yourself (that is, learn self-defense); Get a cryonic suspension contract.[92]

2.7   The Pillars of Transhumanism

Being skeptics, often atheists, a thousand miles away from paranormal, New Age, ufological cults, and so on, Transhumanists can only lean on the one thing capable of resisting the secularization that they themselves promote, that is, contemporary science. And so, passionate about natural science and being *techno-geeks* to the nth degree, the Transhumanists follow, with extreme attention, the frontier of scientific research, in particular, everything concerning biomedical disciplines. Nanotechnology, gene therapy, stem cells, neural chips, tissue engineering: all

---

[92] http://transtopia.tripod.com/5things.htm

broth in the cauldron in which Transhumanists try to prepare their own contemporary version of the Elixir of Eternal Youth. However, there is a set of proposals and disciplinary areas that, although not fully accepted in official scientific research, are also not excluded a priori, and that constitute the *specific* contribution of Transhumanism in terms of ideas and planning. These are approaches and proposals that are found, in some sense, in the antechamber of official science: only the future will tell us, eventually, whether they will be admitted or not. We have compiled a list, without any pretension towards completeness – indeed, those who want to correct and integrate our list, please do so, we will be grateful. These approaches, which we chose to baptize "the pillars of Transhumanism," are: life extension; cryonics; human enhancement, i.e., the enhancement of human physical, psychological and mental abilities, through every possible technological measure, from genetic manipulation to neural implants; nanotechnologies, or, more specifically, nano-machines; mind-uploading, that is, the transfer of human consciousness into a form of non-biological support; the Technological Singularity. The pillars of Transhumanism will be at the center of many of the following chapters; in the meantime, we will limit ourselves to making an observation regarding the nature of scientific research and the excess of optimism with which – in our humble opinion – some Transhumanists approach fringe science.

One thing that, beyond the scope of research – even among culturally prepared people – is not clear is the fundamental difference between acquired science and research in progress. We have cited nanotechnologies, and so now we use them to give an example. Eric Drexler – the best known theorist in this field – supports the possibility of realizing the so-called *molecular assembler*, that is, precisely the nano-machines mentioned above. Richard Smalley – Nobel laureate for chemistry, who died a few years ago – argues instead that nano-machines are not possible, that they do not stand up, because they would violate certain fundamental principles. The debate between the two remained very vague, partly because it is a simple possibility that we are talking about here. Who is right? At the moment, we cannot know whether creating assemblers would turn out to be possible or impossible. Only the research will settle the matter. So: will investing more in the research for the development of nano-machines make it more likely to succeed or not? The answer is yes, in the sense that, if it is possible, then this will make it more likely to succeed. And the answer is also no, in the sense that, if it is not possible, we can invest as much as we want, and we will still never succeed. In practice, we can only speak of a very general and unquantifiable probability – that is, we cannot say: "if we invest *a certain amount*, it is very likely that, within *a certain period*, the nano-machines will be built"; one can only make a generic, non-numerical estimate of common sense, such as: "if it is possible, then investing as much as possible will increase the probability of successfully creating nano-machines." Nano-machines, however, could be impossible to create – a situation similar to the attempts to build the perpetual motion machine, abandoned only after the physical impossibility of realizing it was discovered. So, with such a degree of uncertainty, why invest? Answer: because, by doing research towards a certain goal – as in the case of nano-machines – you could still discover a lot of interesting things. It is the classic example of the search for space travel – fortunately, a successful venture – which also produced a lot of "collateral" discoveries that we still use in our daily life, from paraplegic chairs to nylon.

Certainly, not all examples are as clear, linear and "extreme" as the case of nano-machines or artificial intelligence. Clearly, gene therapy and nanotechnology for cancer treatment are less "critical" areas than nano-machines. However, the reasoning is valid for all of the research related to "fringy" areas, which are very innovative and problematic: you cannot know a priori whether a certain result will be achieved or not, you cannot formulate probabilistic estimates of any type; one can only say, very generically and using common sense, that if and only if something is possible – which often cannot be known *a priori* – then the more we invest, the more likely we are to achieve the result. Investing is still worth it, if only to know the limits of the possible.

And then there is the problem of the frequent "artisanal nature" of certain top research. For example, there are actually cases in which this or that scientist has developed a specific nanotechnology device – such as a nanotech clamp or the like. However, these are *one-off* tools, artisanal in a certain sense, and it is not known whether they can be inserted into a more structured nano-machine, or whether they can be produced in series and at little cost.

Mindful of the famous saying from *Hamlet* – "there are more things in heaven and earth …" and so on – we realize that everything is possible in this world, and that the forecasts of the Transhumanists could actually occur according to the modalities and timing established by them. But also maybe not.

Case closed? Not really: we promised to say something nice in favor of Transhumanism and we are willing to do it. In this case, we want to cite – slightly contradicting the realism that we wanted to demonstrate up until now – a pair of pessimistic technological prophecies proved false. In our case, we are dealing with aviation and astronautics: it is known that, until the beginning of the Twentieth Century – in practice, almost right up to the first official flight by the Wright brothers, in 1903 – the possibility of flying aircraft heavier than air was denied. Likewise, until the launch of the first Sputnik – in 1958 – the feasibility of space travel was denied. As they say: never say never.

2.8 Critique of Transhumanist Reason

Could a movement with such radical ideas not end up under the fire of, let's see, everyone? The answer is obviously no, it could not. In fact, the criticisms have come pretty steadily from all sides, starting with the accusations of insanity, immaturity, ignorance, *hubris*, danger, and so forth. So, let's look at them one by one.

Practical Impossibility

The first criticism that is leveled at the Transhumanists is that of not being in contact with the reality of scientific research. Which may be partly true, given that many Transhumanists are theoretical scholars and often do not belong to the world of biology or the medical sciences, but rather to computer science. In practice – say the accusers – the Transhumanists would be guilty of a form of reductionism, treating very complicated living systems as if they were computer systems. This critique is associated more generally with the idea of the non-feasibility of Transhumanist projects: physical immortality would not really be at hand, the Technological Singularity would be a simplistic or unattainable idea, and so on. However, we will address these issues in the following chapters; for now, we will deal only with ethical and political aspects.

Hubris

This is the classical religious position, which sees, in the projects of the Transhumanists, the desire to "play God"; manipulating our genes, trying to achieve immortality here on Earth, and striving to overcome our limits would be a challenge to morality, to divine authority or to a generic "natural order." The *hubris* doctrine also has a secular counterpart, embodied in Jeremy Rifkin's ideas. For the American essayist and economist, the Transhumanist ideal of "perfecting" through genetic engineering – which he named "algeny," along the lines of alchemy[93] – would clash with the irreducible complexity of living organisms and with the unpredictability of the results of such manipulations. While, as Transhumanists are mostly secular, the objection from traditional religions is largely ignored – apart from some attempts to show that the Transhumanist enterprise would not contradict the dictates of classical beliefs – the Rifkinian one is not so easily eluded – being, in essence, founded. However, this does not scare Transhumanists, and, noting that every innovative research presents a certain degree of risk, they try to develop research protocols that minimize dangers. For example, James Hughes proposes using computer models to prevent the possible dangers to as great a degree as possible. In general, from the point of view of the Transhumanists, the debate revolves around the right to dispose of one's own body and genome, the right to try to self-improve one's own nature, and the right of parents to have all of the possible genetic improvements for their children, in the event that these procedures prove to be safe and advantageous.

---

[93] Jeremy Rifkin, *Algeny, A New Word – A New World*, Viking New York 1983.

Banalization of Human Existence

Another strong point of the anti-Transhumanists. For the American ecological thinker Bill McKibben, author of *Enough: Staying Human in an Engineered Age*[94] – it would be morally wrong to fiddle with genetic engineering in order to overcome human limitations, since eliminating the latter would remove all of the points of reference necessary to live a meaningful life – a characteristic that arises precisely from human finitude. We can speak of all of this, of course; but it must be noted that McKibben's position consists in "playing defense," an option that has always historically been defeated: if any new technical-scientific progress actually improves the duration and quality of life, it will end up being adopted. This is the case with anesthesia, vaccinations and all of the medical practices that have led to our unnaturally long life expectancy – we recall that, in a truly natural environment, most of us would not survive to the end of the year. Moreover, it must be said that the radical strengthening and extension of life would hardly represent the end of every problem and every challenge: it is probable that, in the eventuality that the Transhumanist projects become reality, our descendants will find themselves facing new – and perhaps even more interesting – challenges, which we cannot imagine at the moment.

A Danger to Democracy

This too is an "evergreen"; in this case, we have to mention the German philosopher Jürgen Habermas and the American thinker Francis Fukuyama. In particular, Fukuyama, better known for his theory of the "end of history", has, in recent years, specifically attacked Transhumanism, dedicating an entire book to the subject.[95] Forgetful perhaps of the damage caused by Al-Qaeda, the American philosopher and political scientist has branded Transhumanist thought as "the most dangerous idea in the world," as it would undermine our democratic institutions, founded on human nature. The fear here – summarized by the omnipresent example of the *Brave New World* described by Huxley – is that genetic manipulation can lead to a dystopian scenario, characterized by a division into classes on a biological basis, which would make the division into castes of Indian society seem like a joke. Even Leon Kass, American physician and intellectual, shares the opinion that genetic manipulation would represent a challenge to the social order. In this regard, Kass, who was the resident bioethicist for George W. Bush, wrote a well-known article a few years ago, which garnered a whole host of reactions,

---

[94] Bill McKibben, *Enough: Staying Human in an Engineered Age*, St. Martin's Griffin, New York 2003.

[95] Francis Fukuyama, *Our Posthuman Future: Consequences of the Biotechnology Revolution*, Farrar, Straus & Giroux, New York 2002.

including criticism, in which he proposes, as a criterion for establishing a limit to genetic experimentation, the repulsion that the layman feels against certain themes.[96] It is therefore a criterion based on intuition, which lends itself to all of the criticism and accusations of irrationalism normally launched towards this type of argument.

Here, too, the question can easily be overturned: guaranteeing – and this can only be the result of a conscious political will – that everyone will be granted access to the technologies dreamed of by the Transhumanists, not only would it not reach a dystopian scenario, but, on the contrary, it would further favor the process of democratization; This is, for example, the proposal of James Hughes, who would like the transition to post-humanity to occur in the more general context of democratization and universalization of health care services – still a big problem in the US. It must be said, however, that, in fact, the risk that genetic enhancement – provided that such a thing is possible and at hand – would lead to social inequality on a biological basis, and therefore to the end of democracy, does exist, and therefore should not be underestimated. But this is a criticism that only partially coincides with that relating to the danger to democracy.

Genetic Division

One dystopia-connected scenario is, in fact, that in which democracy remains valid or intact, but individual differences worsen the divide that separates those who can afford the enhancing technologies and those who cannot. The scenario in question has been renamed "Gattaca," after the title of the beautiful science fiction movie by Andrew Niccol, which painted a picture of a society in which those who received genetic improvements inevitably ended up having a much higher – and more satisfying – life path than those who did not receive them. In both scenarios, Hughes' proposal probably remains the most reasonable.

Contempt for the Body

Criticism like this would bring the Transhumanists close to the ascetics of all religions, as asceticism is historically linked to a form of contempt not only for the world and its riches, but also for the pleasures of the flesh and the body in general, which is seen as something fragile and subject to degradation. Here, the target of the critics is the aspiration of some Transhumanists – like Hans Moravec or Ray Kurzweil – to transfer their consciousness into a machine, thus abandoning the fragility of our biological substratum in favor of the solidity of metals and carbon resistance. The critique in question originally arose in relation to the precursors of

---

[96] L. Kass, *The Wisdom of Repugnance*, in "The New Republic", Vol. 22, n. 216, June 1997, pp. 17–26, http://www.catholiceducation.org/articles/medical_ethics/me0006.html

Transhumanism, that is, Haldane and his contemporaries; in particular, a few years ago, the British philosopher Mary Midgley classified the aspirations of these visionaries of the '30s – substantially identical to those of Moravec and colleagues – as para-scientific fantasies connoted by a desire to escape from the body.[97] What to say? In the case of some Transhumanists, it is certainly true, even if – since we are often in the presence of patented atheists – the spiritual dimension is absent, replaced by an otherworldly dimension of the "Second Life" version. It must be said that not all Transhumanists think in this way, and that they did not invent contempt for the body. If this were not enough, Bostrom adds that the desire of a body invulnerable to aging and damage – and therefore structurally different from the "despicable" flesh of which we are made – is an ancient dream present in all cultures, hardly capable of being reduced to the classic contempt for the body cultivated by the ascetics of each country; in short, disdain for the flesh and desire for a super-body are not the same thing.

Fourth Reich

The accusation of wanting to create a new order based on a superior race that enslaves and exterminates the others could hardly be expected to be absent, and, in fact, it has already arrived – as we saw in the first chapter. For their part, the Transhumanists, very careful to avoid any kind of juxtaposition with the nefariousness of National Socialism, have always stressed that their work aims to provide individuals with *more opportunities*, which can be freely chosen and rejected. It is not, therefore, a question of creating a superior race, but of offering to all human beings, regardless of race or culture, social class or gender, the possibility of improving and leading longer, freer and happier lives. On the other hand, the overwhelming majority of Transhumanists – although not all of them – adhere to a democratic and tendentiously leftist political vision. No Fourth Reich, therefore, but rather a free and democratic debate about what is and isn't worth doing with the forthcoming technologies.

Dehumanization

And what about a scenario in which human beings all end up being homologated, losing their individuality in favor of canons of improvement decided from above, socially accepted, and that would therefore make us "less human"? We could respond in various ways, such as: what are things truly like now? Are we not already all slaves of fashion? Or: basing the manipulations of our bodies on

---

[97] Mary Midgley, *Science as Salvation: A Modern Myth and its Meaning*, Routledge, Abingdon-on-Thames 1994.

individual preferences, would we not end up cultivating an even greater variety, going precisely against the human tendency toward homologation?

Existential Risks

Then, there are the existential risks theorized by Nick Bostrom himself. In particular, the risks mentioned by Transhumanism's critics are not those generically linked to natural or artificial "mainstream" catastrophes, but precisely those hypothesized by the Transhumanists themselves, which we addressed a while ago.

Dan Brown, take this

Last but not *least*, conspiracy theories must be addressed; if, in fact, the degree of success of a human person or group is judged in part by the occult powers attributed to her/him/it and by the plots in which he/she/it is considered to be involved, then we can say, without fear of contradiction, that Transhumanism has "made it." In this regard, a special mention goes – if only for the rather complex title – to *Pandemonium's Engine: How the End of the Church Age, the Rise of Transhumanism, and the Coming of the Ubermensch (Overman) Herald Satan's Imminent and Final Assault on the Creation of God,*[98] edited by Thomas Horn; the book – as you can guess – underlines how, among the Transhumanist circles, there's a strong smell of sulfur. Horn also gives us another text of similar depth, *Forbidden Gates: How Genetics, Robotics, Artificial Intelligence, Synthetic Biology, Nanotechnology, and Human Enhancement Herald The Dawn Of Techno-Dimensional Spiritual Warfare.*[99] The book by Joseph P. Farrell and Scott D. de Hart, *Transhumanism: A Grimoire of Alchemical Agendas*, is certainly more interesting. The first a theologian, the second a writer and teacher of English literature, the two deal with "crypto-history" or "alternative history," mixing everything with alchemy, hermeticism and Masonry. Erudite and extravagant, their book seeks to establish not merely an ideal link, but a direct affiliation between contemporary Transhumanism and all of the doctrines in question – those attempting to realize the transmutation of man, the creation of life through the alchemical figure of the *homunculus*, and so on – going back to the most ancient human civilizations.[100]

---

[98] Thomas Horn (ed), *Pandemonium's Engine: The Rise of Transhumanism, and the Coming of the Ubermensch (Overman) Herald Satan's Imminent and Final Assault on the Creation of God*, Defender Publishing, Crane 2011.

[99] Thomas Horn, *Forbidden Gates: How Genetics, Robotics, Artificial Intelligence, Synthetic Biology, Nanotechnology, and Human Enhancement Herald The Dawn of Techno-Dimensional Spiritual Warfare*, Defender Publishing, Crane 2011.

[100] See Joseph P. Farrell and Scott D. de Hart, *Transhumanism: a grimoire of alchemical agendas*, Feral House, Port Townsend 2011.

2.9 The Knot of Post-Humanism

Among the various misunderstandings that arise when dealing with Transhumanism, there is that relating to its possible relationship with another philosophical movement: post-humanism. With this term, we indicate a current of thought – or, more precisely, a family of conceptions – that roughly refers to Nietzsche and to the "dissolution of the subject" he advocated, that is, to the idea that the Cartesian subject – the 'I' that is aware, transparent to itself, free, unitary and autonomous – is essentially an illusion. In place of the classically understood individual, there would be a set of instinctual and contradictory forces. To this idea, in the Twentieth Century, is added a further idea, that in which the boundary between human and non-human – that is, the animal, the mechanical, and so on – would be labile, nuanced. The conception that Man as a well-identified subject and a unitary object of a founded knowledge is only a historical construct was launched by the French philosopher Michel Foucault, who has become another of the *tutelary deities* of Post-humanism.

Post-humanism therefore recognizes the inner disunity and non-perfectibility of man, the impossibility of reconciling the heterogeneous individual perspectives on the world and the fluidity of identity. One of the first intellectuals to use the term Post-humanism was the theorist of American literature Ihab Hassan, in the 1977 article *Prometheus as Performer. Towards a Posthumanist Culture?* The discourse on the cyborg set up by Donna Haraway is thus one of the incarnations of Post-humanist thought, particularly in relation to the hybrid nature of man, to its mixing and remixing with the non-human and, in particular, with the technological.[101] In 1988, Steve Nichols published the *Post-Human Manifesto*, in which he argues that, in relation to the past, men of today can already be considered posthuman. The text by Katherine Hayles, *How We Became Posthuman*, is also important. Published in 1999, it represents a direct critique of Transhumanist thought, which, for Hayles, has remained trapped in the classic rationalistic and dualistic ontology of the West, with its clear distinction between mind and body, subject and object, and so on. In 2003, in *Cyberfeminism and Artificial Life*, Sarah Kembler includes, in the Post-humanist discourse, studies on artificial intelligence and the discipline of "artificial life" – that is, the attempt to build computer simulations that reproduce the logic of life and evolution – developed in the '80s. Also to be remembered is the work by the Englishman Robert Pepperell, *The Posthuman Condition*, released in 1995. In Italy, post-human themes – especially in relation

---

[101] See D. Haraway, *A Cyborg Manifesto: Science, Technology, and Socialist-Feminism in the Late Twentieth Century*, in: *Simians, Cyborgs and Women: The Reinvention of Nature*, Routledge, New York 1991.

to man-machine hybridization – have had a lot of success, through the works of authors such as Antonio Caronia,[102] Giuseppe O. Longo,[103] Roberto Marchesini,[104] Teresa Macrì,[105] Francesca Ferrando,[106] and others.

So, different phenomena, Transhumanism and Post-humanism. This does not mean that no one has ever tried to mix them. This is the case, for example, of Stefan Lorenz Sorgner, the German Nietzschean philosopher who, together with the artist and performer Jaime del Val, has launched "Metahumanism" – which represents a philosophical hybrid between the two currents, and of which there is also a special *Metahumanist Manifesto*.[107]

2.10   The Religious Question

At a glance, a group aiming for earthly immortality and the assumption of a growing power over nature through technology should see religion – any kind of religion – as a blight. And, indeed, it is often the case: the average Transhumanist is, in fact, a hyper-rational, atheistic person, with a strong propensity for the natural sciences and technology. However, this rule presents, as often happens, some exceptions. And this not only because individual Transhumanists, here and there, sometimes claim to believe in this or that traditional religion, but also because there are currents or movements within Transhumanism that cultivate a close relationship with traditional religious forms. This is the case with the MTA, the Mormon Transhumanist Association.[108] Founded in 2006, the MTA has no official structural relations with the "Church of Latter-day Saints" – as the Mormons call themselves – but it does have such with the WTA – the World Transhumanist Association. To understand how such a link is possible, one must at least take a quick look at the religion of the Mormons.

According to the Mormons, the famous lost tribes of Israel ended up in America, and, among their ranks, there was the prophet Mormon, who lived in the Fourth Century, and who recorded their deeds on some tablets. The latter were then found, in 1830, by the founder of Mormonism, Joseph Smith, led in said discovery by an angel. The Mormon doctrine is quite complex, and we will not dwell on it here.

---

[102] A. Caronia, *Il cyborg. Saggio sull'Uomo Artificiale*, Theoria, Rome-Naples 1985.

[103] See G. O. Longo, *Il simbionte. Prove di Umanità futura*, Meltemi, Rome 2003, and ibid., *Homo technologicus*, Meltemi, Rome 2005.

[104] R. Marchesini, *Post-Human. Verso Nuovi Modelli di Esistenza*, Bollati Boringhieri, Turin 2002.

[105] T. Macrì, *Il corpo post-organico*, Costa & Nolan, Milan 1996.

[106] F. Ferrando, *Philosophical Posthumanism*, Bloomsbury Academic, London, Forthcoming. By the same author: *Il postumanesimo filosofico e le sue alterità*, ETS Edizioni, Pisa 2016.

[107] http://www.metahumanism.eu/

[108] See http://transfigurism.org/

Suffice to say that the Mormons await the so-called "transfiguration", at the end of time, when the dead will not only be resurrected, but will be entitled to a bonus: precisely, a transfiguration, i.e., elevation to the divine level. Deification is a theme that was also present in some sectors of early Christianity, so that, in this, the Mormons are not saying anything new or particularly eccentric. In any case, this doctrine explains the interest of some Mormons in Transhumanism, as well as the name of the official journal of the MTA, *Transfigurist*.

Another religious movement that might recall Transhumanism – but which, as far as we know, has nothing to do with it – is that of the Raëlians, a ufological cult founded in 1973 by the Frenchman Claude Vorilhon. The Raëlians await the advent of the Elhoim, an alien people who created humanity with genetic engineering, and who are destined to return to bestow upon all of us a technological form of immortality. The aspiring Transhumanists who read us should not alarmed, however: in reality, Transhumanism, though it may seem (or even be) radical, is a rationalist movement of thought, and does not cultivate ties either with the UFO-like sub-culture, or with the paranormal world in general. Even the links with religious aesthetics are relatively superficial, and are often relegated to individual religiosity. Here, we find, in particular, the cases of James Hughes, who is close to Buddhism, and Giulio Prisco, who coagulated his personal vision of the world in his Turing Church.[109]

For the sake of completeness, we cite the Church of Venturism – now the "Society for Venturism." Founded by David Pizer, the Venturists are a religious congregation whose starting point is one of the most Transhumanist – and apparently least religious – of practices: cryonics.[110]

However, we reiterate that the bulk of the Transhumanists is atheist, agnostic or, more generally, secular. If there is a theme that undoubtedly interests the Transhumanists in general terms, it is that of "neurotheology," understood as a disciplinary field that aims to study the neurological bases of religious experiences, highlighting how human religiosity could have evolutionary roots. Though the term was invented by Aldous Huxley, in a 1962 novel, *The Island*, the book that popularized it was from 1994, *Neurotheology: Virtual Religion in the twenty-first Century*, written by Laurence O. McKinney.[111] This is a general readership text, but has acted as a trailblazer. In addition, the American neuroscientist Andrew Newberg has published – along with other books on the subject – a theoretical work, *Principles of Neurotheology*.[112]

---

[109] http://turingchurch.com/

[110] http://www.venturist.info/

[111] Laurence O. McKinney, *Neurology: Virtual Religion in the twenty-first Century*, American Institute for Mindfulness (Harvard University), Cambridge 1994.

[112] Andrew Newberg, *Principles of Neurology*, Ashgate, Farnham 2010.

And neurotheology certainly does not lack for an experimental side: we refer, in particular, to controversial studies carried out since the '80s by the well-known American neuroscientist Michael Persinger, who, with his famous "God helmet" – a device similar to the one used for transcranial magnetic stimulation – has produced in his experimental subjects the sensation of a "presence," or mystical and religious feelings at any rate.

Well, this type of research is certainly the typical Transhumanist approach to religion, and there is no doubt that some Transhumanists would like to play around with these devices, in order to obtain mystical experiences on command, to reach a spiritual conscience, a deeper interior life or, contrastingly, to follow Trotsky in his lack of need for God.

The love of the Transhumanists for "fringy" ideas has led some of them to welcome, or at least to study with interest, the ideas of Frank Tipler, a mathematical physicist and cosmologist at the University of Tulane, known more for his attempt to justify, on a cosmological basis, the Christian faith in the resurrection of the dead at the end of time.[113]

In reality, the question of the relationship between religion and Transhumanism is rather delicate; certainly, its members – all or in part – nourish religious or quasi-religious sentiments towards their ideas, and Transhumanism certainly offers to those who follow it some psychological compensation that religions and traditional ideologies are no longer able to. And there are also those who maintain that Transhumanist ideology should be seen as a sort of secular faith. But it is precisely here that the question becomes thorny, because other ideologies – such as Marxism – have striven to cover the existential needs of their members. In short, we enter the field of philosophy and history of religion, and of the psychology of worldviews, for which self-perception of the members of a certain movement cannot but influence the analyses made by external scholars – especially in a case like Transhumanism, in which several of its members are respectable academics, and can thus easily intervene in the hermeneutical debates that concern them. To put it in a simpler way: we have no idea how to classify Transhumanism – Religion? Political movement? Ideology? Sect? Philosophical current? Bunch of friends? – and we will gladly leave the question to others.[114]

And anyway, if you are atheist, if you are not satisfied with the limits that nature imposes upon you, if you think that techno-science is the only way, Transhumanism is the maximum – in terms of hopes and consolatory fantasies – that Western rationality can offer you.

---

[113] Cf. Frank Tipler, *The physics of immortality*, Anchor, New York 1997.

[114] Since 2009, the American Academy of Religion has organized an annual symposium dedicated to "Transhumanism and Religion," in which several academics address the identification and analysis of any religious beliefs implicit in Transhumanist thought. See http://papers.aar-web.org/content/transhumanism-and-religion-group

3

Living Forever

3.1 The Keystone

As this is a chapter about death and the human search for immortality, I was tempted to begin it with some kitschy rhetorical statement about "the dream of eternal life that has always motivated human beings," the Tree of Life in the Garden of Eden, the Fountain of Youth, and so on. Nothing you don't already know; that's exactly why I'd rather begin with something different, that is, a concept not very well known to the masses: TMT, the Terror Management Theory.

Our story begins in 1973, with the publication of a book, *The Denial of Death*, authored by an American anthropologist, Ernest Becker, who won a Pulitzer for it.[1] The basic idea expressed by Becker in the book is rather simple: the better part of human action aims to forget or sidestep the only unavoidable thing in life, besides taxes: death. This despite the fact that, unconsciously, the terror of our absolute annihilation resides in every human being, to a point that we spend our entire lives trying to find/give it meaning. Basically, Becker substitutes sexuality – Freud's favorite thing – with mortality, and uses the latter to explain everything else; in other words, the purpose of all of the symbolic universes that we human beings build – from laws to religions – is to reassure us, make us feel safe, and give a meaning to our life and our death.

Becker's idea is simple, maybe even a little bit simplistic, but not without appeal, and with some good points, so much so that it has inspired three American

The original version of this chapter was revised with the late corrections from the author.
The correction to this chapter can be found at https://doi.org/10.1007/978-3-030-04958-4_11

[1] E. Becker, *The Denial of Death*, Simon & Schuster, New York 1973.

social psychologists – Jeff Greenberg, Sheldon Solomon and Tom Pyszczynski – to create, starting in the '80s, the so-called "Terror Management Theory."[2] This is an articulated approach, which begins with the identification of what the three psychologists consider more or less the keystone of the human psyche: a fundamental psychological conflict derived from the encounter between the willingness to live and the sharp awareness of the unavoidability of death. This conflict doesn't just cause fear, but terror, a trait typical of human beings. And the solution is typical of human beings as well: culture. Human cultures are symbolic systems aimed at attaching meaning and value to human life. Therefore, if we perceive our life as meaningful, death – which we don't know anything about, besides the fact that it hits randomly and is unavoidable – becomes less scary. More specifically, this goal is achieved through self-esteem, which is bound to the culture we belong to; that is, if we adhere to the values of our cultural context, this confers self-assurance upon us, and, somehow, a sense of "invulnerability" against death. It is a strong approach, and, scholarly, a very fertile one, but it is not without criticism[3]; the general model per se seems to hold, so much so that it originated a program of investigation that is still very much active.

However, the TMT approach dictates that the most obvious examples of values useful in managing the terror of mortality are the ones that seem to guarantee actual immortality – for example, faith in an afterlife and similar solutions. But there are other values, apparently not connected to our mortality, that, according to Greenberg and colleagues, seem to exert the same function: for example, the value of national identity, which makes us feel part of a higher collective identity, able to resist to death, our sense of superiority to animals, our desire to have kids to whom to transmit our cultural heritage, and so forth.

Essentially, our value systems offer us, one way or another, a kind of symbolic immortality. To sum it up: like the acquisition of bipedalism, with its advantages and disadvantages, like back pain, so the development of intelligence and self-awareness provided us with ambiguous benefits. On one side, it provided us with a powerful tool for survival, on the other, it gave us the awareness of our mortality, the fix for which is our invention of culture.

At this point, I would like to ask: is it possible to better classify, at least a little bit, the strategies adopted by mankind in order to confront death? Of course it is,

---

[2] Cf. J. Greenberg, T. Pyszczynski, S. Solomon, *The causes and consequences of a need for self-esteem: A terror management theory*, in: F. Baumeister (ed.), *Public self and private self*. Springer-Verlag, New York 1986, pp. 189–212. Ibid., *A terror management theory of social behavior: The psychological functions of self-esteem and cultural worldviews*, in: «Advances in experimental social psychology», 24(93), Springer-Verlag, New York 1991, p 159.

[3] For example, we might object, along with Abraham Maslow, that human individuality fully manifests itself in the very act of resisting the process of enculturation thrust upon us by our socio-cultural context. Cf. A. H. Maslow, *Toward a Psychology of Being*, John Wiley & Sons, New York 1998 (first published 1961).

and we don't even need to work that much, as most of the work has already been done for us by Stephen Cave. A British writer and thinker, Cave has done exactly that. In his book *Immortality: The Quest to Live Forever and How It Drives Civilization*,[4] the writer – who believes that immortality cannot be achieved under any form, not even symbolically – classifies five "narratives" that every human culture has used to deal with or to face death. Actually, Cave lists four strategies, and adds a fifth one, presenting it not as a strategy per se, but as "the way" to deal with mortality. I don't want to follow him down this path, and I prefer to consider his proposal as just another strategy among the others. Here you have them, with their related "mottos": "Stay alive," "resurrection," "the soul," "legacy" and "wisdom and acceptance."

The first is obvious: trying not to die, developing technologies and procedures in order to keep the Grim Reaper under control. From this strategy, we humans derived all of our medical knowledge, from the prehistoric techniques for extracting arrow tips from the body up to contemporary gene therapy; this strategy is not only about medicine, as it entails every technology aimed at growing and conserving food, or the practices of public health, the use of artificial lights, and every other measure of public security adopted in every historical and social context.

Most scientific and technological progress can be classified as part of this strategy, including all of the attempts – so many, during the course of human history – to obtain immortality, from the gymnastic practices of the Taoists to the Middle Ages-era alchemic search for the Elixir of Life. The "narrative" of resurrection begins with Christianity – even though the myth of the deity that dies and comes back from the dead is much older – and takes unexpected shapes, like in the case of Mary Shelley's *Frankenstein*, which represents a very human attempt to resurrect the dead. The narrative of the soul regards the idea that, inside us, there is an eternal element, which can escape death, and maybe enjoy the blessings of Heaven; an alternative to this option is the belief in reincarnation.

Then, we have the strategy of legacy, which takes different shapes and entails the "self-actualization" of the individual through his/her offspring, but also through whatever can be left as heritage for posterity: political and military undertakings, artistic and literary works, maybe even a simple contribution to a super-individual entity, like dying for one's country or other more or less worn-out ideals. This means essentially staying alive in the memory of the future generations.

In any case, these are strategies that cannot work: the search for immortality has not worked (so far), same for the resurrection strategy, the concept of the soul has been weakened by contemporary neurosciences and the philosophy of the mind,

---

[4] S. Cave, *Immortality: The Quest to Live Forever and How It Drives Civilization*, Crown Publishers, New York 2012.

and, finally, the idea of legacy doesn't work that well – either because we will not be there to enjoy it, or because the collective memory is transient, and, in the best-case scenario, you are going to end up in some history book that students will hate. Cave's solution resides in the fifth strategy – reprised from the thought of Stoics and Epicureans, and from Bertrand Russell – which consists in the serene acceptance of the limits of human existence. Never mind the effectiveness of these five strategies – besides, the one proposed by Cave is not very convincing, as, in following it, we wouldn't have even invented agriculture and we would still be in the caves[5] – it is worth noting that every human society basically represents a different mix of these five strategies, including our own contemporary culture.

3.2   One More Sip of Life

What's the point of all of this? Well, the point is that, far from being an oddity devoid of common sense and scientific rationality, Transhumanism represents just a simple incarnation of the first narrative; one just a little bit ahead of the curve with respect to contemporary bio-medical research – or so Transhumanists like to think. We might even classify them as eccentric people, but their ideas are spreading through the mainstream culture. Moreover, Transhumanist thinking works both sides of the street, as, on one side, it supports the prolongation of life by any possible means, and, on the other side, following the resurrection narrative, it kickstarted the cryonics movement. But do not think that the desire to use science to live a very long life is a contemporary trend; here, you have some examples of the contrary. Let's start by mentioning the work of Gerald Gruman, a well-known American historian of medicine; in his 1966 classic *A History of Ideas About the Prolongation of Life*,[6] he retraces the history of the search for immortality – or at least extreme longevity – until the Nineteenth Century. According to his work, the desire to find a way to prolong life seems to have been quite widespread in the Ancient World – and especially in the Ancient East. Gruman divides those who worked in this field into two differing groups, the "pro-longevists" and the "apologetics" – the latter representing those who were in favor of the acceptance of mortality, a trend more common in the West. In Asia, the search for immortality has never been a taboo, as proven by the long list of characters – mythological and real – who have tried to cheat death. For example, the great Chinese alchemist Ko Hung, who lived in the Third Century C.E. and was author of several treatises on how to achieve a long life through human means. Before him, we have Qin Shi

---

[5] I hope Cave forgives this very bad pun.

[6] Gerald Joseph Gruman, *A History of Ideas About the Prolongation of Life*, Springer Publishing Company, New York 2003.

Huang, the first historical Emperor of China, known for his obsessive search for the elixir of immortality. Moving into the modern and contemporary age, we can mention characters like Metchnikoff, Brown-Séquard and Steinach. Do they ring a bell? These were all scientists who lived between the Nineteenth and Twentieth Centuries, well-respected in their own fields of investigation and all interested in finding a way to prolong life and youth's vigor using scientific means. Élie Metchnikoff was a Russian biologist and zoologist; remembered for his work on the immune system, in 1908, he shared the Nobel Prize for Medicine with the microbiologist Paul Ehrlich. He was the first – in 1903 – to use the term "gerontology"; besides that, he also developed a theory of aging based on the idea that this process is caused by toxic bacteria in the guts and that lactic acid could prolong life – and, for this reason, Metchnikoff got into the habit of drinking yogurt every day. He wrote a book on this topic, *The Prolongation of Life: Optimistic Studies*,[7] which inspired subsequent research on probiotic bacteria.

Charles-Édouard Brown-Séquard, British physiologist and neurologist, became very famous in 1889 – at 72 years old – for his experiments on rejuvenation based on the injection of a liquid extracted from the testicles of dogs and other animals; the results of his experiments were shown in public demonstrations, during which Séquard used to state that the therapy made him feel thirty years younger. In spite of the many criticisms, during the following years, thousands of physicians started to use Séquard's "fluid"; despite the fact that, later on, these miraculous effects were explained by the placebo effect, his research brought about the birth of endocrinology, a discipline considered quackery at the beginning, which then turned into a respected science.

Another pioneer of endocrinology was the Austrian physiologist Eugen Steinach. In 1912, he implanted into a female Guinea pig the testicles removed from a male of the same species; the substance produced by the testicles, known as testosterone, pushed the female to display male sexual behavior. This persuaded him that those glands were related to sexuality. Later on, he developed a technique – actually ineffective – called the "Steinach operation," basically, a partial vasectomy, which was supposed to reduce the symptoms of aging and reinvigorate the patients; this procedure became very popular among artists, actors and VIPs.

To sum it up: even in contemporary science, we can find characters who considered it at least reasonable to try to fight the effects of the aging process, in the hope of incrementing our life expectancy beyond the known limits. Transhumanists are just a little bit more extreme, but they are made from the same mold. Before evaluating, from a scientific viewpoint, the Transhumanist take on aging, we have to answer another question: what does contemporary science think about the aging process?

---

[7] https://archive.org/stream/prolongationofli00metciala#page/n5/mode/2up

3.3   A Few Answers from Biology

To put it simply: why do we have to grow old and, after a certain amount of years, die? Obviously, biology has already asked this question, and the theories developed over the years to explain aging and death are numerous and have been very widely articulated; generally speaking, we can divide them into two groups: theories that explain these phenomena through a genetic program established by our DNA, and theories that explain them through the accumulation in our organism of cellular damage and random mutations. Back in 1889, August Weismann, a German evolutionary biologist and follower of Darwin, proposed the idea that aging evolved in order to make room for the future generations, a kind of evolutionarily necessary cleansing act; it is a finalistic theory, though it assumes – consciously or not – that nature can act wittingly, showing "intentions." The first modern theory of aging was formulated by the British biologist Peter Medawar; in 1952, the scientist hypothesized that aging was a simple matter of carelessness. The natural world is characterized by merciless competition, and almost every animal dies before having enough time to grow old. There is no need for nature to invest energy and resources into the goal of preserving the organism's efficiency until old age; or, in less finalistic terms, there's no selective pressure that favors genetic traits suitable for keeping people alive and in good health until an age at which they would be dead anyway, because of incidents, diseases or predators. Medawar's theory is known as the "theory of mutation accumulation" and concerns the exact accumulation – during the evolutionary process – of negative mutations whose effects manifest themselves only at a certain age, and so cannot be erased by natural selection. The American evolutionary biologist George C. Williams suggested, in 1957, the theory of "antagonist pleiotropy," which states that there are some genes that exert two different effects on the organism, one positive, which manifests itself during youth, and one negative, which manifests itself only after a certain age. If this is true, altering the aging process is basically impossible, as it would imply the alteration of a fine-tuned equilibrium on which life is based. Implicit in this theory is the fact that genetically manipulating an organism in order to lengthen its life would necessarily impair its fertility; in spite of this, in 1994, the American biologist Michael R. Rose selected a group of fruit flies that were especially long-lived, noticing that, contrary to Williams' theory, they were even more fertile than their normal-lived counterparts. More generally, Williams' approach has been judged "too rigid," as it implies that nature cannot sidestep – if adequately "motivated" – the abovementioned rule. In 1977, Thomas Kirkwood, a British biologist, proposed the theory of "disposable soma," which implies that the body – that is, the soma – has a limited quantity of resources available, which it has to distribute between metabolism, reproduction, reparation and maintenance. Because of these limits, the body is forced to settle for one or more compromises, and, because of this, after a while, it begins to deteriorate – right after the age of reproduction, that is, when it is not useful anymore.

Of course, these are just a few examples, and we can find many other theories. Right now, under different shapes, the dominant paradigm seems to be the one of damage accumulation, that is, the idea that aging and death are just the side effect of a nature not very interested in the fate of the organism after the latter has had its chance to reproduce.

Truth is, we are very far from owning a complete knowledge of the metabolic processes underlying aging. That said, we can't avoid noticing that aging and death are not universal phenomena, not at all: for example, there are several species of reptile and fish that do not manifest any functional or reproductive decline; that is, they don't age. And this trait is often connected to a life expectancy way longer than ours – as in the case of the Aldabra Giant Tortoise of the Seychelles Islands, which can reach 255 years. The ocean quahog – *Arctica islandica* – is a marine bivalve mollusk in the family of Arcticidae; it lives in the North-Atlantic Ocean and can reach 507 years. The Hydra – a very simple marine animal – does not manifest any sign of aging and it seems not to die of old age.

And let's now introduce another interesting concept, the one of cellular senescence. Our story begins with Alexis Carrel, a French biologist and surgeon, who, in 1912, won the Nobel Prize in Medicine; during his research, he concluded that, if adequately nourished, cells cultivated on a petri dish could keep multiplying indefinitely. Sadly – for him – his theory was debunked in 1961 by the American biologist Leonard Hayflick, who discovered that, under such conditions, cells could only keep multiplying for a limited number of times – more or less fifty – before becoming senescent. This phenomenon – known as the "Hayflick limit" – is related to the telomers. These are "caps" that cover the extremities of the chromosomes: they play a protective role, and they get shorter and shorter as the cells reproduce, until they reach a limit that signals the beginning of cellular senescence. A specific enzyme, called telomerases, can lengthen the telomeres again; in cancerous cells, this compound is produced continuously, causing the proliferation of the tumor. According to supporters of the idea of a genetically programmed aging and death, the Hayflick limit could prove the existence of a biological "internal clock" that controls our organism.

Bottom line: aging is still a mysterious and poorly understood phenomenon; most theories attribute it to Mother Nature's negligence, and the fact is that we living organisms are not equal: some age, some don't; some live long lives, some very short. As things stand, does the fact that some people are trying to alter the aging path come as a surprise?

3.4 Prometheus Unchained

Back in 1965, sociologist Robert Fulton used to say that, in America, death was treated as if it was a contagious disease, as the cause of personal carelessness and not a trait of the human condition[8]; today, this vision is even more widespread, and it is actually part of our common cultural background, made of prevention, campaigns against this or that disease, against this or that bad habit. And it is exactly this mentality that brought about the birth of what we can consider an informal international movement, that of the longevists – actually an "umbrella" under which we can find a very heterogeneous ensemble of groups, associations, clinics, and enterprises, more or less serious, all interested in prolonging their stay in this world as much as possible. The first on our list – primarily because of their incredible dedication – are the supporters of so-called "caloric restriction." This is a nutritional regime impossible to understand for us normal people, and it consists in obtaining all of the fundamental nutrients – and especially vitamins, minerals, and so on –, but keeping the caloric intake *a little bit below the minimum*. The basic idea dates back to 1934, when two Cornell University researchers, Mary Crowell and Clive McCay, put a few mice through a very harsh version of this regime. As incredible as it sounds, the life expectancy of the mice increased way beyond the norm, and this discovery has been confirmed by many studies conducted on several species of animals, primates included. More specifically, a study conducted since 1987 by the National Institute of Aging on some rhesus monkeys has shown that caloric restriction provides several health benefits, but it hasn't shown that the effects seen on mice – a form of extreme longevity – work on bigger mammals too. Only time will tell. Obviously, this is true for human beings as well, on which nobody has ever conducted a specific clinical investigation on the effects of caloric restriction. Anyway, the possible effects include benefits for the circulatory system and memory; the side effects include a possibly reduction in libido, skeleton and muscle mass loss, and so forth. Even though the effects of caloric restriction have been known for almost eighty years, the underlying mechanism is still unclear; one of the possible hypotheses mentions so-called "hormesis,"[9] a biological phenomenon in which, at least in some cases, a little quantity of stress – and surely caloric restriction is stressful – could stimulate the organism, activating its defenses. This could push it to optimize its resources, and put its immune system in motion. Basically, the "perception" of what is, in fact, a condition of famine might stimulate body and mind – likely, this mechanism dates

---

[8] Cit. in: B. Appleyard, *How to live forever or die trying*, Simon & Schuster, London 2007, p. 108.

[9] A phenomenon studied and named for the first time by the American researchers Chester Southam and John Ehrlich in 1943.

back to our prehistoric life, when three meals a day were a utopist dream. Another hypothesis regards our reproductive system: maybe, in "perceiving" a condition of chronic famine, the organism might "decide" to postpone reproduction, and, in doing so, it might try to keep the organism healthy and to lengthen its lifespan, empowering some mechanism of reparation – like so-called "autophagy," a process in which cells recycle their content, for example, selectively eliminating damaged organelles. The most famous researcher to work on caloric restriction – and a superstar to all Transhumanists – was Roy Lee Walford. American, born in 1924, during the '80s, Walford conducted – together with Richard Weindruch – several experiments of caloric restriction on mice, cutting down caloric intake to around 50 percent and almost doubling their life expectancy. Far from limiting his work to pure research, Walford has published several general readership books, in which he promotes caloric restriction as a lifestyle, like *Maximum Life Span* (1983) and *The 120-Year Diet* (1986). Sadly, in spite of his regime of caloric restriction, Walford died in 2004 at just 79, of amyotrophic lateral sclerosis. A connection with caloric restriction is unlikely – it would be the only known case –, but it is possible that the progress of the disease could have been hastened by this nutritional regime, and that's why caloric restriction is not recommended for people with this pathology. Anyway, Walford launched a real movement, with many followers – mostly Transhumanists willing to gain some more time. And, in 1994, Walford created – together with his daughter Lisa and Brian M. Delaney – a dedicated organization, called the CR Society International, which sponsors devoted conferences, raises funds for anti-aging research and provides practical information to its members – all very skinny, if we are to judge from the pictures on their website.[10]

Inside the longevist movement, someone decided to go beyond this, that is, to propose a global approach against the aging process. Which suits the Transhumanists very well, as busy as they are preparing themselves – rather maniacally, I would say – for the world of tomorrow. I am talking of so-called anti-aging medicine, a very disputed medical practice, which aims to treat senescence as a disease like all of the others. This approach is promoted by the American Academy of Anti-Aging Medicine[11] – codename: A4M –, a nonprofit organization that holds seminars on this topic and releases licenses for physicians. Founded in 1993 by two osteopaths, Robert Goldman and Ronald Klatz, A4M has 26,000 member in 110 countries; their anti-aging program includes some recommendations that are scientifically sound – eat properly, exercise, and so on – and some that have been quite criticized by official medicine – like taking hormones and avoiding tap water, which is said to be polluted with dangerous chemicals. The official mission of this

---

[10] http://www.crsociety.org/

[11] http://www.a4m.com/

organization is "the advancement of tools, technology, and transformations in healthcare that can detect, treat, and prevent diseases associated with aging." Good, isn't it? According to A4M, many of the disabilities associated with the aging process are caused by physiological dysfunctions that can be improved by medical means, and this can lengthen our life expectancy. Or, to quote Klatz, "we're not about growing old gracefully, we're about never growing old."[12] Well, it looks like a promising program, and our hopes become elevated even more if we take a look at their code word, as well as their final goal, that is, *The Ageless Society*, which will end the "apocalypse of aging" – a good slogan indeed. Applying the procedures recommended by A4M should bring us straight into immortality. "Fishing" in the Transhumanist background and in the vocabulary of cutting-edge research, Klatz speaks of "emerging technologies," like nanotechnology and stem cell-based therapies, which will increase our life expectancy well beyond our present limitations.[13]

Just one question, though: what does the scientific community think of this "anti-aging medicine"? The majority of researchers working in the field of gerontology separate themselves from the ideas promoted by A4M, considering them pseudoscientific and stressing the strong economic interests of the organization. For example, Stephen Coles, physician and researcher at the UCLA Medical School, who has studied centenarians for many years, has stated that "there is no such thing as anti-aging medicine."[14] Many other specialists also frown upon Klatz's organization and consider it a mix of medical practice and commercial interests. The American Board of Medical Specialties – a nationally recognized nonprofit organization that evaluates and certifies the level of professionalism of every medical procedure and organization – does not recognize A4M as a professional organization. Even the official journal of A4M, the *International Journal of Anti-Aging Medicine*, has been under attack: in a letter published in *Science* in 2002, Aubrey De Grey himself declared that the contents of that journal represent no more than an advertisement in favor of the pseudoscientific anti-aging industry. A cold shower indeed, and from a prominent Transhumanist. *Science* has also published an article – signed by De Grey, Coles, S. Jay Olshansky and other prominent bio-gerontologists – branding anti-aging medicine as pseudoscience.[15]

---

[12] Cfr. http://www.nytimes.com/1998/04/12/style/anti-aging-potion-or-poison.html

[13] http://www.worldhealth.net/news/forever_young_the_scientific_fountain_of/

[14] http://articles.latimes.com/2004/jan/12/health/he-antiaging12. "The good news is scientists may achieve real breakthroughs that could lead to longer and healthier life spans – in 20 to 30 years."

[15] AA. VV., *Antiaging Technology and Pseudoscience*, in «Science», New Series, n, 296, 26 April 2002.

No hope, then? Actually, as we mentioned before, mainstream official science doesn't rule out the possibility of radically prolonging our life expectancy, and doing so in decades, not centuries. Among the supporters of this hypothesis is a well-known American biologist, William Haseltine. Scholar, entrepreneur and philanthropist, Haseltine is a VIP of American biotechnology; after working on AIDS and the human genome, he created several biotech companies, among which is Human Genome Sciences, an enterprise aimed at applying genomics to medical science. Just to get an idea of his position, in 2001, *Time* magazine named him one of the twenty-five most influential businessmen in the world.[16] In 1999, during a conference at Lake Como (Italy), Haseltine coined a term hopefully destined for a bright future, "regenerative medicine." His idea is that technologies still in their infancy, like gene therapy, stem cells and tissue engineering, could one day restore normal functionality in cells, tissues and whole organs, regardless of whether the damage has been caused by pathologies, injuries or simply the aging process. After co-founding the *Journal of Regenerative Medicine* and the Society for Regenerative Medicine, Haseltine explained this new discipline's general principles in several papers. Not satisfied, he has even coined another new term, "rejuvenating medicine," to describe the next revolution that could one day open the path toward immortality. In a 2002 interview for *Life Extension Magazine*,[17] Haseltine stated that

> "I agree that many medical treatments might be made obsolete by some general and systematic solution to aging. Many of the conditions that we are working on are consequences of aging. Should the fundamental aging clock be stopped, or indeed reversed, the need for most of these medications would evaporate. That would be a happy day indeed.
>
> (…)
>
> "In the past few years it has become possible for the first time to construct a scenario in which humans may become immortal: by the systematic replacement of stem cells."[18]

---

[16] An interesting presentation of the life and work of Haseltine can be found in: Brian Alexander, *Rapture. A raucous tour of cloning, transhumanism, and the new era of immortality*, Basic Books, New York 2003.

[17] The Life Extension Foundation (LEF) is a non-profit organization in Fort Lauderdale, Florida, founded in 1980 by two Transhumanists, Saul Kent and William Faloon, with the goal of promoting research and information about longevism, preventive medicine and sports performance. Besides selling vitamins and supplements, LEF publishes *Life Extension Magazine*. Cfr. http://www.lifeextensionfoundation.org/ and http://www.lef.org/index.htm

[18] http://www.lef.org/magazine/mag2002/jul2002_report_haseltine_01.html

At the time of this writing, 2018 is almost over, and we are still waiting for the results of regenerative medicine; meanwhile, we take comfort from another character in between mainstream science and Transhumanism, Michael D. West. And while Haseltine lives and works comfortably in that high-level, financially well-endowed social environment at the crossroads between biotechnology, business and politics, West is an old acquaintance of the Transhumanists: he attends their meetings, he knows them personally – basically, he is one of them.

West's path is rather unique: initially a supporter of creationism and a strong opponent of Darwinian evolutionism, his father's death pushed him to the side of scientific investigation, infusing him with the desire to find a solution to the problem of aging and death. With this purpose in mind, he created, in 1990, the Geron Corporation[19] – headquartered at Menlo Park, and still active –, the first biotech company officially aimed at finding a "cure" for aging. West's strategy revolves around telomers and telomerase, the latter, again, being, the protein that can regenerate the former, re-lengthening them; of course, the company hopes to use telomerase to modify the aging process. In 1998, West became CEO of Advanced Cell Technology,[20] a biotech company in Santa Monica, California, specializing in therapeutic cloning and regenerative medicine. The better part of West's work at Advanced Cell Technology regards the possibility of bringing human cells back in time, biologically speaking. West's musings begin with the classic distinction between germline and soma. The first term indicates all of the germ cells you can find in an organism: that is, sperm and eggs, and the cells they are derived from, spermatocytes and spermatogonia, as well as the cells at the "bottom" of the germline, gametocytes and gametogonia. All of these cells form, metaphorically and chronologically, a line: in fact, we are the product of our parents' germ cells, and they derive from our grandparents, and so on; same thing for our offspring and our descendants. So, it's possible to talk in terms of a germ "relay" that dates back to the dawn of life on this planet.

The soma – the ensemble of every other cell that composes our body – is a different story. The soma gets sick, ages, and dies; the germline is, potentially, forever. Somehow suggestively, West talks about his work as an attempt to get close – and to place the soma near – the germline, the true source of immortality. From a practical viewpoint, all of this means working on an important expression of the germline, the stem cells. As you probably know already, embryonal stem cells are not-yet-differentiated cells, cells that can turn themselves into any kind of tissue. If we can manage to master this process, we would be able to cure a huge number of different degenerative pathologies, from Alzheimer's to muscular dystrophy. And, as these cells haven't yet decided if they want to be germ cells or

---

[19] www.geron.com
[20] http://www.advancedcell.com/

soma cells, they are not yet prisoners of soma's mortality, and actually enjoy the germline's immortality, meaning that they can proliferate indefinitely. According to West, theoretically, the aging process could mostly be treated with stem cells; ideally, we could re-create almost all of our organs in the lab, and replace the originals with them, one at a time or all together. In the simpler cases, we wouldn't even need to do new original research: the procedures of rejuvenation would simply follow the ones that we would use to treat muscular dystrophy or other pathologies. And it was West himself who made public, in 2001 – so, during the Bush II era – his intention to carry out the first procedure of therapeutic human cloning. In case you don't know, this procedure consists in inserting a whole nucleus of a somatic cell – which has all of its 46 chromosomes – into a non-fertilized egg; from that, a human embryo genetically identical to the donor is obtained, and, from the embryo, you can harvest all of the stem cells you need, ready to be cultivated in the lab. Charged – of course – with the accusation of "playing God," West became the center of an intense debate, in which a journalist even compared him to bin Laden. After all of this, Bush urged the Senate to ban both reproductive and therapeutic cloning, a bill previously already approved by the House of Representatives.

In 2003, West published an autobiographical history of his research, *The Immortal Cell.*[21] Right now, he is CEO of BioTime,[22] a biotech company in Alameda, California, which covers R&D in the field of stem cells.

Besides all of this, Michael West has played another important role, that is, role model and inspiration for the most important theoretician and promoter of scientific immortality in the contemporary world: Aubrey David Nicholas Jasper de Grey.

3.5 The Phoenix 2.0

De Grey has developed the most complete system for curing aging ever imagined, a set of therapies that – if applied on regular bases – will allow human beings to regenerate themselves periodically, metaphorically arising from their own ashes – just like the mythological creature of the Near East. If it works, of course, and if someone decides to finance it generously, defining every detail and carrying it out in every one of its parts. About the man De Grey, we know everything we need[23];

---

[21] Michael D. West, *The Immortal Cell: One Scientist's Quest to Solve the Mystery of Human Aging*, Doubleday, New York 2003.

[22] http://www.biotimeinc.com/

[23] A good portrait of De Grey can be found in: Johnathan Weiner, *Long for this world. The strange science of immortality*, Harper Collins, New York 2010.

now it's time to take a closer look at his plan, to savor the details, to see if, scientifically speaking, it can stand. De Grey adheres to the dominant paradigm of biogerontology, that is, the idea that aging and psycho-physical decline are produced not by inherent genetic program, but by the simple carelessness that Mother Nature reserves for us after we become too old to be useful for reproduction. The approach proposed by De Grey looks very reasonable, really; he recognizes that human metabolism is poorly understood, and that we cannot really get rid of aging, or even slow it down, as the knowledge necessary to re-design our body is far from sufficient. Moreover, we cannot really wait to have all of the answers in order to begin to intervene. The questions that need to be answered are way too numerous. For example: which metabolic alterations cause aging, and which ones are just consequences that would disappear if the deeper causes were removed? Or: are we sure that any metabolic alteration wouldn't have a cascade of unforeseen and dangerous side effects, just like the famous butterfly that, flapping its wings in Brazil, can cause a thunderstorm in Japan? Furthermore, De Grey adds that the border between disease and aging is actually illusory. This distinction apparently makes sense: diseases are not universal, aging is. But – De Grey stresses – age-related diseases are actually – surprise surprise – related to age. They appear at an advanced age because they are mostly consequences of that process; to put it another way, aging can be seen as an initial *collective* phase of the many age-related diseases we suffer. Aging has always been considered something very mysterious, something qualitatively different from everything else and, for this reason, intractable; it is, in reality, just a "downward spiral." There is not – says De Grey – a "biological clock of aging," a program, because, if this was the case, our aging should be as ordered as our development, and we know that it is actually messy, disordered and very subjective. And, from another viewpoint, age-related diseases are really universal, because, sooner or later, we all catch one, if we haven't already caught another type of age-related disease earlier. There is no "time-bomb," just a slow accumulation of damage. What matters – stresses De Grey – is the fact that, compared to people in their twenties, people in their forties can expect fewer years of life and health, and this depends on the molecular and cellular differences between the two groups, not on the mechanisms that produced them. What matters are the actual differences. In order to sidestep what looks like an insurmountable obstacle, De Grey promotes an "engineering" approach. Why don't we just classify the kinds of damage inflicted on our organism by the aging process – asks the scientist – and try to develop methods of restoration to be applied time and time again, as soon as a new form of damage manifests itself? After all, if we have to do maintenance work on a house or a car, we don't really need to have the blueprints of the architect who designed the house or the engineer who created the car; we just need a little bit of knowledge, to take a look at what's wrong and fix it. And this is true for engineering in general; for engineers, it is normal to design new technologies without a perfect comprehension of the underlying physics – electricity, nuclear fission and super-conductor magnets have all been used for a long time without a complete theoretical understanding of the phenomena at play. And let's not forget that the same happens all of the time in medicine; for example, acetylsalicylic acid was in use long before a full chemical understanding of it was realized. Bottom line: it is not necessary to have all that many details to get things in motion and *begin* to develop this or that therapy. You just have to begin, and that's it. Now, if we began to develop concrete solutions, we could repair aging damage, without any worry about metabolism, the Butterfly Effect, and so on. You see the damage, you remove it with an *ad hoc* therapy, and everybody is happy. So, the solution would be sufficiently thorough and frequent maintenance. De Grey's work doesn't stop here; in fact, the pioneer of scientific immortality has prepared a list – apparently complete, but you never know – of the kinds of damage that the aging process inflicts on us on a daily basis. Let's take a look.

The first type of damage is that related to mutations in the genome and epigenome – the latter is the ensemble of mechanisms that regulate the activation of this or that gene. Basically, we are talking about the mutations that hit the DNA and the proteins synthesized from it.

De Grey's hypothesis – quite a "strong" one – is that, globally, the aging process is *not* determined by genomic or epigenomic mutations, that is, these mutations can cause only those pathologies that are quite visible, like cancer. There are further mutations in the DNA, silent ones, although they might, just might, cause damage only if we lived way longer than we do. Anyway, it's a complex topic, and for all of the details, I would like to refer you to the book published by De Grey in 2007.[24]

Then, we have the mutations that affect the mitochondrial genes. Mitochondria are organelles – "organs" of the cell – that work as a kind of cellular power-plant. They produce energy and, at the very same time, they release the infamous free radicals, very reactive molecules, able to damage our organism in various ways. Mitochondria have their own genetic material, and the related mutations can damage their ability to work properly; indirectly, this can speed up the aging process. At least, this is De Grey's theory – one that earned him a PhD at the University of Cambridge. According to the biogerontologist, generally speaking, the mitochondria are not culprits *per se* of the oxidative stress that hits the body; the real culprits are those few mitochondria that enter a maladaptive state, provoking and spreading oxidative stress toward other cells.

---

[24]Aubrey de Grey, Michael Rose, *Ending Aging. The Rejuvenation Breakthroughs That Could Reverse Human Aging in Our Lifetime*, St. Martin Press, New York 2007.

The third type of damage is the "junk" inside of the cells – the so-called intracellular aggregates, a mix of different compounds collectively known as "lipofuscin." Our cells constantly metabolize proteins and other molecules that are no longer useful or that might even be dangerous. It's a lot of work, and the molecules that cannot be "digested" are allowed to accumulate as junk inside of the cell. Atherosclerosis, macular degeneration and Alzheimer's are somehow related to this issue.

And here we have the fourth kind of damage, another type of "junk," the so-called extracellular aggregates, which accumulate outside of the cells and can have a role in certain diseases – let's just think of the amyloid plaques, which accumulate in the brain and are related to Alzheimer's disease.

The fifth problem is "cellular death." Some of our cells cannot be replaced, while others can only be replaced very slowly. For example, with the passing years, the numerical decrease in these cells can weaken our heart, our immune system or our brain – sometimes, the loss of neurons can have dramatic effects, as in the case of Parkinson's disease.

And then we have the issue of "cellular senescence." In truth, some cells stop dividing, but neither do they die nor do they allow other cells to reproduce; moreover, they can secrete dangerous proteins.

Last, but not least, the AGE, which stands for "advanced glycation endproducts": a random process in which sugars bind to certain groups of proteins – including structural proteins like collagen or elastin, which have a central role in the architecture of our tissues. In more technical terms, glycation – also known as non-enzymatic glycosylation – is the product of the fusion between a sugar, like glucose, and a protein or a lipid – without the mediation of an enzyme. Glycation is a basically random process that reduces or stops the functioning of biomolecules. Accumulating AGE can damage organ functions, stiffen muscles, thicken arteries, cause wrinkles, and manifest other symptoms of biological aging. Something similar happens when you cook food, and so we can say that, metaphorically, aging is also similar to a process of cooking over a low heat.

And there you have them, the seven capital sins of aging: cancerous nuclear mutations, mitochondrial mutations, intracellular junk, extracellular junk, cellular loss, cellular senescence, and AGE. All waiting for the engineering solution promoted by De Grey. Not an easy task, both because of the tremendous complexity of the body and because, as we didn't design it, we have to arrange a colossal amount of retro-engineering in order to complete the job. At the end of the day, the trick consists in using or manipulating our natural systems of defense, to enhance them, using specific therapies. Sure, but which ones?

De Grey has in mind an articulated plan, composed of interventions that – he says – are already under development, or very near, or at least conceivable. Ladies and Gentlemen, meet the SENS, which stands for "Strategies for Engineered

Negligible Senescence." With a complex – but already imaginable – gene therapy, we could break the tie between aging and the free radicals that are freely released by malfunctioning mitochondria without interfering in the activity of these organelles. For De Grey, toying with the mitochondria *might* – and I stress *might* – decrease the aging process by 50 per cent. The trouble with the mitochondria is the following: in these organelles, we can find about one thousand different proteins, and only thirteen of them are synthesized by genes residing inside the mitochondria themselves. All of the others are synthesized from regular DNA and then transferred into the mitochondrion. The reason for this is pretty simple: the nucleus is protected against external mutation-causing agents way better than the mitochondria, and the only proteins synthesized *in loco* – that is, inside the mitochondrion – are those that, for structural reasons, cannot be transferred from the nucleus to the mitochondrion. It's thirteen proteins, and they are hydrophobic – that is, they cannot be transferred through the aqueous content of the cell. Sadly, mitochondria are less well protected than the nucleus, which means that the genes/proteins inside them suffer way more mutations than the nucleus. They are easier to damage. How do you fix this? You would need to find a way to create a back-up copy of these thirteen genes inside of the nucleus, and find a way to transport the thirteen proteins through the aqueous liquid. Is it possible? Yes, according to De Grey. Or, at least, there are already people working on that. What De Grey is referring to here is the so-called mitochondrial diseases, very rare pathologies that can take different forms and affect different tissues. These pathologies usually provoke heavy disabilities, and several researchers are trying to develop an adequate gene therapy to cure them. And, of course, they had to face the very same problem of the hydrophobic proteins. Some solutions are possible, though, and De Grey mentions the work of Michael P. King, an expert on mitochondrial diseases at Thomas Jefferson University, Philadelphia. In this case, King discovered – during the '90s – that six out of the thirteen proteins have variants – to be found in green algae – which are actually codified by the nucleus. So, at least theoretically, it is possible to imagine – and many researchers are working on that – a gene therapy that transfers these thirteen genes into the nucleus in order to provide a backup for the malfunctioning mitochondria.

One down, six more to go. Now it's intracellular junk's turn, also known as lipofuscin, because it is composed mostly of lipids. Here, De Grey's proposal is quite fascinating, as it represents a cross-over between two different fields, genetic engineering and bio-remediation. The latter is a very interesting field that consists in the creation of genetically modified bacteria able to digest this or that toxic compound, restoring the environment.

To elaborate this strategy, De Grey has enlisted the help of John Archer, Cambridge researcher and internationally recognized authority in the field of bioremediation. The idea elaborated here by De Grey is intellectually brave – and I hope it can work, of course. Now, lipofuscin is fluorescent; so why don't dead bodies glow in the dark? They should, as decomposition should free lipofuscin for everybody to see. The answer is the so-called geo-bacteria, typical organisms that you can find in soil, and which – at least some of them – produce enzymes that can digest lipofuscin. So, here's the plan: why not take the genes that codify these enzymes and transfer them into us? We *might* make our lysosomes – organelles that work as waste disposal plants – able to digest lipofuscin.

Two down. And now, about the extracellular junk. In this case, we have different types of debris, like the beta-amyloid plaques, related to Alzheimer's, transthyretin, a molecule produced by the liver, and so on. Here, De Grey proposes using genetic manipulation to turn our immune system against this type of junk, a phenomenon known as "phagocytosis." And there is already research in this direction going on; for example, the California biotech company Elan[25] is working on a vaccine able to push the immune systems of mice against amyloid plaques. This system might be used to fight against cellular senescence, too; senescent cells express specific proteins, which could be used as a target of choice by a specific immunologic therapy. In this case, the strategy would consist in tricking these cells into committing apoptosis – that is, cellular suicide.

To fight the AGE, we can imagine a pharmacologic therapy. This was the case of Alagebrium – codename: ALT-711 –, a drug researched by the Alteon Corporation that is able to dissolve – with some limitations – the AGE, bringing the situation back to square one. In spite of the initial interesting results, Alteon closed down, and the study has been stopped. In the future, someone might reprise the research, though, and, according to De Grey, we should look for specific enzymes able to dissolve the AGE.

The issue of cellular death – neurons, cardiac cells, and so on – could be fixed using adult stem cells, to be extracted from the patient, multiplied in the lab, genetically manipulated and reinserted into the patient. Which is exactly the regenerative medicine promoted by Haseltine and West.

Well now, the only thing we have left to do is to defeat cancer. What could we say about this condition that hasn't been said by the main experts of this field? The war on cancer is the widest sector of bio-medicine, well-financed and provided with well-prepared and very motivated researchers. De Grey's proposal doesn't want to replace the actual therapies or the ones in development; in fact, it aims to prevent cancer from being born in the first place. Just to be clear: De Grey's proposal against cancer represents the most controversial part of his plan. The researcher has christened his strategy WILT – Whole-body Interdiction of Lengthening of Telomeres. The trick here would be to find a therapy that doesn't depend on some factor that cancer might side-step through a mutation in its genic

---

[25] http://elan.com/

expression. Therefore, we need to deprive cancer of something that it absolutely needs to thrive and spread. The "tool" should be removed in such a way that cancer cannot do anything to get it back, and the "tool" has to be such and such that healthy tissues can do without it. So, why not – asks De Grey – erase the gene that produces telomerases? Cancer cells are immortal because they have telomers that can regenerate themselves continuously, thanks to an always active gene for telomerases. And what about the healthy cells that use and need telomerases? Here's De Grey's take on this issue. The whole body is not composed of cells that reproduce on regular bases; and, among them, we have the ones that compose the alveoli in the lungs, the bone marrow, part of the bowel, and so on. These are parts of the body – says the scientist – easily accessible by medical tools, and that could be periodically regenerated with stem cells expressly cultivated for this task. It is not very clear how we should erase the gene for telomerases and the mechanism connected to cancer growth[26] – probably through gene therapy. There is another critical aspect of this proposal, that is, to cause permanent damage to the organism and then repair it periodically using stem cell-based interventions. Maybe we could consider WILT as a temporary solution, while we wait for a new, still unknown, biotech intervention; in spite of this, the critical aspects remain.

Very good, anyway. Everything seems to work – at least on paper. What's the next step? The RMR, of course, which stands for 'Robust Mouse Rejuvenation' – basically, taking a mouse and rejuvenating it *a lot*. Our goal will be to take twenty common mice –*Mus musculus* – and make them live beyond their normal limits; in other words, we should take two-year old mice – which have a life expectancy of three more years – and make them live *five* more years. Such a result would cause a huge international debate, because it would persuade basically everyone that defying aging is not an impossible goal. Protests would probably follow – and maybe some riots, but RMR might convince politicians, entrepreneurs and scientists to kickstart a "War on Aging" that could make the "War on Cancer" launched by Richard Nixon at the beginning of the '70s look lame. Aging has always been considered an immutable part of the human condition, but if we show that it can be at least partly repaired, this would probably kickstart a whole new bunch of studies and experiments that would be more and more effective.

Of course, says De Grey, the therapies that we will test first on mice and then on humans in a decade or so will be far from perfect; no matter how good they are, the body will still accumulate some residual damage that they won't be able to treat; even though we will apply them frequently and thoroughly, in the end, the aging process will win. For every kind of damage, we will find aspects that are easy to treat and aspects that are difficult – and the latter will keep accumulating

---

[26]The issue with telomerases is not the only one: De Grey speaks of a further mechanism used by cancer to reproduce, a mechanism that could be blocked with a specific gene therapy as well. For details, we refer you to his book.

and carrying out their effects. As the seven types of damage are removed, it is likely that an eighth type will emerge, and a ninth, a tenth, and so on. Furthermore, it is also likely that the seven types of damage will reveal themselves as being composed of different sub-categories, more or less difficult to treat. For example, some AGE might turn out to be treatable only with new and more powerful chemical agents, still to be developed.

To tell the truth, these therapies don't have to be perfect: it's enough that they make middle-aged persons – like De Grey himself – live a few more decades in good health, decades that can be spent in further investigation of even more powerful therapies. The basic idea is one of a competition between us and the aging process: that is, trying to reach a rhythm of scientific development so quick that it allows us to gain more years than the aging process and entropy can take away. With the final goal of being able to repair not all of the damage produced by aging, but the minimum necessary to keep us alive and well – no need to go beyond that minimum –, avoiding physical and mental senescence indefinitely. This is what De Grey calls "Longevity Escape Velocity": the scientist did the math for us and, according to his calculations, if we double the efficacy of our therapies every forty years, we should be able to beat aging permanently, reaching a lifespan of 5000 years, a number estimated by De Grey using actuarial statistics, in which you calculate the probability that a certain person will be involved in an accident. So, according to the scientist's guesswork, with the Longevity Escape Velocity, you can reach 5000 years, after which you have 100% chance of getting into an accident.

Everything alright, then? We follow this plan and we become immortal? Mainstream science has nothing to add? Actually, yes. Despite the fact that several pieces of the SENS program belong completely to official science – for example, the research on stem cells, on the genetics of cancer, and so on –, the entire program has faced severe criticism from several members of the scientific community. The main charge is that De Grey's proposal is "eccentric" and, as our knowledge of the aging process is still incomplete, it will take a lot of time to become actually viable. Although legitimate, De Grey's theory, according to which the only DNA damage that counts is that conductive to cancer, is not universally accepted.[27]

Moreover, in 2005, the *MIT Technological Review*[28] – a magazine of "high" popular science, published by the Massachusetts Institute of Technology in Boston – issued an article in which the authors branded SENS as clearly

---

[27] The "theory of aging provoked by DNA damage" has its followers as well, and, during recent years, growing experimental evidence has been gathered. Cf. for example, C. Bernstein e H. Bernstein, *Aging, Sex, and DNA Repair*, Academic Press, San Diego 1991.

[28] http://www.technologyreview.com/

unrealizable and offered a not-so-nice portrait of De Grey.[29] The article inspired many comments, and the intervention of De Grey himself, notoriously very argumentative. The debate – or, better yet, the clash – spurred the "SENS Challenge" – promoted by the *MIT Technological Review* and the Methuselah Foundation –, a competition awarding 20,000 dollars to anyone who could demonstrate that SENS "is so wrong that it doesn't even deserve a detailed debate." For the jury, the magazine picked Craig Venter – the famous American biologist who challenged the Human Genome Project in the race to decode the human genome –, Nathan Myhrvold – mathematician, entrepreneur, expert in technological innovation –, Anita Goel – specialist in nano-biotechnologies –, Vikram Kumar – pathologist at the Brigham and Women's Hospital, Boston – and Rodney Brooks, expert in artificial intelligence and computer science at MIT. The magazine received five articles authored by as many scientists or groups of researchers aiming to debunk De Grey's work; two were rejected, as they didn't satisfy the previously established criteria. The Cambridge biogerontologist was allowed to answer the papers, and their authors were then allowed to reply. In the end, the jury made its decision: no article managed to show that SENS was "so wrong that doesn't even deserve a detailed debate." Neither did De Grey manage to cogently show the validity of his project. An impartial decision was handed down, with the prize split into two parts. Myhrvold stated that SENS makes many unfounded statements and surely is not scientifically proven. He also pointed out that Estep and colleagues[30] didn't prove SENS wrong, as this would require more research. Speaking about SENS's critics, Rodney Brooks has stated that he thinks that they don't understand engineering, and that their criticisms about a legitimate engineering project are weak. Succinctly, Venter ruled that neither the critics nor the supporters of SENS had proven their respective points. All of the members of the jury agreed that De Grey's proposal is somehow "proto-science," that it is in a kind of scientific "waiting room," that is, waiting for an independent verification or rejection.[31] De Grey took this positively, admitting, without any problem, that his work is unavoidably speculative, but worthy of consideration.

In 2005 as well, *EMBO Reports* – the journal of the *Nature* group specializing in molecular biology – published an article signed by Huber Warner, Julie Anderson, Steven Austad and 25 other scientists, *Science fact and the SENS*

---

[29] http://www.technologyreview.com/featuredstory/403654/do-you-want-to-live-forever/

[30] One of the competing groups was led by Preston W. Estep, an American biologist working in the field of aging processes. His team presented a very articulated critique of De Grey's project, but it was very argumentative as well, criticizing the jury's decision very harshly.

[31] http://www2.technologyreview.com/sens/

agenda,[32] in which they state that De Grey forgot to mention that his methods never managed to extend the lifespan of *any* organism. Which is totally true, as admitted by De Grey himself.

At this point, I would like to ask: is De Grey's project really so weird, so eccentric? After all, the Modern Age – starting around the Eighteenth Century – introduced the concept of the "deconstruction of mortality," the idea that death should be deconstructed and divided into individual causes, to be fought one by one. As it stands, De Grey is just continuing the very same project that has been keeping our society busy for centuries.

Let's take a look at the official mission of the SENS Research Foundation:

Aging is unarguably the most prevalent medically-relevant phenomenon in the modern world and the primary ultimate target of biomedical research. (…) when it is developed, this panel of therapies may provide many years, even decades, of additional youthful life to countless millions of people. Those extra years will be free of all age-related diseases, as well as the frailty and susceptibility to infections and falls that the elderly also experience. The alleviation of suffering that will result, and the resulting economic benefits of maintained productivity of the population, are almost incalculable. In our capacity as the overseers of SRF's research strategy, we urge you to do all you can to help SENS Research Foundation carry out this mission with maximum speed.[33]

24 scientists signed this statement, among which were Anthony Atala, director of the Wake Forest Institute for Regenerative Medicine; Maria A. Blasco, a specialist in telomers and director of the program of molecular oncology of the Spanish National Center for Cancer Research; Judith Campisi, famed biogerontologist and member of the Lawrence Berkeley National Laboratory; George Church, geneticist at Harvard Medical School; Irina Conboy, a bio-engineer at Berkeley University. And, of course, Michael West and – surprise surprise – William Haseltine. He might look a little bit weird in this company, this self-taught British biogerontologist; in fact, he managed to assemble a team of respected researchers. And here you have it: maybe De Grey – an eccentric and a little bit of a neurotic genius – has achieved a result for sure. He has offered us, even if just for one second, the opportunity to nurture the sweetest of all delusions: the possibility – at last – to cheat death.

---

[32] http://www.ncbi.nlm.nih.gov/pmc/articles/PMC1371037/

[33] http://www.sens.org/about/leadership/research-advisory-board

3.6 Three Bridges to Eternity

Okay, De Grey might make us immortal; so what? Should we just hang out without a care, and wait for the first results, keeping our fingers crossed? Not really. Or · at least not according to Ray Kurzweil. In fact, our "super-baby boomer" has made a decision: to take his biological destiny into his own hands, developing a very personal plan in order to maximize his chances of living forever. A plan that you can adopt, too, by reading the book he wrote in 2004 with Terry Grossman, *Fantastic Voyage: Live Long Enough to Live Forever*, a title that exemplifies the philosophy of its two authors.[34] In 2009, the two published a follow-up, *Transcend: Nine Steps to Living Well Forever*, in which the authors translate the content of the first book into an easy-to-remember framework.[35]

The starting point is quite simple. If – like the authors – you are middle-aged persons, if you try to stay in shape and to take care of your health the best possible way, if you manage to reach ninety – or, even better, one hundred and twenty –, well, you have a good chance of not having to die anymore. And all of this because we will be able to take advantage of the genetic therapies under development right now first, and of nanotechnologies second. It's the Kurzweilian theory of the Three Bridges, or, better yet, "a Bridge to a Bridge to a Bridge." Bridge One consists of nothing more than our present medical and scientific background about health, mostly in regard to prevention. It is the most substantial part of the plan, the one explicitly developed by Kurzweil and Grossman themselves. The instructions provided to us by the two consider the main pathological conditions of the modern world, from diabetes to cancer, from heart disease to Alzheimer's. The proposed solutions combine ideas that come from both official medicine and – alas – from alternative medicine. The plan includes physical activity, caloric restriction, low-glycemic index foods – so no carbs and no sugars –, and ionized water – a pseudo-scientific practice. Not satisfied by this punishing regime – 1500 calories per day, and just 80 grams of carbs, basically, bye bye pasta and pizza –, Kurzweil is undergoing a regime that is more rigid than the one recommended to his readers: which means weekly injections of vitamins – a method *not* endorsed by official science – and chelation therapy. The latter is a therapy that was developed for the first time during the First World War, and consists in injections of chemical agents into patients intoxicated by heavy metals in order to remove them; right now, it is offered by alternative practitioners, who state that some conditions, like heart disease and autism, might be caused by a silent – and unproven – intoxication of metals. Rejected by the FDA, chelation therapy looks good to

---

[34] R. Kurzweil, T. Grossman, *Fantastic Voyage: Live Long Enough to Live Forever*, Rodale Books, Emmaus 2004.

[35] R. Kurzweil, T. Grossman, *Transcend: Nine Steps to Living Well Forever*, Rodale Books, Emmaus 2004.

Kurzweil. And this is not even the most radical thing that he does: his supplementation regime is even more radical. Kurzweil used to ingest 250 pills per day – now 150 – of several types of supplement: vitamins, minerals, herbs, even aspirin, with the final goal of "reprogramming his own biochemistry," even though it's not clear if he means that metaphorically or literally – the second one, probably. Some of these supplements do work, some others are dubious, for example, a Chinese herb, gingko biloba. He also uses tools to limit his exposure to magnetic fields and an air purifier in his bedroom. And let's not forget the benefits the inventor gets from meditation and lucid dreaming.[36] Even Kurzweil's diet *might* be scientifically sound: caloric restriction has been researched, and the benefits of a low-glycemic index diet have not been proven, but they might be in the future. It is promoted by Cynthia Kenyon, an American biogerontologist who studied the aging process in *Caenorhabditis elegans*, a little roundworm; the scientist discovered that feeding these animals with sugar would shorten their lifespan. And that's why Kenyon decided to get rid of all high glycemic index foods, pasta, bread, pizza, sweets, potatoes. As rigorous as Kurzweil, then; for his part, Kurzweil recommends – without laughing – that his readers wait a couple of decades in order to get an ice cream. That is, to wait for the therapies of Bridge Two.

There you go: if you do the homework, you might reach Bridge Two, which is basically everything medical science is working on right now, and the results of which we should be able to see – according to Kurzweil – in a couple of decades. The list of therapies compiled by Kurzweil and Grossman in their book is quite long: from therapeutic cloning – using stem cells to cure degenerative pathologies – to gene therapy – inserting new genes into the body using genetically modified viruses as vectors – to genetic testing – knowing our predisposition to this or that disease – to neural implants – easing the symptoms of severe pathologies like Parkinson's or epilepsy – and much more. These are technologies that have been under development for many years now; for example, the first intervention of gene therapy – an attempt to cure a girl suffering from the ADA syndrome, a severe malfunction of the immune system – was performed for the first time in the US in 1990. Time passed, and progress has been made; it is still difficult to say when gene therapy will really revolutionize our approach to health and disease. What matters is that these are promising therapies, aggressively researched; the only issue I have with Kurzweil's approach is that he presents Bridge Two as a set of

---

[36] Lucid dreams are dreams in which dreamers are aware of dreaming and can actually guide their dreams, using them for different purposes. After a few anecdotal reports, the existence of lucid dreams was scientifically proven for the first time, in the '70s, by Stephen LaBerge, a Stanford University researcher, who also developed specific techniques to facilitate the fruition of this oneiric phenomena. Cf. S. LaBerge e H. Rheingold, *Exploring the World of Lucid Dreaming*, Ballantine Books, New York 1991.

developments that will come at us quickly – actually, according to Kurzweil, all progress, not just that of medicine, is undergoing a process of acceleration. I am not very persuaded by this idea; anyone who works in biotech knows that it is a complex, uneven terrain, and that it's difficult to make predictions. But, as usual, I might be wrong.

Bridge Three – which refers to the possibility of curing and regenerating our body using nano-machines as small as viruses and guided by nano-computers – is the most controversial; here, Kurzweil's predictions enter *terra incognita*, as we are going to see when we talk about nanotechnologies.

Anyway, Kurzweil's optimism knows no bounds; while he has predicted the arrival of immortality by the year 2045, recently, the inventor joined another Transhumanist foundation, the Maximum Life Foundation, which has the mission of reversing the aging process by 2033 – or 2029, according to the specific web page that you visit.[37]

One more musing. It has been a number of years now since science writers and scientists began to consider the idea that aging and death are two processes that could be manipulated using science and technology.[38] In recent years, this idea has been spreading more and more, that is, death and aging have now lost the "sacred aura" conferred upon them by us humans since the dawn of time. What separates Transhumanists and the mainstream scientific community is more about the timing, rather than the feasibility. What matters is that the time in which looking for the Fountain of Youth[39] was a forbidden dream is over. Now we can talk freely about immortality and rejuvenation. Bottom line, if you are contemplating this perspective, be advised: you might find a lot of people to talk to.

3.7   Fighting Aging Today

De Grey and Kurzweil are the harbingers of a wider movement, which, in recent years, has gathered momentum as well as an increasing number of representatives. Let's take a look, naming some of the groups, enterprises, researchers and activists that want to cheat death, or at least to live a very long life.

Liz Parrish is a controversial American life-extensionist and CEO of the Seattle-based biotech company BioViva. In September 2015, she made headlines for

---

[37] http://www.maxlife.org/

[38] Cfr. B. Bova, *Immortality: How Science Is Extending Your Life Span – and Changing The World*, Avon Books, New York 1998.

[39] As historically done by the Spaniard Ponce de León in 1513, an enterprise that ended with the discovery of Florida.

flying to Colombia in order to experiment on herself; more specifically, she received two experimental gene-therapies, a telomerase gene therapy – supposedly able to lengthen telomers – and a myostatin inhibitor – a compound that is being tested in the treatment of the muscle loss caused by a few diseases and by the aging process itself. Criticized by the scientific community, Parrish – who claims that her approach did work – has been strongly supported by the Transhumanist community.

Betterhumans started as an educational website in 2001, as a resource for all things Transhumanist. It ceased such activities in 2008, becoming a Florida-based nonprofit Transhumanist bio-medical research organization. Among its advisors, Betterhumans has no less than the famed Harvard Medical School geneticist George Church. The research carried out at the organization includes a study on supercentenarians – people who have reached or surpassed 110 years – aiming to unlock the genetic secrets of their "performance" – and the study of a few senolytics – compounds that might protect the body from the damage caused by aging.

Another Transhumanist scientist, Greg Fahy, is working on the rejuvenation of the thymus, which is an essential part of our immune system – it produces T cells – and begins its decline in our 50s, a process known as thymic involution. Fahy's firm, Intervene Immune, is trying to rejuvenate the thymus using HGH – human growth hormone.

This is nothing compared to Ira Pastor and his BioQuark Inc., which aims to bring the dead back to life. More specifically, using stem cells, lasers, nerve stimulation and peptide injections (peptides are short chains of amino acids), the company is trying to promote brain repair in the case of patients who have been declared clinically dead, mimicking some amphibians and fish known for their ability to regrow parts of their bodies and even of their brains.

The Movement for Indefinite Life Extension (MILE) is a loose grouping of associations aimed at reaching indefinite life extension as soon as possible; it was founded by Eric Schulke and, among its delegates, lists the Nevada Transhumanist Gennady Stolyarov II. Stolyarov also developed Stolyarov's Wager, a philosophical argument in favor of life extension that goes like this:

> If you believe in human life extension and are right, you have everything to win – a happy, prosperous, indefinite life that you can be sure of in this world.
> If you believe in human life extension and are wrong, you cease to exist.
> If you do not believe in human life extension and are right, you cease to exist.

If you do not believe in human life extension and are wrong, it may be that the effort that you did not put in to promoting the idea was just enough for the possibility not to come to pass. Then, you will cease to exist.[40]

The Life Extension Advocacy Foundation[41] is very active and aims to crowdfund longevity research; among its members, I'd like to mention Keith Comito, Oliver Medvedik and Elena Milova.

Marios Kyriazis is a gerontologist and life extensionist; his approach entails, among other things, fighting the informational entropy that causes aging – that is, the idea that aging represents a loss of information and order – through a stressful and constantly stimulating lifestyle.[42]

Among the more active life extensionists, we have to mention the Israeli scientist Ilia Stambler, author of the highly recommended "A History of Life-extensionism in the Twentieth Century" and founder of the international life extension advocacy group Longevity For All and the umbrella organization Longevity Alliance[43]; he also founded an international organization, the Longevity Party, together with the Russian researcher Maria Konovalenko, who describes their movement as "100% Transhumanist." Among other things, Konovalenko is working on a Longevity Cookbook.

Let's also mention another prominent longevist and Transhumanist, the Portuguese scientist João Pedro de Magalhães, who, besides doing research, runs an interesting website, senescence.info.[44]

A veteran of longevism worth mentioning is Bill Andrews, an American molecular biologist whose research revolves around finding a cure for aging. Andrews founded and is the president of the biotech company Sierra Science. In 1997, while at the Geron Corporation, he successfully identified human telomerase.

It is important to note that the people whom we have just mentioned are – or at some point were – considered in some way to be outsiders, but now, the anti-aging movement has finally gone mainstream, with at least two companies, Google's

---

[40] G. Stolyarov, *An Atheist's Response to Pascal's Wager*, The Rational Argumentator (Issue 137), December 2007. http://rationalargumentator.com/issue137/pascalswager.html. Stolyarov also wrote a children's book on immortalism, *Death is Wrong*, Rational Argumentator Press, 2013.

[41] https://www.lifespan.io/

[42] M. Kyriazis, *Reversal of Informational Entropy and the Acquisition of Germ-like Immortality by Somatic Cells*, https://arxiv.org/abs/1306.2734

[43] http://www.longevityalliance.org/

[44] www.senescence.info/

Calico[45] – California Life Company – and Craig Venter's Human Longevity Inc.,[46] both aiming to hack the aging process. It's seems like the beginning of an important and momentous revolution, but, as they say, time will tell.

There is one last point I would like to make, which involves the infamous "Murphy's Law." Just think about this: you are a prospective immortal, you follow Kurzweil's plan thoroughly, you diet, take supplements, meditate, and so on. And still, you die. It could happen. After all, it happens all of the time in this grim world. Maybe the ultimate technology, the technology that would have given you immortality – what computer scientists like to call the "killer-app" – will be launched the week, the day or the second after your death. That would definitely suck, for every aspirant immortal Transhumanist. Do you really think Transhumanists haven't thought of that? Of course they have. And that's exactly why they prepared a specific "Plan B."

---

[45] http://calicolabs.com/
[46] https://www.humanlongevity.com/

4

Plan B

4.1 Upside Down, in Liquid Nitrogen

Transhumanists are incurable optimists, no doubt about that. They do strongly believe in techno-scientific progress, and they hope that physical immortality is really around the corner, that they are going to enjoy a limitless life. I feel nothing but sympathy for these nerds, and, in spite of my skeptic background, I admit that these are normally very educated and smart people. That's why they know that the transition from human to post-human – a package that includes physical immortality – might take some time, that it might take too many decades or too many centuries. What can they do then? How can they cheat death? They need a "Plan B," and the solution is a magic word: "cryonics."

Usually, we hear about cryonics sporadically, when some television program covers this topic in a superficial way, often paying attention to its most colorful aspects. To cut it short, cryonics is the practice of freezing the corpses of recently deceased people, hoping that, in a more or less far future, science will be able to bring them back to life, making them healthy and young again as well. A very good program, even though, right now, it qualifies as more of a weak hope than a scientifically sound enterprise. This hasn't prevented its supporters from creating several organizations that offer to whoever wants it – and is willing to pay a decent amount of money for it – the opportunity to get frozen at the moment of clinical death. And maybe bringing along an already deceased pet. Just to be clear: the process of freezing – or, as cryonicists say in their peculiar jargon, the "cryonic suspension" – is allowed only after the customer has been declared clinically dead by competent authorities; otherwise, it would be a homicide or, in the best-case

© Springer Nature Switzerland AG 2019
R. Manzocco, *Transhumanism - Engineering the Human Condition*,
Springer Praxis Books, https://doi.org/10.1007/978-3-030-04958-4_4

scenario, an assisted suicide. In an ideal scenario, the process of cryonic suspension should begin within a few minutes of the cardiac arrest that killed the patient; promptly alerted, the members of the cryonic organization the patient signed up with would then be able to quickly saturate the organism with anti-coagulants and anti-freeze compounds, through a heart-lung bypass machine. At the same time, the operators would lower the temperature of the corpse down to −79 degrees Celsius, using dry ice, that is, frozen carbon dioxide. This stage, called "stabilization," is followed by the actual suspension, in which the body is stored in a specific tank – called a Dewar tank – at −196 degrees Celsius, which should allow for the indefinite preservation of the "patient." The choice to position the bodies upside-down is due to the fact that, should something go wrong – if, for some reason, the patients begin thawing unexpectedly – the brain, the location of our memories, our most important part, would be the last to thaw. Of course – and very honestly – cryonicists admit that they cannot pledge that their project will work, that their organization would actually be able to "reanimate" its hosts; the only thing that they can guarantee is that they will do their best to preserve their patients impeccably, in the hope that, sooner or later, science will be able to do something. And if you cannot afford a complete cryonic suspension, you can opt for a more affordable "neuro-suspension," which consists in solely freezing the head – in this case, the waiting time for resurrection would become much longer, as the already burdened science of the future would also need to find a way to regrow whole bodies. Neuro-suspension is probably the most gruesome aspect of the whole cryonic movement, one that caused the media to brand the whole enterprise as the "gathering and freezing of severed heads."

Let's review the list of cryonic organizations on the market. Starting in the '60s – when cryonics was born – several providers followed each other, going from an unsophisticated stage to a more professional one. Right now, cryonicists throughout the whole world can count on four main organizations, three American – Alcor,[1] the Cryonics Institute[2] and TransTime[3] – and one Russian – KrioRus.[4]

To this list, we have to add those organizations that don't manage facilities or containers, but simply work in the field of cryonics promotion or offer further

---

[1] http://www.alcor.org/. Alcor owes its name to one of the stars in Ursa Major, whose ability to be seen is made difficult by the near and brighter Mizar. Because of this, the Ancient Egyptians used Alcor to verify their youngsters' sight, in order to determine whether they could become good bowmen and hunters. The symbolic meaning of this choice is thus clear: Alcor – the organization – wants to indicate the far-sightedness of its members and founders.

[2] http://www.cryonics.org/

[3] http://www.transtime.com/

[4] http://www.kriorus.ru/en

services, like the American Cryonics Society[5] and Suspended Animation Inc.[6] – which organizes the procedures of stabilization and transportation to the final location. Let's start with the biggest facility, then, Alcor's, which is headquartered in Scottsdale, a lovely Arizona town known for its nightclubs. As we speak, Alcor counts 968 members and 111 patients preserved in liquid nitrogen – two-thirds of whom have opted for neuro-suspension; to this count, we have to add about 30 pets. Legally speaking, Alcor is a nonprofit organization, while, from the viewpoint of its patients' cryopreservation, it applies the local Arizona laws about organ donation – that is, legally speaking, the cryo-suspension of the members is interpreted as "donation of the body to science." Alcor was founded in 1972 by two California cryonicists, Frederick and Linda Chamberlain, and the first cryonic suspension – Frederick's father – was performed in 1977. Alcor has a Scientific Advisory Board, which includes Aubrey De Grey, nanotechnology theoreticians Robert Freitas and Ralph Merkle, Marvin Minsky, and Bart Kosko, a scholar specializing in "fuzzy logic."[7] Among the most famous subscribers, I'd like to recall baseball star Ted Williams and his son John Henry – both already suspended – while, among the active members, I'd mention Charles Matthau – son of the actor Walter –, Ray Kurzweil and Eric Drexler. Originally founded in California, Alcor has since been transferred to Arizona, in order to avoid the seismic risks of the original location – from this viewpoint, we can say that cryonicists really have tried to consider any possibility.

Let's talk now about the Cryonics Institute, headquartered in Clinton Township, Michigan. The organization has 938 members, of which 488 have already signed a contract for cryonic suspension; right now, the facility hosts 111 patients and 80 pets. Moreover, the opportunity to preserve a sample of your DNA is offered. As we said before, the Cryonics Institute was founded in 1976 by Robert Ettinger, who died at 92 in July 2011 and was promptly subjected to the cryostasis procedures he promoted – which at least proves how seriously he took the cryonic enterprise.

Then, we have Trans Time, in San Leandro, California. One of the few for-profit cryonic organizations still in existence, it doesn't share statistics about its patients – but, as far as we know, it has suspended three persons.

The last organization is KrioRus, founded in 2005 by a group of cryonicists in Alabushevo, thirty kilometers from Moscow; right now, its facilities host seventeen patients and six pets.

---

[5] http://www.americancryonics.org/

[6] http://www.suspendedinc.com/

[7] An approach that privileges approximation over the dichotomist choices of classic logic.

4.2   How Do You Freeze – *pardon me*, "Suspend" – A Patient

Now let's take a look at the protocols followed by cryonicists to perform a cryonic suspension. Our story begins with the development, during the Twentieth Century, of increasingly sophisticated procedures for cardiopulmonary resuscitation, using the cardiopulmonary bypass machine, invented in 1931 by the American surgeon John Gibbon, and the defibrillator, invented at the end of the Nineteenth Century by two Genevan researchers, Frederic Batelli and Jean-Louis Prévost. These developments inside medical science have saved a huge number of lives, but they have also changed the way in which we die and have transformed death from an instantaneous event to a process prolonged over time and divisible into various stages. In previous ages, if a person's heart stopped, there was not much that could be done. Now, with the abovementioned technologies and procedures, it is literally possible to reanimate patients who are otherwise doomed. Of course, these developments have spurred a lively debate inside the scientific community about the criteria we should use in order to define the death of a human being. And so, during the '60s, we saw the birth of a few important distinctions, like the one between clinical death – which consists in the arrest of the heart – and cerebral death – the complete and irreversible cessation of any electrical activity in the brain, a definition coined for the first time in 1968 by a specific Harvard Medical School committee. It's a very complicated topic, especially for its ties to the topics of organ donation, euthanasia and assisted suicide; for our purposes, what we need to remember is that the progressive medicalization of human life has turned death into an articulated process, prolonged in time, thus opening a space in which cryonics could find a role.

   The main assumption of cryonicists is that, if medical science has developed subtler and subtler definitions between clinical death, respiratory death, cerebral death, and so on, it is possible that further investigations will bring about even more refined definitions. And so – the theoreticians of cryonics think – as the abovementioned technologies have persuaded us that clinical death might not be a "final" event, so, in the future, even cerebral death might become a reversible phenomenon, losing its status as "ultimate death."[8] The cryonicists' bet – very risky indeed – is the following: if we freeze a patient who is just deceased, the biochemical degradation of the brain will also stop; plus, the almighty science of the future will be able to repair the damage the brain underwent because of cerebral death and the freezing process. And the procedure would literally end with the reanimation of what – from the viewpoint of our far-future descendants – has never been a simple frozen corpse, but rather a patient waiting for adequate

---

[8] For further information, read the article by Benjamin P. Best, president and CEO of the Cryonics Institute: B. Best, *Scientific Justification of Cryonics Practice*, «Rejuvenation Research», Vol. 11, n. 2, 1988, http://www.cryonics.org/reports/Scientific_Justification.pdf

treatment. In order to make their approach scientifically sound, cryonicists have introduced a new and, according to them, more accurate definition of death, that is, "information-theoretic death." A few years ago, the Californian nanotechnologist and cryonicist Ralph Merkle[9] coined a definition of information-theoretic death based on the idea that the individual is essentially made up of her/his memories and behavioral patterns – that is, his memory and her character. While our short-term memory might be tied to the dynamic activity of the brain – and then disappear when the latter ceases –, the long-term memory might depend on microscopic structural changes in our nervous architecture – an idea that belongs to neurosciences *tout court* as well. To Merkle, the obvious consequence of this is that, even a few hours after the cerebral death, our long-term memory could be substantially intact, and a prompt cryonic suspension – in spite of the damage caused by freezing – might allow us to preserve it. A sufficiently advanced technology – by this, Merkle means the nanotechnologies of the near future – could fix, at a molecular level, the body and the brain of the patient in cryostasis, bringing her/him back to life with an intact memory and personality.[10] As an alternative to nanotechnologies, we can opt for mind-uploading, which consists in the very speculative possibility of transferring the content of the mind into a non-biological substrate – basically, a very powerful computer. Quite a leap, I have to say; anyway, being aware of the technical difficulties that the future scientists will face in the process of reanimation, cryonicists are always trying to develop procedures for cryonic suspension that are as little detrimental as possible. Which means: to prevent as many of the types of damage caused by the freezing process as possible – contrastingly, those caused by the process of thawing are none of our business, as they will be a concern for our descendants. So, this is the essence of cryonics: that the real death is not the cerebral one, but the information-theoretical; as long as this doesn't occur, a person cannot be pronounced completely dead. Then, if we intervene promptly, we can preserve memories indefinitely, solidly nestled in the cerebral microstructure, while we wait for medical science to restore our patient.[11]

Now let's take a detailed look at how things work for the customers of these cryonic organizations. Let's imagine that a person has signed a contract for cryonic suspension – for example, with Alcor –, following all of the advice given by the organization's consultants. And so, this person constantly wears the stainless steel bracelet and collar provided by the organization – which reports the phone numbers to call in case of an emergency and the protocols to follow by whoever

---

[9] Scientifically speaking, Merkle has contributed to the field of public-key cryptography, a sophisticated system used in the field of data protection.

[10] Cf. R. Merkle, *The Molecular Repair of the Brain*, http://www.merkle.com/cryo/techFeas.html

[11] The American press has already christened these people in cryostasis *human popsicles*, or *corpsicles*.

finds the body of this person –; has informed his/her spouse and his/her relatives and has made them understand his/her motives[12]; always brings with him/her a document stating his opposition – for religious reasons – to any form of autopsy,[13] and so on. And then, death – in the form of a terminal illness – hits this person. Contrary to how it may seem, an incurable disease is an ideal scenario, as it allows the patient to move to an American hospital that has an agreement with Alcor. The patient is constantly watched by the nurses, and the cryonic suspension team is ready to intervene. As soon as the physician pronounces the patient dead, the team starts the procedure; the body and the head are covered in dry ice, the patient is connected to the heart-lung bypass machine and two venous catheters are inserted into the body. Following this protocol, the team can inject anti-coagulants into the body – in order to avoid the formation of blood-clots in the brain – and to remove the blood, perfusing the whole body with anti-freeze compounds – mostly glycerol. This latter compound is toxic, but this is a problem that our descendants will fix, right? The most imminent issue is to prevent the formation of ice. As is known, the main side effect of freezing is the considerable damage to the cellular structure of the patient; in particular, if this process is very fast, the water cannot exit the cells, and its transformation into ice breaks them down. If, on the contrary, the process is slow, the water can exit the cells, but the latter end up with a too-high concentration of potentially toxic chemicals. During recent years, the cryonic organizations have begun to adopt a new procedure based on the massive use of anti-freeze compounds, which manages to vitrify the corpse – and especially the brain –, that is, the water inside the body acquires a glass-like texture, without turning into ice. Anyway, our patient undergoes a further lowering of its temperature – down to −79 degrees Celsius; then, the body is transferred to the Alcor facility and immerged in liquid nitrogen, until the temperature reaches −196 degrees Celsius. Theoretically, at this temperature, the body chemistry should undergo a huge slowdown. There is another problem, though: no matter how carefully you suspend a patient, the body will spontaneously undergo micro-fractures – to which there is no solution, you can just leave it to the science of the future. Anyway, the patients get immerged into a container full of liquid nitrogen, which is not based on an electric cooling system – in order to avoid any risk connected to a possible blackout –, but is constantly replenished with fresh liquid nitrogen. So, as we can see, the damages suffered by the organism – which, according to our

---

[12] Sometimes, the family members do oppose cryonic suspension, and it happens fairly often that some relative blocks the procedure of cryostasis. As the customers are often men, the cryonic community speaks half ironically of the "syndrome of the hostile wife," – and Ettinger himself recommends divorcing any partner who opposes cryonic suspension.

[13] It's a stratagem used by cryonic organizations to avoid – when possible – the requisition of the body by health authorities and the damaging of the brain by the coroner's intervention.

criteria, is totally and definitely dead – are consistent, and are related to a long series of chemical and physiological processes that we are going to spare you. Essentially, cryonics finds solace in the fact that, as far as we know, the possible reanimation of these "corpsicles" does not violate any physical law, and that the science of the future may be able to manipulate matter at the molecular and atomic level, allowing us to do practically anything.

4.3 Cryonics' Prices, Not for the Faint of Heart

Now let's talk money, that is, how much does it cost to get frozen – or vitrified? One of the stereotypes – an incorrect one – about cryonics is that this practice is for rich people. Actually, the better part of its customers is represented by middle-class folks, and they have different payment methods available to them. And what about the prices? Their range goes from ten thousand to two hundred thousand dollars, according to the organization we consult and the kind of suspension we want. And so, Alcor charges two hundred thousand dollars for a whole-body sus-pension and eighty thousand for neuro-suspension. The Cryonics Institute charges much less, twenty-eight thousand for members and thirty-five thousand for non-members – that is, those who decide at the very last moment to get frozen. To these costs, we have to add other extra expenses, for example, the cost of the procedure of vitrification and stabilization with dry ice, transportation, funerary expenses, the extra-price for the last-minute option, and so on. Basically, if you pick the Cryonics Institute, you have to use the services of Suspended Animation Inc., which range from eighty-eight to ninety-five thousand dollars, plus extra if you need aerial transportation. The most convenient organization for your one-way trip to the future is KrioRus, which charges thirty-thousand for a whole-body suspension and ten-thousand for a neuro-suspension. If the costs look excessive to you, keep in mind that they are comparable to the average cost of American health care – notoriously very expensive, mostly because of its private nature. Don't worry, though. You can finance your cryostasis through a specific life insurance that names your cryonic organization of choice as beneficiary; if you are young and relatively healthy, you can get away with a limited monthly fee – more or less fifty dollars per month. In truth, finding an insurance company willing to fund your cryonic suspension is not easy, and that's why, connected to Alcor, there's a financial planning agency that offers this service.[14]

These are the duties of the customers. And what about those of the cryonic organizations? How do we force them to comply with their promises? Clearly,

---

[14] If you are seriously considering this option, you can contact the American agency that offers this kind of insurance plan, Hoffman Planning: http://www.rudihoffman.com/

they cannot assure us that they will bring us back to life when the right technologies are ready; to tell the truth, they cannot even assure us that they will still exist and be operative on that far day. This is explicitly stated by these organizations themselves; more specifically, they inform us that they cannot assure us that those technologies will exist, and neither that the people of tomorrow will want to invest energy in re-animating us. That said, it is definitely possible that our descendants will have unlimited riches and will be curious to speak with the people of the Twenty and Twenty-First Centuries; this is nothing more than a possibility, though. To offer a remedy to this uncertainty, as early as the '80s, Frederick and Linda Chamberlain launched the "Lifepact." The purpose of this initiative is to push cryonicists to sign actual contracts with each other in which they commit to do their best to reanimate the contracting party suspended before them. Essentially, the contracts invented by the Chamberlains are just a starting point for an organization – still developing – in which the members commit to paying the costs of reanimation and the care of the other members, and to helping them to resurrect whoever is left. The Lifepact is based on one of the mantras of cryonics, that is, "last in first out": as the suspension technologies of today are better than those of yesterday, and those of tomorrow will be better than those of today, the future "corpsicles" will be in a better condition than those of the past and the present, thus their reanimation will be easier. In other words, they will be the first to get out of the suspension tanks, and will be forced by the organization's Lifepact – basically a "relay" among generations – to reanimate the patients belonging to the previous generation. These will have to reanimate the previous one, and so on, until they get to the first people who were frozen. Basically, the Lifepact method should infuse the cryonicist movement a kind of "sense of community" that makes them feel compelled to work for each other. And it's not over yet: there is the possibility of signing personalized contracts, in which the patient can list the conditions under which he/she wants to be reanimated. So, you can choose whether you want your mind transferred to a non-biological substratum, or if you want to wait for the advent of nanotechnologies, if you want to be reanimated even though your memories got lost, and so on. The contract includes the necessity that the contracting parts make as complete a list as possible of the main aspects of their life, just in case those memories get lost and as an "instruction manual" for those who will resurrect them. Plus, the Lifepact includes the option of storing, in specific facilities, documents and objects of any kind, in order to help the recovery of memories and adaptation to the new life. Finally, the Lifepact will work as a group for psychological support, just in case the resurrected are haunted by the memory of those that they cannot recover.

Now let's talk money again, and let's make a hypothesis. Let's pretend for one second that cryonics can work, and that the science of the future brings us back to life. Maybe the world of tomorrow will be so rich that we won't need any money, and neither will we need to work; maybe money will be just a bad memory of a dark age long gone – like in "Star Trek," in which they don't have money anymore. But it's possible that money will keep existing and playing a central role in everyday life for a very long time. In that case, the resurrected will be without any money, and maybe an eternal life of misery in an unknown and incomprehensible world will await them. No worries: cryonics has – almost – taken care of this, too. In particular, the Cryonics Institute has instituted, in Lichtenstein, the Reanimation Foundation, an investment fund that will manage the money that the members of this organization wish to deposit, in the hope of recovering it after reanimation – the minimum deposit allowed is twenty-five thousand dollars. We have to say that the State of Lichtenstein allowed the operation, but didn't like it that much – maybe because cryonicists are seen as quite eccentric people. Right now, in the US, cryonicists are working on alternative solutions, maybe picking states like Alaska or South Dakota, where the internal legal systems are not hostile to this kind of initiative.

4.4   The Almighty Science of the Future

Here we are, covering the most controversial part of cryonics: reanimation. It's completely useless to sign a contract for cryonic suspension, commit to a Lifepact or pour money into the Reanimation Foundation, if our chances of being brought back to life are near to zero. That's why cryonics supporters have tried, since the beginning of the movement, to imagine which systems or technologies our descendants would use to bring them back to life, that is, to prove that the cryonics enterprise is plausible. The first one to raise the issue was, in 1961, Robert Ettinger himself, who settled for "super-robot surgeons able to work for years or centuries." In 1969, another cryonicist, Jerome White, spoke of "specifically programmed viruses."

Not satisfied by this vagueness, and by the excessive dimensions of Ettinger's robot surgeons, in the '70s, Michael Darwin[15] – president of Alcor between 1983 and 1988 – devised the anabolocytes, imaginary biotech organisms able to move autonomously and repair all of the damage suffered by the patients in cryostasis. It would have been enough to flood their bodies with millions or billions of anabolocytes and everything would have been fine. Let's face it: it's a slapdash solution, as nobody knows how to create a micro-organism able to invade a host, feed,

---

[15] Real name Michael Federowicz, while "Darwin" is the nickname awarded to him by his classmates because of his strong opposition to creationism and his defense of Darwinian evolutionism.

maybe reproduce, and, in addition, precisely fix any kind of cellular damage – produced either by aging, disease, death, freezing or anti-freezing compounds. More recently, cryonics has gotten some theoretical help from Eric Drexler, his *Engines of Creation: The Coming Era of Nanotechnology* and the nano-machines hypothesized by the author. In spite of the theoretical and technical issues raised by nano-machines, between cryonics and nano-assemblers, it was love at first sight; thanks to Drexlerian nanotechnology, it would be possible to repair the patients' bodies, molecule by molecule, with no limit. Even better: Merkle and Freitas have developed a detailed scenario regarding the way in which this could happen.[16]

Almost the entire process should happen before the thawing of the patient, in order to avoid the damage this could provoke. To begin with, the nano-machines would free the circulatory system, from the main vessels to the capillaries; that action should look like a super-precise digging operation – that is, the vessels should be treated like tunnels completely obstructed by super-compact debris. The nano-assemblers' work would be controlled by very sophisticated computers located outside of the organism and able to follow the whole procedure step by step, through systems of imaging of the future. Once granted access to the various tissues, the nano-machines would fix the micro-fractures, "tying" the different parts together through some biologically inert material. At this point, it should be possible to slowly start the procedure of warming up, until the temperature allows for the presence of liquid compounds, which would allow the nano-machines to intervene on the molecular chemistry, fixing the damage suffered by nucleic acids, proteins and the cells in general. Once that is done, the metabolism could be kick-started again, and the patient reawakened. No matter how unlikely and fanciful, there's no doubt that these "rational fantasies" elaborated by cryonicists are quite fascinating; they must be considered part of the history of science, even though their role – pioneering or marginal – will be ruled only *a posteriori*. For my part, I will now try to outline a general history of the cryonic movement from its origins until today.

4.5   A History of Cryonics, from Benjamin Franklin to Nano-Machines

Let's reprise the history of cryonics that we briefly sketched in the second chapter. Searching for an inspirational figure, cryonicists often look to Benjamin Franklin, who, in a 1773 letter to the scientist Jacques Dubourg, confessed his desire to come back to life in the far future, so that he could observe the destiny of the

---

[16] Cf. R. Merkle e R. Freitas, *A Cryopreservation Revival Scenario using MNT*, in: «Cryonics», October/November 2008. MNT stands for *molecular nanotechnology*.

United States; the half-serious method proposed by the American politician and scientist consisted in immerging himself and a bunch of his friends in a barrel of Madeira, ending up somehow embalmed, so as to later be revived by the Sun's rays.[17]

Besides science fiction, the first real systematic proposal of cryonic suspension comes from a book published in 1961 by the first official cryonicist, Evan Cooper, who published, at his own expense, a book, *Immortality: Physically, Scientifically, Now*. Afterwards, in 1964, Cooper founded the Life Extension Society, for the purpose of promoting these ideas.

Later on, in 1962, Robert Ettinger published – at his own expense, too – *The Prospect of Immortality*, a book in which the author independently developed the same ideas as Cooper. The paternity of cryonics is normally attributed to Ettinger; this stems from the fact that his book got republished – with great success – by an important American publisher, Doubleday, thanks, above all, to the recommendation of two famous science fiction writers, Frederick Pohl and Isaac Asimov. So, the cultural background of cryonics is apparently the world of science fiction literature, which was quite popular in the US in those years.

Starting with Cooper and Ettinger, the cryonics movement slowly began to spread, finding proselytes everywhere. The name "cryonics" was invented in 1965 by cryonicist Karl Werner; in the same year, Curtis Henderson and Saul Kent founded the Cryonics Society of New York. The following years saw the birth of the Cryonics Society of Michigan, the Cryonics Society of California, and the Bay Area Cryonics Society – later re-named the American Cryonics Society. In 1967, Ettinger himself joined the party, founding the Cryonics Institute.

The first person to be cryo-suspended was, in 1967, James Bedford, a 73-year-old professor of psychology – among those frozen during those pioneering years, Bedford is the only one still preserved, at the Alcor facilities.

The first, hard blow to the cryonics movement was an event that the cryonicists remember as the "Chatsworth disaster," named after the location at which Robert Nelson – a TV repair technician and president of the Cryonics Society of California – had stored a number of bodies entrusted to him. All of a sudden, it was discovered that nine of them had thawed, because of a lack of funding – the families of the "patients" had paid a certain amount of money and, when it ran out, in some cases, Nelson used his own money and kept a few of his corpsicles frozen for one and a half more years. Some of the bodies had thawed years earlier, and Nelson hadn't alerted the families. The scandal ended up in court, and cryonics received really bad press coverage as a result.

---

[17] Quoted in: E. Drexler, *Engines of Creation: The Coming Era of Nanotechnology*, on-line edition, http://e-drexler.com/d/06/00/EOC/EOC_Chapter_9.html

The '70s saw the birth of the biggest cryonic organization in the world, Alcor, founded by the Chamberlains in 1972 – initially, the name was the Alcor Society for Solid State Hypothermia, changed in 1977 to the Alcor Life Extension Foundation.

A further technical breakthrough was brought about between the end of the '70s and the beginning of the '80s, when Jerry Leaf – former University of California researcher in cardiothoracic surgery – introduced the heart lung bypass machine and other technologies into the cryonic practice, making the protocol of cryonic suspension more systematic. Together with Michael Darwin, Leaf launched the procedure of standby, which consists in assisting – when possible – the patients in their last minutes of life, in order to begin stabilization promptly.

Besides the advent of Drexler's theories about nanotechnology, the '80s saw the birth of CryoNet, a mailing list that allowed for the consolidation and expansion of the cryonics community. It's also worth mentioning the case of the actor and TV producer Dick Clair; terminally ill with AIDS, Clair sued the State of California and the hospital in which he had been admitted – and won – so that he might be cryonically suspended at the moment of his death – which happened in 1988. In the same year, Alcor was investigated by the local authorities because of the suspension of Dora Kent. Mother of Saul Kent, then a member of that cryonic organization, the woman died in 1987, and was cryo-suspended by her son – actually neuro-suspended. The death was not witnessed by a public officer – so nobody could pronounce her dead. Moreover, the autopsy of the body revealed the presence of the substances used by the suspension, which – from the viewpoint of the coroner – could also have been used to kill her. Alcor's properties were seized, and many members of the organization were arrested, but not before they managed to hide the cryo-suspended head of Dora Kent. The story was widely covered by the media, and, after that, Alcor was cleared of any wrongdoing and received a big compensation.

The year 1993 saw a schism inside Alcor, and a group of activists – among them, Michael Darwin – left and founded the CryoCare Foundation; afterwards, two for-profit companies joined the list, CryoSpan Inc. and BioPreservation Inc. The "rebels" added further innovations, including using bigger quantities of glycerol, thus promoting a less damaging process of suspension. The schism didn't last long, though, and, by 1999, CryoCare had been absorbed by Alcor and CryoSpan by the Cryonics Institute.

Essentially, the for-profit companies didn't have any success and, besides a couple of exceptions, the non-profits ruled. So, right now in the US, we can find Alcor, the Cryonics Institute and Trans-Time – which have the facilities to store patients –, and Suspended Animation Inc. and the American Cryonics Society – which offer only the procedures of stabilization and transportation. The latter also works at the spreading of the cryonicist philosophy. Same for the Immortalist

Society, an organization connected to the Cryonics Institute and devoted to education and cultural promotion.

Inspired by mainstream medical research, in 2001, Alcor began to use a mix of anti-freeze compounds able to induce the vitrification of the patient, that is, freezing without the formation of ice.

The '90s and the '00s saw the birth of several groups of cryonicists around the world – in Italy[18] and several other European countries, in the UK, in Canada and Australia – all more or less connected to the American organizations. Besides KrioRus, there are no other facilities able to host patients outside of the United States, even though the Australian cryonicists have been trying for a while to create one.[19] And let's not forget the providers of cryonic services headquartered in Europe, like Cryonics UK[20] and the Portuguese Eucrio,[21] which allow Europeans to get frozen and shipped to the US facilities.

The most recent cryonic initiative is the Timeship, planned by the well-known architect Stephen Valentine. His plans entail the construction – which he has been working at for several years now – of a facility devoted to cryonics and life extension. What Valentine is planning to do is to build a kind of "Fort Knox" to preserve biological materials – DNA, tissues, organs and whole human beings –, and, for this purpose, Valentine has designed it so that it might survive blackouts, hurricanes, terrorist attacks, natural catastrophes, and even the sea level rising. The Timeship should also conserve the DNA of species at risk of extinction. Basically, the Timeship aims to become the biggest facility of cryonic preservation ever created,[22] and, for this purpose, Valentine has already bought a wide plot of land in Comfort, near San Antonio, Texas.

Recently, the super-popular American TV anchorman Larry King has confessed, in an interview, his deep fear of death and his desire to undergo cryonic suspension,[23] joining the short list of celebrities who have manifested the same intention. Among them, we should mention the publisher and pornographer Larry Flynt – member of Alcor since 1986 –, the Silicon Valley venture capitalist Peter Thiel – signed up for many years –, the singer Britney Spears and Seth MacFarlane – the "Family Guy" creator.[24]

---

[18] http://www.estropico.com/id298.htm

[19] Cfr. http://www.cryonics.org.au/

[20] http://www.cryonics.uk.com/index.html

[21] http://www.eucrio.eu/

[22] Cfr. http://www.timeship.org/

[23] http://www.newsmaxhealth.com/headline_health/Larry_King_freeze/2011/12/07/421500.html

[24] To be clear: Walt Disney was never frozen; that's just an urban legend.

The last chapter – for now – of our history of cryonics is the case of Kim Suozzi, a 23-year-old girl from St. Louis, diagnosed in 2012 with a very aggressive glioblastoma multiforme – a very severe brain tumor. Against her family's wishes, Suozzi launched a campaign on Reddit – a social news web-site – to collect the seventy thousand dollars necessary for her cryonic neuro-suspension, which got successfully funded. Dead on January 17, 2013, Kim passed her last two weeks of life in a palliative care facility in Scottsdale, near the headquarters of Alcor.[25]

So, just to crack a bad joke, the world of cryonics is alive and vital; all of this despite the fact that this practice both does and doesn't have a scientific basis. Skipping the most obvious considerations – like the one that cryonics has nothing to do with the phenomenon of hibernation[26] –, what is the opinion of mainstream science about cryonics?

4.6 Science's Take

Cryobiology is the discipline of studying the effects of low temperatures on tissues and living organisms; academically respected, it has inspired several national and international scientific societies, among them the Society for Cryobiology. This latter society has openly distanced itself from cryonics, stressing its belief, as early as 1982, that the latter practice doesn't constitute real science. Even the American skeptic organizations – the associations of researchers that debunk the paranormal and pseudoscience, like CSICOP – have taken a rather negative stance toward cryonics.[27] For example, the well-known skeptic Michael Shermer has compared cryonics to a form of faith, even though it is secular.[28] Finally, official American institutions interpret cryonics as a form of burial, albeit an intricate one. But do not think that science *in toto* rejects cryonics; on the contrary, we can find a few figures that have actually advocated for it. For example, this is the case with Arthur C. Clarke, who once declared that "although no one can quantify the

---

[25] http://www.dailymail.co.uk/news/article-2268011/Kim-Suozzi-23-head-cryogenically-frozen-reborn-cure-brain-cancer-found.html

[26] In fact, hibernation is a natural physiological process during which some animal species reduce their vitals to the minimum, in order to pass the winter season in sleep. Another very common mistake is the confusion between cryonic suspension and suspended animation: with the latter term, we indicate a yet-to-be-developed technology able to induce hibernation in human beings. In this case, the slowing down should be such that the hibernating humans would age more slowly than normal, in order to be able to deal with very long and boring journeys into the deep space.

[27] http://www.skepdic.com/cryonics.html

[28] http://www.michaelshermer.com/2001/09/nano-nonsense-and-cryonics/

probability of cryonics working, I estimate it is at least 90% – and certainly nobody can say it is zero!"[29]

Concretely, which technical problems are cryonicists unable to tackle in a persuasive way? First of all, there is the issue of long-term memory. As we said, their idea is that, if we vitrify the patient's brain quickly, we can preserve the cerebral structures underlying memory and personality; which is possible, but not certain (and definitely unproven). It's just wishful thinking. Moreover, the idea of information-theoretical death doesn't erase the fact that, for now, cerebral death is still considered to be final. It is possible that future research will push forward the boundaries between life and death, forcing us to classify cerebral death as a simple stage, maybe a reversible one. But this also might not happen, and cerebral death might really be the last stop of human existence. To tell the truth, cryonicists do recognize these issues and, in spite of their enthusiasm for what they call "the second cycle of life" – the infinite existence that will begin after their reanimation –, they openly declare that their enterprise does not belong to any empirically consolidated official science and that they cannot promise anything to anybody.

Then, we have the issue of cerebral damage: even a limited amount of damage can erase whole portions of our memory, or impair this or that cerebral function; you can just imagine the conditions of a brain dead and frozen for decades or centuries.

Not to mention the purely technical issues – recognized by the cryonicists as well –, like the fractures provoked by the temperature of the liquid nitrogen and other forms of severe damage that, before we can fix them, we will need to understand scientifically – which is not a given, as scientific progress is unpredictable, and it could end up taking paths completely different from the ones cryonicists hope for.

But the biggest difficulty of all is the technological means that cryonicists expect to use in order to bring life back to their patients. The miracle solution they are waiting for, nano-machines, is, for now, just a slogan. And let's not forget all of the conceptual issues related to so-called mind-uploading, that is, the possibility of transferring the human mind into a non-organic substratum: maybe it is not feasible; even if it was, we have no idea as to how to do it; neither do we know if frozen brains are "readable," that is, if the patients' memories can be extracted – if they are still there. And, of course, right now, we have no idea as to how to read them.

Dulcis in fundo, we have the problem of the awakening. Let's suppose that our molecular reparation procedure works: the super-surgeons of the future have in their hands a perfectly healthy corpse. How do you kickstart it back to life, putting

---

[29] http://www.alcor.org/Library/html/clarke.html. As far as we know, Clarke never thought about getting a cryonic suspension contract for himself.

in motion, all at the same time and systematically, the metabolism, the heartbeat, the respiration, and so on? In other words, how do you bring life back to a corpse?

Now let's take a look at the management problems. The two American cryonic facilities are small and, from a financial viewpoint, relatively unstable – even though it seems that the situation has greatly improved in recent years, especially for Alcor. They are owned by nonprofit organizations, managed by enthusiastic volunteers, and they always need money – so much so that Alcor has sometimes only been able to replenish its finances thanks to generous donations from rich supporters. In other words, the cryonic organizations might not last long, reanimation might be too expensive or, simply put, the people of tomorrow might not be interested in bringing a handful of people from a time long ago back to life.

4.7   Anthropology of the Average Cryonicist

At this point, let's try to sum things up, and let's ask: what kind of people are the cryonicists? Are they crazy, scammers or very forward-looking pioneers? First of all, notice that I carefully avoided calling them ignorant or stupid: in fact, on average, they are smart people, and with a good scientific background; it's enough to read their articles or to look at their blogs to figure out that, from the viewpoint of their debating style, they are closer to philosophers and scientists than charlatans. If we really have to label them, we should opt for "nerd," just like every other Transhumanist – and, in truth, Transhumanism *tout court* and cryonics cross paths fairly often. It is also interesting to take a look at cryonicists' typical manner of thinking, because it allows us to analyze their self-perception. Its supporters call cryonics "medical time travel," stressing the fact that it's a one-way trip; the cryostasis tanks are compared to ambulances that are bringing their patients to the medical facilities of the end of Twenty-First Century and the beginning of the Twenty-Second. Moreover, cryonicists have developed a quasi-medical jargon that they constantly use – for example, the habit of defining as "patients" what are, in fact, corpses.

Despite stressing the purely speculative nature of their enterprise, cryonicists try, at the same time, to look as scientific as possible, to the point that they draw a scientific manner of speaking from mainstream science, by which they would like to be seen as a kind of avant-garde.

I also don't think that cryonics is a scam: reading their books and their web sites, I had the strong impression that these people are "true believers," and that, when they speak about their magnificent high-tech future, they really mean it.

This almost religious nature of cryonics is proven indirectly by the reformulation of the famed "Pascal's wager" by Ralph Merkle. Wishing to persuade himself of the truthfulness of the Christian faith, the French philosopher tried to prove that faith came with at least a few advantages. Pascal's thinking goes more or less like this: if we believe in God and He exists, then we will gain eternal life; if we have faith and he doesn't exist, at least we can live a peaceful life freed from the fear of death. Contrastingly, if we don't have faith, and God doesn't exist, we will end up in nothingness – plus we will be obsessed by the fear of death –, while if God exists, eternal damnation awaits us. Likewise, if we sign up for cryonic suspension and it works, we will get a new life, possibly infinite; if cryonics doesn't work, we lose only the money we have invested – which we would have lost anyway by dying. Contrastingly, if we don't sign up for cryonic suspension, we will end up rotting, no matter what. Unlike classical faiths, cryonics tends to downplay the religious elements, and it prefers to consider itself simply as a very long-term experiment; joking about it, Merkle said that cryonics is an experiment in which the control group is not doing very well.[30]

To sum up, what is cryonics? In my opinion, it is not – yet? – a scientific practice, but it does present elements taken from mainstream science, and particularly from cryobiology. It has religious aspects as well, which can be connected to very ancient religious practices, like the mummification process practiced by the ancient Egyptians – which aimed to confer eternal life upon the Pharaoh. Given almost every cryonicist's lack of religious faith, and given the many concepts taken from orthodox science, mixed with rational speculations,[31] I might classify cryonics as an ideology. In fact, ideologies – for example, Marxism – tend to begin with empirical facts, but, after that, they build theoretical constructions detached from reality, which people end up believing in religiously – and, in fact, sometimes, in light of their enthusiasm, cryonicists do look like a small cult.

To this, we have to add three facts. First of all, there are several cryonicists who don't believe that cryonics will work; it's just that people are afraid of dying – well, they are not wrong, death is scary – and are willing to do something to avoid their complete erasure. Secondly, cryonics is an ideology; true, but it is not that rigid. Inside it, there are different opinions on many topics – it's not like the Soviet "Diamat," which couldn't be criticized at all. Thirdly, in my opinion, cryonics

---

[30] Medical methodology is based on so-called "control groups." When researchers have to test the efficacy of a certain therapy, they compare the test-subjects with another group of people who don't receive any treatment. The reason for this is to verify that the therapy is more effective than a placebo, and more effective than already existing therapies, otherwise, there is no point to keeping the research going. So, in Merkle's joke, the control group is everybody who didn't decide to join the cryonic enterprise. More seriously, cryonicists present cryonics to their customers as an experiment that they can join after paying a certain amount of money.

[31] Maybe it's better to say "ratiomorphic," that is, built on unproven assumptions, but developed in a logical and consistent way.

represents an emanation of the classic "American Dream," with its idea that America is the land of opportunities, where no idea or initiative is too crazy. From this viewpoint, death is just one more frontier to defeat. And it is not a coincidence that, when they speak about cryonics, its supporters use a very eloquent expression: "conquering death."

5

Nanometer Cornucopia

5.1 The ABCs of the Nano-world

In the course of this book, we have often encountered nano-machines and nano-technologies, and the time has come to clarify some things about them. First of all, a dictionary definition: nanotechnology consists in the manipulation of matter at the atomic and molecular levels. In particular, the extended version of this concept speaks of *precise* atomic and molecular manipulation, with the aim of producing objects for everyday use – a hypothetical technology also known as "molecular nanotechnology." The National Nanotechnology Initiative, the American federal program for the development of nanotechnologies, defines nanotechnology as the manipulation of matter in which at least one of the dimensions, i.e., length, width or height, is between one and one hundred nanometers. A nanometer corresponds to a billionth of a meter – more or less the order of magnitude of molecules, so to speak. You can already imagine how the term "nanotechnology" is currently an umbrella under which very different ideas and practices are included. So, we ask: what are nanotechnologies made of now? What fields do they include, what devices do they use, in short, what do they do and what do they produce? Let's take a brief look at this.

Fullerenes

A fullerene is a molecule composed entirely of carbon and shaped like a hollow sphere, an ellipsoid or a tube; spherical versions are also called *buckyballs*, while cylindrical versions are called nanotubes. The first fullerene or buckminsterfuller-ene molecule was prepared in 1985 by Richard Smalley, Robert Curl and Harold Kroto at Rice University, an achievement that earned the three the Nobel Prize in

© Springer Nature Switzerland AG 2019
R. Manzocco, *Transhumanism - Engineering the Human Condition*,
Springer Praxis Books, https://doi.org/10.1007/978-3-030-04958-4_5

Chemistry in 1996. The name is a tribute to architect Buckminster Fuller, author of the design of geodesic domes – architectural structures formed by triangular elements. Fullerenes have various properties; for example, nanotubes are among the strongest and most resistant materials in circulation – both mechanical and heat-resistant – and these properties are useful in many technological fields, for example, for improving the mechanical and thermal properties of this or that product.

Nanoparticles

These are the result of relatively traditional techniques, that is, they merely constitute an extension of the current science of materials. Nanoparticles are obtained by transforming this or that material into ultrafine powders, practically dividing it into pieces of nanometric dimensions, and thus succeeding in obtaining better mixtures of different materials, ones with properties – of various kinds, depending on the mixture – superior to the classic ones.

Nanomanipulation

Probe scanning microscopy is a branch of microscopy that is based on the production of surface images by means of a probe – a physical object – that scans the sample to be visualized. The apex of the probe must be extremely sharp, as it is the feature that contributes most to the resolution of the microscope. In the context of this technique, of particular importance is the atomic force microscope, which has a very high resolution – in terms of fractions of a nanometer. The first prototype of this microscope was developed in 1986 by a Euro-American team composed of Gerd Binnig, Calvin Quate and Christoph Gerber. The atomic force microscope is one of the most important tools for the manipulation of matter at the nanometric level; visualization is obtained by making the probe tip and the object being explored interact – in practice, the probe "feels" the surface with which it comes into contact. Piezoelectric elements are included in the device – that is, they are composed of materials that can vary slightly in size when crossed by a slight electric current – which allow for greater accuracy. That is, in being able to perform very small movements, thanks precisely to the above variations controllable by the operator, you can get a very high resolution. What I am offering you is, of course, only a simplification, and what matters is that, currently, thanks to this type of microscope, it is possible to manipulate and move single atoms in an ultra-precise way.

Wet Nanotechnology

And then there are the biotechnologies that, if desired, can be considered a particular form of nano-manipulation, in this case, of organic molecules immersed in saline solutions – hence the use of the adjective *wet*. Alongside this, we must say that, over the years, we have witnessed a growing convergence between biotechnologies and nanotechnologies, to the point that this has led to the emergence of a specific discipline, bionanotechnology. After all, it is a question of the simple application of nanotechnology devices and procedures of various kinds to classical biological research. DNA nanotechnology deserves special mention; it concerns the creation of nano-structures starting from nucleic acids. In this case, the synthetic nucleic acids thus obtained are used, given their versatility, as a nanotech construction material, rather than for the normal transmission of information in cells. At the beginning of the '80s, the American scientist Nadrian Seeman launched the nanotechnology of DNA, but this field of research didn't start to arouse widespread interest until the mid-2000s, in regard to, for example, the possibility of using DNA for computation. In short, we can speak of a convergence between biotech and nanotech, and of a mutual contamination, with biotechnologies trying to imitate the "industrial" methods of nanotechnology and the latter trying to use the knowledge gathered by the former.

Nanomedicine

This is also a term that acts as an umbrella for a wide range of nanotechnology applications. Nanomedicine includes the medical use of nanomaterials and nano-electronic biosensors and, more generally, the development of new therapies, new methods of administration, and new drugs. A 2006 report by *Nature Materials* found that there were one hundred and thirty drugs and systems of administration based on nanotechnologies under development in the world.[1] The turnover of nanomedicine is growing, with hundreds of companies, numerous products and sales amounting to several billions of dollars. One example of the medical use of nano-materials is that of nano-shells of gold that can be used – for example, by connecting specific antigens – to diagnose and attack a tumor; another nanotech product is liposomes, lipid spheres that can be used to administer drugs. In essence, nanomedicine aims, among other things, to develop methods for introducing drugs into the body cell by cell, thereby increasing its effectiveness and reducing the necessary doses and side effects. Then, there is a field known as "nanobiopharmaceutics," which is based on the therapeutic use of proteins and peptides – short sequences of amino acids – capable of exerting biological actions on the human

---

[1] See http://www.nature.com/nmat/journal/v5/n4/full/nmat1625.html

body and administered with appropriate nano-particles. Another field of application is that of visualization: for example, you can hook "quantum dots" – that is, in short, "artificial atoms" containing only electrons – to specific antigens; the former can then be made fluorescent, thus allowing us to see, for example, the growth or shrinkage of a tumor. Finally, we can mention tissue engineering, a very futuristic research field, which aims to literally "grow" organs suitable for transplantation in the laboratory. Here, nanotechnologies are also making their contribution, in particular, offering scaffolds composed of nano-materials and promoting cell proliferation – that is, the numerical growth of cells that will form the finished organ.

If we wish, we could go on to mention micro- and nano-lithography, which is a set of technologies used to produce electronic microchips; or nano-electronics, which instead works on the production of transistors. Essentially, we could go on for a long time, but we want to spare you other boring technicalities; what matters is that you get an idea of the degree to which the nanotechnology world is articulated. And, most of all, that you realize that, although the Transhumanists undoubtedly appreciate all of the progress that nanotechnologies have made, this is not what they mean when they talk about such topics. In particular, their vision revolves around the concept of the "nano-machine," of the *replicator* and, above all, of the *assembler*. What is all of this about?

The precise definition that Drexler gives is that of "a hypothetical device capable of guiding chemical reactions through the positioning of reactive molecules with atomic precision." In other words, it is a type of molecular machine. And what are molecular machines, then? A molecular machine is, by definition, any discrete set of molecular components that produces quasi-mechanical movements in response to a specific stimulus. It is an expression that applies to molecules that perform actions resembling mechanical actions that take place at the macroscopic level. In short, it is a term that was born in the bio-chemical field and that can be applied to all structures and reactions that have the above characteristics: in practice, molecules that act as gears or mechanisms. And chemists have a lot of fun with these things, so much so that, over time, they have been able to synthesize many of these molecules. Sometimes, they are single molecules, sometimes more complex structures, composed of several mechanically interconnected molecules in a rather complex way. We could mention, for example, the rotaxanes[2] or the catenanes,[3] but we would need to become unnecessarily technical; on the other hand, it is better to limit ourselves to a classification by function of the molecular machines produced up to now. We have "molecular motors," i.e., molecules capable of one-way rotation following a pulse; the "molecular helices," which can

---

[2] https://en.wikipedia.org/wiki/Rotaxane

[3] https://en.wikipedia.org/wiki/Catenane

rotate and push fluids; the "molecular switches," which can reversibly assume two equally stable states; the "molecular shuttles," capable of transporting other molecules from one place to another; and the "molecular logic gates," which can be used to compute. Then, we have the "molecular sensors," which interact with specific molecules, producing a perceptible change, and the "molecular pincers," structures capable of supporting other structures – as if they were pincers, in fact. There is a lot of stuff, in short, of which one would not even normally suspect its existence. Obviously, the most complex molecular machines are those produced by nature: we have, for example, myosin, a protein engine that produces, among other things, muscle contraction; kinesin, another protein engine, which, in this case, moves materials along the microtubules (internal channels to the cells); and so on. And, from a theoretical viewpoint, there are obviously those who work on the construction of even more complex molecular machines, although, in some cases, they are destined to remain on paper, given that adequate construction methods are not yet available. Here, then, is the starting point that allows us to stress that, however fanciful, Drexler's visions did not materialize from nothing. And about the assemblers: do they exist in nature? Since they receive instructions from DNA, from which they synthesize proteins, ribosomes can certainly be considered natural assemblers.

At the terminological level, there is some confusion, in the sense that nano-machines can be seen as nanoscale-sized machines capable of moving single molecules or single atoms; to this could also be added, although not necessarily, the ability to self-replicate. The production of nano-machines could then be delegated to the appropriate "replicators," nano-factories that act as assembly lines for the assemblers; the possibilities are limitless, and nanotechnologists have indulged in imagining many different nano-machines.

5.2 Nano-mania, or the Two Souls of Nanotechnology

There has been much talk of nanotechnology in recent years; one could even argue in favor of the existence of a form of transversal "nano-mania." It is not just industry, the armed forces, university research centers, and non-profit organizations who are for or against the nano-world. Popular culture – and, in particular, science fiction – has also welcomed the nanotechnology lexicon with open arms. And so, if the protagonists of *Star Trek* have had to face the "nanites" – microscopic and malicious nano-machines – on several occasions, so, too, was the Incredible Hulk, in the film version given to us in 2003 by Ang Lee, only able to become the green monster we all know thanks to nanotechnologies. And these are just two examples of penetration of the nano-world at the level of public and "pop" perception. In short, if they mention "nanotechnologies" to us, what comes to mind are, most likely, machines as small as viruses and capable of doing a lot of things: taking care of our bodies, building objects of all kinds, and much more. And this is precisely the version of the facts favored by the Transhumanists.

But nano-machines do not exist yet, nor do we have any way of knowing whether they will ever exist, or even if they are possible in principle. In the meantime, what do all of those who recognize themselves in the term "nanotechnology" do? A lot of things actually, as we have just seen, because, at the present time, this label can be used by all of those operating – regardless of the disciplinary area – at the nanometric level. For example, in the US, the university laboratories that deal with MEMS[4] promptly renamed their sector "nano-science" in order to attract the rich funding provided for nanotechnology. Often, what is presented today as nanotechnology is actually materials science, and the activities of hundreds of nano-tech companies are often the result of classic chemical engineering associated with our new ability to move at the nanometric level. Furthermore, the experts in mainstream nanotechnology generally consider Drexler's theories to be science fiction or, even worse, pseudoscience. However, nanotechnology, in the broad sense, is or may soon become a GPT – *General Purpose Technology*. According to Elhanan Helpman,[5] GPT technologies are defined by characteristics such as: the possibility of further improvement and a progressive decrease in costs; the great variety of uses that can be made of them; the progressive "invasion" of economic activities, in the sense that, over time, an increasing percentage of economic activities makes use of it; and the ability to associate with existing technologies, acting as a complement. The steam engine, electricity, and the railways can all be considered GPTs, and they have all been the basis of great economic revolutions; in this regard, some argue that the impact of nanotechnology could be similar to that of plastic. In the 2000s, some applications of nanotechnologies, such as titanium dioxide and zinc oxide nanoparticles in sunscreens, cosmetics and some foods, began to appear on the market; as did silver nanoparticles in clothes, disinfectants and food packaging and nanotubes in stain-resistant fabrics. Nanotechnologists promise new and more effective drugs, cleaner and more efficient manufacturing systems, better information storage systems, and more. Then, there are the potential risks, ranging from the effects – currently unknown – of nano-materials on human health to those on the environment.

---

[4] Micro Electro-Mechanical Systems, electronic devices whose components oscillate between one and a hundred micrometers – one micrometer is one millionth of a meter. We are talking about devices composed of a processor surrounded by components that interact with the surrounding environment, such as sensors. By pushing its focus "down" even further – to the nanometric level – MEMS users have been able to include the term "nano" in their repertoire.

[5] E. Helpman (ed.), *General Purpose Technologies and Economic Growth*, MIT Press, Cambridge 1998.

The expansion of this sector was made possible by the National Nanotechnology Initiative, formally proposed to Bill Clinton in 1999 by Mihail Roco, well-known nanotechnology expert, a leading exponent of the National Science Foundation and one of the main architects of the subsequent development of the initiative. The NNI is essentially a program for the promotion of research and development of all nanotech fields and for the use of this knowledge in the public and private sectors. Convinced of the validity of the initiative, Clinton then officially promoted it with a speech held on January 21, 2000, at Caltech in Pasadena, in which the then-president cited the famous speech by Feynman, expressly naming the possibility of "arranging the atoms" one by one in the way that we want. Apparently, Drexler's vision had no detectable effect on Clinton's speech. However, when we talk about such initiatives, we talk about money – *a lot* of money – and we will add that, in conjunction with this declaration, the Clinton administration allocated five hundred million dollars to finance this research. George W. Bush later re-launched it, signing the "21st Century Nanotechnology Research and Development Act" in 2003, which authorized $3.63 billion over 4 years for five of the agencies participating in the program. And this is only the beginning. It is a lot of money, as you can see, none of which could escape the logic – sometimes perverse – that governs state funding for research.

When it came to "selling" nanotechnology to Clinton, there was no lack of hyperbole among those directly involved; when it came to the matter of fundamentally assigning the funds, the Drexlerian visions were completely ignored, and it was decided that everyone needed to "get real." Why? For various reasons. First of all, because of the need to convince the federal agencies to finance this or that research, presenting credible and well-detailed projects. Secondly, for fear of arousing the anxiety of public opinion – it is easy for people to get nervous, especially if you talk about nano-machines capable of eating the planet – and thus causing the blocking of research, along the lines of what happened with stem cells during the Bush II era.[6] Add to this the fact that, in a hyper-competitive context such as that of state funding for research, describing one's work as "nanotechnology" is a good way to get it taken into consideration by those in charge – as well as to attract brighter students. So, in the end, many researchers have adopted the term "nanotechnology" popularized by Drexler because it evokes very powerful future technologies and revolutionary capabilities, but what they really work on is, in essence, chemistry – or, in any case, more traditional disciplines. And, then, it must be said that many of the insiders do not take Drexler's theories seriously. So,

---

[6] For the interesting background to this question, see the following, excellent work: D. M. Berube, *Nano-Hype. The truth behind the nanotechnology buzz*, Prometheus Books, Amherst 2006.

"nanotechnology" has, in fact, two meanings. The first concerns all of those technologies – and there are many – that have to do with dimensions less than one hundred nanometers. We can also call it "nanometric scale technology." The second is the original meaning, that of precise technology at the atomic and molecular levels; basically, actual nano-machines.

To summarize, at present, the direction in which nanotechnologies are going is not clear, that is, we do not know whether this interdisciplinary field is destined to go towards nanometric chemosynthesis – in practice, a more sophisticated form of chemical engineering than the current one -, towards the creation of a nano-robotic handyman, or rather towards a level intermediate to these two. Certainly, there is a lot of enthusiasm around – some would probably speak of *hype* – but, at the same time, there is also the tendency to keep a low profile, a schizophrenic condition dictated by the above reasons. But now, we would like to know another thing: how did all of this come about?

5.3 There's Plenty of Room at the Bottom

The process of penetration into the nano-world began on December 29, 1959, with a speech given by the famous physicist Richard Feynman at a meeting of the American Physical Society, at Caltech in Pasadena. During the speech, entitled *There's Plenty of Room at the Bottom*,[7] Feynman took into consideration the possibility of directly manipulating individual atoms, thus giving rise to a series of chemical synthesis techniques that would be much more powerful than those currently available. The scientist also hypothesizes the realization of nanometer-sized machines that are able to "arrange atoms in the way we want," through mechanical manipulation. Feynman's plan is to construct machines of macroscopic dimensions, which will then build similar machines of considerably smaller dimensions – a quarter of the original, to be precise – which, in turn, should build even smaller machines, and so on, down into the nano-world. The physicist obviously also considers the need to redesign, as the "descent" proceeds, some tools and parts for the machines, as the relative intensity of the various physical forces involved changes. Thus, while gravity would become less important, Van del Waals' forces – attractive or repulsive forces between molecules – would be more so, and so on. Practical applications of this method would go from developing denser computer circuits to more powerful microscopes and more. Another

---

[7] R. Feynman, *There's a Plenty of Room at the Bottom.* http://www.zyvex.com/nanotech/feynman.html

possibility examined is that of "swallowing the doctor"; it consists in building a small surgeon robot controllable by an external operator. At the end of the intervention, Feynman offered a thousand dollars to the first person who could create a functioning electric motor no bigger than a 64th of an inch[8] and another one thousand dollars to whoever was able to restrict a page of text to a twenty-five thousandth of its dimensions. The first result was achieved as early as the following year by a craftsman, William McLellan, using conventional tools. For the second result, we had to wait until 1985, when Tom Newman, a Stanford graduate, managed to shrink the first paragraph of a novel by Charles Dickens, *A Tale of Two Cities*, by a factor of twenty-five thousand times.

Feynman's speech had a follow-up, delivered on February 23, 1983, in front of a large audience of scientists and engineers at the Jet Propulsion Lab in Pasadena. Title: *There's Plenty of Room at the Bottom, revisited*. In it, the physicist claimed that the small machines mentioned in the previous speech did not have any particular utility, and that, since then, there had not been much progress in that direction. It also speculated on the particular uses of these machines as drills and mills; finally, it eliminated some theoretical problems related to their construction and use, such as the source of energy, control, movement and the problem of friction.

A final consideration: despite all of the above, Feynman's influence on nanotechnology development should not be exaggerated, as evidenced by the fact that many of those who have actually developed nanometric manipulation technologies never heard of the speech in question before the beginning of the "fashionable" period of nanotechnologies, and, in particular, before the famous speech with which Bill Clinton launched the National Nanotechnology Initiative.

In addition, on the subject of "founding fathers," we also have to mention the German-American scientist Arthur R. von Hippel and the Japanese scholar Norio Taniguchi. The first coined the term *"molecular engineering"* in 1959,[9] while the second coined, in 1974, the definition of "nanotechnology." Taniguchi, a researcher at the University of Science of Tokyo, defines this term this way: "Nanotechnology mainly consists of the processing of separation, consolidation, and deformation of materials by one atom or one molecule."[10]

---

[8] An inch corresponds to 2.54 centimeters.

[9] A. R. V. Hippel, *Molecular Science and Molecular Engineering*, MIT Press, New York 1959.

[10] N. Taniguchi, *On the Basic Concept of 'Nano-Technology,'* Proc. Intl. Conf. Prod. Eng. Tokyo, Part II, Japan Society of Precision Engineering, 1974.

5.4   The Engines of Creation

The reference point of all nanotechnologists in the Transhumanist area, however, is the super-classic *Engines of Creation. The Coming Era of Nanotechnology*,[11] published in 1986 by Drexler, with a preface by Marvin Minsky. In this veritable milestone of Transhumanism (which, at the time, generated much talk about itself), Drexler starts from a distinction: that between the two different styles or ways in which technology acts, i.e., the "ancient" way and the next one. The ancient technology – in practice, all that we humans have done to date – is defined by Drexler as *bulk technology*. It consists basically in handling large quantities of atoms, relatively precisely, which is what we do today, in the factory and beyond, through actions such as cutting, welding, melting, assembling, and so on. This old method is opposed by *molecular technology*, a procedure that consists in manufacturing what we need by assembling it from the bottom upwards, arranging all of the necessary atoms in an ultra-precise way. Molecular technology – guarantees Drexler – will change our world in more ways than we can imagine. It will lead to a "new industrial revolution" and an era of extreme abundance, in which the very concept of "scarcity" – at the base of the economy as we know it – will lose its meaning. Awesome, the techno-utopia of the Transhumanist scientist. How to get there? But that is obvious! Through nano-machines, or *assemblers*, i.e., machines of nano-metric dimensions capable of manipulating, with absolute precision, the matter at the atomic and molecular levels. In practice, these are devices, all yet to be built, capable not only of arranging atoms and molecules according to a defined scheme – guided in this by special "nano-computers" – but also of producing other nano-machines at will, almost like real life forms do. According to Drexler's calculations, "on the bottom," there would be enough space to make all of the necessary structures fit to make a nano-machine function. The proof that these assemblers are possible would come from life itself: are not viruses, bacteria, living cells or even mere proteins already perfectly functioning nano-machines? Regardless of where they are located – inside or outside of a cell, for example – or the elements they are made of, whether biological or not, nano-machines obey the laws of physics; ordinary chemical bonds hold atoms together, and ordinary chemical reactions – guided by other nano-machines – assemble them. The first stage of nanotechnology is therefore, for Drexler, tied to biotechnology, i.e., the manipulation of organic molecules, proteins, genes, enzymes, and so on. The biochemists of the future will be able, for example, to use proteins to build motors and moving parts, so as to build nano-arms and nano-hands capable of handling single molecules. It is something similar to what we can see today in so-called "synthetic

---

[11] K. Eric Drexler, *Engines of Creation. The Coming Era of Nanotechnology*, Anchor Books, New York 1986.

biology," a new bioengineering discipline that aims to isolate parts of living systems, such as viruses or bacteria, redesign and optimize them, learn how to produce them in series, and, finally, assemble them, creating brand new living forms, perhaps with functions never seen in nature.[12] Nano-biological machines in short, starting from which we could arrive at "second generation nanotechnology," i.e., a type of technology that is no longer based on proteins, but on non-organic materials. Like ribosomes, real nano-machines can work following pre-programmed instructions but, unlike them, they can handle many different types of molecule, assembling them with a greater degree of freedom. In the assemblers themselves will be combined the ability of enzymes to divide and unite molecules with the programmability of ribosomes. The versatility of the nano-machines will allow us to build solid objects, made of metal, ceramic or diamond, all materials that we can assemble atom by atom – in the end, the construction of any type of structure is just a matter of the way in which the atoms are arranged, and, with the assemblers, this procedure is certainly not a problem. Not only in terms of versatility, but also resistance: unlike living organisms, non-organic nano-machines will be able to work in environments of all kinds, very hot or very cold, using as "instruments" almost all types of reactive molecule normally used by chemists, placing atoms in any type of stable pattern and adding them a little at a time until the structure is complete. In short, those that Drexler has baptized *universal assemblers*, and that, according to him, will allow us to build, at high speed, all that is permitted by the laws of nature, as well as – just to add a bit of melodrama – to remake our world or to destroy it.

And how do we build these nano-machines? The approach proposed by Feynman, the *top-down* one, should not be as simple as it seems; otherwise, it would have been in practice for some time. It is already difficult – from a technical-design point of view – to produce machines capable of manufacturing parts of themselves that are as precise as necessary. Imagine making them a quarter as big while maintaining precision – in practice, the machine that the factory should be able to produce has more precise versions of the same parts that produced it. Then, you would need to find a design that is insensitive to the scale, or a way of changing the design as the laws of scale make certain techniques ineffective. The machine should have the ability to self-correct itself, because, at the nanoscopic level, the human operator would have limited intervention power. Instead, Drexler proposes a *bottom-up* approach, that is, creating the right conditions for making simple systems self-assemble, then making them self-assemble into even more complex systems, and so on.

---

[12] For example, think of Craig Venter's attempt to create the first cell with a synthetic genome. Cf.   http://www.newscientist.com/article/dn23266-craig-venter-close-to-creating-synthetic-life.html#.UgjpoaxJxSo

The manipulations at the nanometric level will allow us to progressively reduce the size and costs of the circuits, at the same time increasing the speed, up to the coveted nano-computers, which will allow us to make a qualitative leap in the management of the assemblers. The nano-computers – equipped with special "molecular memory devices" – will provide the latter with a constant flow of instructions necessary for the construction of complex objects. In addition to the assemblers, we will also have *disassemblers*, instruments that, as the name implies, will allow us – thanks always to "molecular memory" – to analyze the objects, if necessary dismantling them one atom at a time. Essentially a sort of "controlled corrosion" for exploration purposes. But it will be a long time before the nano-machines – assembler, disassembler and nano-computer – emerge; for Drexler, however, it is inevitable that we get there, given that, under the heading of "genetic engineering," the first steps have already been taken.

Drexler also enters into much more daring speculations about where the evolution of nanotechnology will take us. In particular, the development of increasingly sophisticated nano-machines will lead us to create assemblers capable – in a completely automatic way – of assembling copies of themselves,[13] to build tools of all kinds, to manage mines and generators of energy and to provide all of this to the other nano-factories. Which, in turn, will end up merging into a single expanding and self-replicating nanotechnology production system – this system can then be made highly eco-compatible, so do not be afraid that we will find ourselves with smoky and polluted industrial complexes. In this regard, it is possible to imagine nano-replicators that will not resemble cells, but rather factories shrinking at the cellular level, and constituted by molecular scaffolds with various nano-machines embedded – nano-assembly lines, in short. The omnipotent nano-machines of the future will be able to build everything, even objects of considerable size – like a skyscraper. After all, the nano-machines produced by nature also built whales, although this took some time. To proceed quickly enough, it will be necessary for many assemblers to work together, but they can be produced by the nano-replicators on the order of tons. We can then imagine a tank full of water in which many nano-machines are dissolved, which aggregate around a nano-computer that contains the construction program – and that therefore acts as a "seed" of the object that we will "grow" in the tank, for example, a rocket or, for that matter, a skyscraper. So: end of factories, end of the economy, end of work. Abundance without limits. If we then add artificial intelligence, there is no limit to what could be done. You can create nano-machines as small as viruses to be used to study the structure and functioning of the brain cell by cell and, if necessary, molecule by

---

[13] In reality, the idea of machines capable of self-reproducing is not credited to Feynman or Drexler, but to John von Neumann. Von Neumann conceptually elaborated these machines – which he named "universal builders" – in the '40s, and the details were then published in 1966 in the posthumous work *Theory of Self-Reproducing Automata*.

molecule. Thus, structures similar to the brain can be constructed, but much more efficient and smaller – according to Drexler's calculations, a complex human brain composed of nano-computers connected by nano-cables should occupy less than one cubic centimeter. Artificial intelligence systems with access to nano-machine networks could perform many experiments very quickly. They could design experimental devices in seconds, and assemble them; they could build without the logistical problems that afflict contemporary research. At the cellular and molecular levels, it will even be possible to conduct one million experiments in parallel in one go. There is no limit to what could be done.

The nano-machines will then allow us to expand into space, reducing the costs of building spaceships and transport between the planets, thus opening the Solar system to human colonization. In this regard, Drexler also imagines a special nanotech spacesuit: padded with nano-machines, a similar suit could be "arranged" in order to be soft, light and transparent; it could allow – through the intermediation of special layers of nano-machines – the perception of tactile sensations coming from outside – in practice, as if we were touching the things with our bare hands. The nano-machines inside the suit could – thanks to sunlight and carbon dioxide – regenerate the air breathed; contrastingly, behaving like a real internal ecosystem, they could decompose waste products – we do not need to elaborate on that – and assemble fresh food. There would be – even if we never mentioned it – some limitations: our wonder suit could not indefinitely resist huge speeds or very low and very high temperatures; it could not allow us to walk through walls; it could not withstand strong explosions.

Let's go back to the economic aspects now. The slogan of Drexlerian nanotechnologies is: abundance. The things that our nanotech utopia will take away are numerous: work; the exhaustion of raw materials (everything will be recyclable or obtainable through nanometric manipulation); the problem of energy (the Sun will be enough); the occupation of land and damage to the landscape (the nano-factories will occupy very little space); waste disposal (just disassemble it); the need to organize work and distribution (nano-machines will do everything, *of course*), and since everyone will have their own nano-home factory, what purpose will there be to going to the supermarket or to Ikea? So, the end of the zero-sum society – in which, if someone wins, someone else must necessarily lose – and the advent of a positive sum society – in which everyone wins. Finally, the costs of production and maintenance of public goods will fall to such an extent that it is almost zero. And so, De Grey having gotten rid of death, Drexler will also free us from taxes.

Let's move on to medicine. Our body is also made of atoms and molecules, and it is obvious to think that, from nanotechnology, a new nanomedicine will ascend, in which, interpreting diseases, traumas and aging in terms of the wrong dispositions of atoms, Drexler's nano-robots can remedy practically anything. So: destroy viruses, bacteria and cancer cells, repair the damage caused by any type of degenerative disease, treat injuries of any kind – including those that would today put us in a coma or a wheelchair. Even here, therefore, there are no limits: thanks to nano-computers, damage can be repaired at the cellular level, restoring youth and prolonging human life indefinitely. Drugs will disappear, replaced by nano-machines that are much more precise and free of side effects. Not only that, it will also completely change the methods that we have used to cure diseases since prehistoric times. Until now, we have limited ourselves to acting in acute conditions, with surgical and pharmacological treatments able to buffer the damages, leaving the rest to our natural repair mechanisms. Instead, nanomedicine will allow us to carry out repairs or maintenance at a molecular level, in a very precise way and by anticipating the repair systems available from nature, which will no longer be strictly necessary. We will not have to do anything but "swallow" a doctor – or, for that matter, an entire hospital – in the form of a small clot of nano-machines and wait quietly for its work to be done. Given the level of molecular control guaranteed by nanotechnology, Drexler also imagines a new type of anesthesia, which he calls "anesthesia plus." It consists in completely blocking the metabolism – no vital chemical reactions of any kind, but neither signs of degradation, as in the case of a death – for minutes, hours, days, years or, for that matter, centuries. This situation of "biostasis" could provide doctors – human or artificial – with a very calm system to work with and astronauts with a way to travel for centuries in space, without consuming resources or getting bored. Biostasis would also be an excellent first aid procedure to protect the patient from death. It can also be considered a cryonics suspension technique more advanced than those available now. And, speaking of this, we can add that it is Drexler himself who suggests the possibility of using nano-machines to repair the molecular damage of cryo-suspended patients, bringing them back to life – this suggestion, and the fact that the nano-technologist explicitly cites Ettinger, have certainly not helped to make him popular among the most orthodox scientists. Not only will nanotechnologies guarantee immortality, they will also allow us to repair all of the damage inflicted upon the ecosystem up to now, so that the ecological impact of our civilization becomes greatly reduced[14] – despite the fact that the population is destined to grow exponentially. Drexler also notes that this last feature would happen in any case, immortality or not – so it is worthless to use the argument of overpopulation against nanotech immortality.

The American engineer then tackles his number one enemy, the one that pushed him on the road to nanotechnology, that is, the discourse on the limits to growth

---

[14]The nano-machines will, in fact, be able to clean the atmosphere, the waters and the land of any type of toxic substance and, in the case of air, of greenhouse gases as well; with them, you can recover all of the radioactive materials scattered in the environment and bury them forever – on Earth or – why not? – on the Moon.

set by the Club of Rome, which resulted, in 1972, in the report *The Limits to Growth.*[15] The nano-world represents a way out of this *impasse,* as it will allow us to reduce production costs, eliminate waste, maximize the raw materials and energy sources available to us and, finally, to expand into space, thus meeting a growth of indefinite duration.

And then there's the *gray goo.* It is a scenario in which some nano-machines go crazy and start to reproduce non-stop, devouring, for this purpose, all of the carbon on which they can put their nano-hands, including us – a possibility christened "ecophagy." This is a purely theoretical possibility, as it is based on technologies that do not yet exist; despite this, Drexler tries to propose a remedy: that we rely on the notion of "redundancy," which, in this case, consists of putting multiple copies of the instructions into every replicator and assembler, so that these nano-machines resist the harmful mutations.

In short, gray goo apart, we will eventually create nano-machines that will take full care of us, from cleaning the house to the refreshment of domestic air, that will eat dirt, that will produce fresh food – real meat, wheat, vegetables, and so on, a bit like the food generator of the spaceship Enterprise. And again: it will be possible to form neural connections – through transducers and electromagnetic signals – between brains, creating telepathy. Nanotechnologies will allow people to change their bodies in ways ranging from trivial to surprising and bizarre. Some people may abandon the human form as the caterpillar transforms into a butterfly; others will limit themselves to achieving a perfect human form.

While *Engines of Creation* continues to arouse enthusiasm and criticism, under the supervision of Minsky, Drexler, in 1991, received his doctorate in molecular nanotechnology at MIT – the first of the kind ever awarded. And, if his first book is considered annoyingly on the border between reality and fantasy, Drexler's doctoral thesis, later published under the title of *Nanosystems,*[16] was much better received; and it wasn't just chemists – Ohio State's Leo Paquette, Cornell's Roald Hoffman, Columbia's Clark Still, and others – who found several of Drexler's ideas interesting; this is essentially a work of science and engineering. There are also various criticisms, relating, above all, to the concrete way in which to implement those systems. For example, Philip Bart of

---

[15] The Club of Rome is a think tank formed of economists, scientists and entrepreneurs, founded in 1968 by the Italian entrepreneur Aurelio Peccei. The report, commissioned by the Club to the MIT in Boston, foresees the very negative consequences that our planet will suffer – and we with it – if the growth of the population and its consumption continues at the same rate as was occurring then. The report – edited by Donella H. Meadows, Dennis L. Meadows, Jørgen Randers and William W. Behrens III – became a book, also published in Italy. Cf. AA.VV., *The limits of growth*, Potomac Associates, Falls Church 1972.

[16] K. E. Drexler, *Nanosystems: molecular machinery, manufacturing, and computation*, John Wiley & Sons, New York 1992.

Hewlett-Packard claims that, in the Drexlerian text, the holes are larger than the substance and that, although there are plausible arguments for everything, there are no details about anything.

However, starting in 1992, Drexler combined the idea of the assembler with *molecular manufacturing*, which he defines as "the programmed chemical synthesis of complex structures through the mechanical positioning of reactive molecules, and not the manipulation of single atoms." From molecular manufacturing then come the "nano-factories," systems of nano-machine organized on several hierarchical levels, and able to assemble objects of all kinds. From a dimensional point of view, a nano-factory could easily sit on a desk, as Drexler himself explains in *Nanosystems* – work that he defines as "exploratory engineering." At this point, other nanotechnologists from the Transhumanist area, such as Ralph Merkle, J. Storrs Hall, Forrest Bishop, Chris Phoenix and Robert Freitas, who, in turn, offer as many versions of the nano-factory, come into play. The buzzword now becomes "mechanosynthesis," which defines any chemical synthesis in which the results of a certain reaction are determined by mechanical constraints that direct the reactive molecules in specific molecular sites. In 2000, Freitas and Merkle launched the Nanofactory Collaboration, a project involving twenty-three researchers with the aim of developing a research program for the development of a nano-factory based on the mechanosynthesis of diamondoids – structures physically similar to diamonds – possibly small enough to fit on a small table.[17] So, no nano-machines for Freitas and Merkle, at least for the moment.

5.5   Nano-schism. The Drexler-Smalley Debate, and Beyond

As we said, the criticism of Drexler was intense, and the most famous debate on the matter was that between him and Richard Smalley, which took place through a series of letters published between 2001 and 2003 in *Scientific American* and *Chemical & Engineering News*. The bone of contention was the feasibility of the assemblers, which, for Smalley, was impeded by some fundamental physical laws and, in particular, by the fact that, atoms being sensitive to each other's presence, they could not be manipulated in the "clean" way desired by Drexler. In particular, to work, the assembler would require several arms, which should consist of several atoms; there would therefore not be enough space for all of the arms that the assembler needs to precisely control a certain reaction – a difficulty that Smalley baptizes the "problem of the fat fingers." In addition, the arms would sometimes

---

[17] See http://www.molecularassembler.com/Nanofactory/

end up adhering to the atoms that they would like to manipulate and could no longer release them – the "problem of the sticky fingers."

In response, Drexler advanced the example of ribosomes, cellular organelles that do exactly what Smalley says is impossible, that is, they produce proteins in a precise way. The "fat fingers" would not be a problem, because, in reality, many reactions would require only two reagents. The "sticky fingers" would also be a problem only for certain reactions, not for all of them.

The subsequent skirmishes between Smalley and Drexler focused on the question of water. It is true that enzymes and ribosomes are at the center of a very complex chemistry, admitted Smalley, but the reactions that concern them are facilitated by the aqueous solutions in which they are immersed. Enzymes and ribosomes thus cannot construct anything that is chemically unstable in water, and therefore could not create the materials that modern technology uses. In practice, for the Drexlerian nano-machines to be possible, there should be a chemistry as complex as the organic one, but not based on water – a chemistry that, in centuries of research, we should have already encountered. Drexler replies that his idea is to "increase" the chemistry based on solutions, starting with the self-assembly of assemblers based on the chemistry of solutions that they would then use to go further, building other more complex assemblers. In the end, the debate – which always had a rather polemical tone, to tell the truth – turned into a sort of "dialogue between the deaf," in which Smalley claimed that the assemblers are impossible because chemical reactions are much subtle than Drexler thinks and because few reactions would satisfy his needs, while Drexler referred him to the principles contained in *Nanosystems*. To this was added the accusation – lobbed by Smalley at Drexler – that he was trying to "scare children" with his stories about gray goo. Smalley's statements go beyond scientific debate, suggesting a strategic nature. At the time, the chemist – now deceased – had big financial interests in nanotechnology; in particular, he created Carbon Nanotechnologies Inc., a world leader in the production of nanotubes and holder of more than a hundred patents. Smalley was also a great promoter of the NNI; this being the case, you understand why he had so little patience with the idea of nano-bots. In fact, if – from the point of view of the Nobel Prize for Chemistry – the Drexlerian fantasies can be labeled as such, there is no reason to fear impossible nano-machines, and the most serious work can continue – generously supported by the quiet American taxpayer.

Smalley was Drexler's most heated critic, but certainly not the only one. Among others, we mention George Whitesides, a well-known Harvard chemist who, among other things, invented a microfabrication technique known as soft lithography. According to him, Drexler's nano-machines could not be fueled and it was uncertain how molecular bonds could be broken with brute force. For him, Drexlerian nano-submarines would need to be so big as to prevent blood flow; it is not, in fact, a coincidence that bacteria are measured in microns. For David

Berube,[18] there are few physical and chemical reasons to say that mechanosynthesis is impossible, but there are still many positive steps that are missing. While it is true that living cells prove that molecular fabrication can take place, the transition from ribosomes and enzymes to nanotechnological molecular fabrication is – at least for now – outside of the realm of chemistry. *Scientific American* contributor Gary Stix said that, in order to give nanotechnology a chance, the nano-bots and the idea of reviving corpses must be abandoned altogether.[19]

It is not just NNI's leadership that has rejected Drexlerism, but also the leaders of the private sector, particularly in the States. Christine Peterson, Drexler's ex-wife and president of the Foresight Institute, attempted to include, in George W. Bush's 21st Century Nanotechnology R&D Act, the authorization to fund a single feasibility study on Drexlerian mechanosynthesis. Apparently, however, at the time, the NanoBusiness Alliance – the main association of nanotech entrepreneurs – approached Senator John McCain about removing this proposal from the final text. In short, funding for the study of molecular manufacturing would have been cut without public debate and without scientific motivation.[20] Drexler then publicly blamed Mark Modzelewski – former executive director of the NanoBusiness Alliance – who denied the allegations but continued to show strong animosity towards supporters of Drexler's nanotechnology – hostility due, according to him, to the hyperbolic statements of the latter.

In 2006, the US National Academy of Sciences published *A Matter of Size:* a report that examined, among other things, the concept of molecular fabrication, also analyzing the content of *Nanosystems* and concluding that, despite the possibility of making multiple theoretical calculations in regard to many aspects of these nano-systems – for example, operating speed, thermodynamic efficiency, and so on – it was impossible to draw definitive conclusions; rather, the report said, we should move on to action, that is, to pursuing experimental research.[21]

But here's a nano-twist. In 2008, the British government decided to finance a five-and-a-half-year research on the mechanosynthesis so dear to Drexler, Freitas and Merkle. In particular, Philip Moriarty, a researcher at the University of Nottingham's Nanoscience Group, would work on mechanosynthesis using tunneling microscopy techniques. The project, entitled *Digital Matter? Towards Mechanised Mechanosynthesis*, was funded by the UK Engineering and Physical Sciences Research Council. The focus of the project was, in particular, the

---

[18] See D. M. Berube, Op. Cit., 2006.

[19] Gary Stix, *Little Big Science*, «Scientific American», September 16, 2001, www.ruf.rice.edu/~rau/phys600/stix.htm

[20] See L. Lessing, *Stamping out good science*, «Wired», July 7, 2004. www.wired/archive/12.07 view,html? Pg = 5.

[21] http://www.nap.edu/catalog.php?record_id=11752

development of new protocols for manipulation with tunnel effect probes – the usual super-microscopes, in practice – capable of assembling one atom at a time automatically, with the aim of arriving at three-dimensional nano-structures. Not bad. Who knows? Maybe now Drexler, with a little luck, will get his nano-machines.

5.6 Flesh and Computronium

We briefly mentioned the concept of "programmable matter," and now we want to say something more. If you have seen *Terminator 2*, you will certainly remember the T-1000, Schwarzenegger's opponent, an android made of a strange "liquid metal" – a substance that allowed him to assume any form and any consistency. And one of the goals of nanotechnologists – the *mainstream* ones, but, above all, the more imaginative ones – is precisely that of producing analogous materials, i.e., highly structured compounds at the molecular level capable of changing physical form and properties, turning into everything we desire. The history of programmable matter began in 1991, when two MIT scholars, Tommaso Toffoli and Norman Margolus, hypothesized the possibility of manipulating matter at the molecular level and making it capable – through specific software – of modifying every physical property on command: density, shape, elasticity, optical properties, and more. Then, the theoretical and experimental studies multiplied; just to give one example, we mention a recent research by a team from MIT and Harvard, directed by Robert Wood and Daniela Rus.[22] These scholars have developed special nanotechnological sheets that can fold automatically, changing their structure and transforming themselves into a paper boat or paper plane. This material consists of small triangular sections, and is crossed by thin strips of "shape memory" metal, a special nickel-titanium alloy that can change shape when it is crossed by a slight electric current. It is only a beginning, but it is the intention of these scholars to progressively miniaturize the triangular sections, up to a nanostructured material capable of taking on many different configurations. In practice, it would almost be a "substance" that behaves as if it were semi-liquid, and with which it will be possible to produce many types of object, from "super-swiss knives" – that is, tools capable of turning into many tools – to dishes and glasses capable of changing their capacity, and more. Before them, in 2002, Seth Goldstein and Todd Mowry – researchers at Carnegie Mellon University – started a new discipline, which they baptized *claytronics* – from *clay* and *electronics*. The goal of "claytronics" is to develop hardware and software necessary to create a completely

---

[22] http://www.zdnet.com/blog/emergingtech/mit-harvard-researchers-create-programmable-self-folding-origami-sheets/2293

programmable object. The final purpose: to create modular robots with sub-millimeter dimensions – renamed "claytronic atoms," or "catoms" – able to join one another in many ways, generating any type of object – self-propelled or not. The perspectives opened by the programmable matter are many, and, in this regard, we mention one of the elements that most struck the imagination of researchers, writers of science fiction and – needless to say – Transhumanists, i.e., computronium, a theoretical material also hypothesized in the '90s by Toffoli and Margolus. It consists of a form of matter whose atoms have been rearranged in order to function as an ideal substrate – and optimized – for any type of computational process. Essentially, this is a possible future computational technology capable of producing an intense and sustained "downward" computation – which basically consists in optimizing matter and using the same atoms – even sub-atomic particles – as elements of the computation process. Everything is essentially about arranging atoms and subatomic particles to provide the maximum amount of computation possible. Each atom is given a logical value – for example, "yes" and "no," as in the case of the binary code, but systems of more complex values can be chosen – and used for computation. No atom is wasted. If, in a more or less distant future, computronium becomes reality, this would lead to the replacement of classical computers with a real form of "intelligent matter." Indeed, in the intentions of the Transhumanists, this matter, adequately programmed, could be used to build anything, so that all of the objects that we deal with would be, in a certain sense, animated. But it is not over yet: since one of the goals of many Transhumanists is to upload their mind into a non-biological support, some of them have contemplated the possibility of synthesizing a beautiful post-human body made of computronium, which would make them intelligent entities up to the atomic level, as well as much more versatile than we traditional human beings, still made of mere flesh.

Now, maybe you'll think, "Enough, right? After becoming computronium, these Transhumanists will finally be satisfied." But no. Not satisfied to have managed to control matter at the molecular and atomic levels, our heroes have decided to *go further*. And what could be further – or, more precisely, further down – than nanotechnology? It is time to get acquainted with "picotechnology" and "femtotechnology." The first works at the level of the picometers – a picometer is one thousandth of a nanometer – and concerns the manipulation of atomic matter; it consists in the alteration of the structure and chemical properties of individual atoms, generally by the manipulation of the energy levels of the electrons. Phenomena of this type occur both in nature and in the laboratory – but obviously with results currently far from the expectations of Transhumanists. The idea of a picotechnology should not be so far-fetched, since a group of

American and Indian researchers has dedicated a specific article to it in *Nature*,[23] exploring its potential.

Further below, there is, as we have said, femtotechnology – a femtometer corresponds to a quadrillionth of a meter; in practice, we would find ourselves at the level of sub-atomic particles. For example, it is possible to imagine – although it is science fiction – self-replicating organisms composed of "molecules" made of protons and neutrons, instead of whole atoms – they are physical entities that some theorists believe are possible under certain conditions. In this regard, the astrophysicist Frank Drake had fun playing with the hypothesis that creatures made of these "molecules" exist on the surface of a super-dense neutron star, and are therefore endowed with physical characteristics that, in theory, would allow for the existence of such entities.[24] As you can see, these are phenomena that, if possible, would require conditions of extreme pressure and gravity, far from those that are ordinary to us. Always eager to find new limits to overcome, Transhumanists have thought considerably of jumping on the femtotechnology bandwagon. And so, in 2011, Hugo de Garis published an article, entitled *Searching for Phenomena in Physics that May Serve as Bases for a Femtometer Scale Technology*,[25] dedicated precisely to this theme. After Drexler conceptually tried to work with atoms in an ultra-precise manner, De Garis then did the same with particles, trying to assemble them into stable configurations – which are not atoms, of course. De Garis clearly has no problem admitting that his attempt was highly speculative, amateurish and probably wrong. His goal is, in essence, to find some loophole that makes femtotechnology possible. In this regard, the Transhumanist spoke about it – at the end of the '90s – including with Murray Gell-Mann – the discoverer of quarks. The latter, even though he did not know how to answer De Garis's questions, told him that he had considered, on one occasion, the possible industrial applications of a particular particle, the kaone. The question now, however, is: what are we to do with femtotechnologies? According to De Garis, they would be a trillion of a trillion times more performative than nanotechnology; an organized "block" of protons, neutrons or "femtotechnologized" quarks would be monstrously denser than a similar atom made of nanotechnologies, and therefore the speed of internal processes would be a million times faster. Who knows, perhaps one of the applications could be a beautiful femto-bomb, much more powerful than the current ones and capable of destroying an entire nation. Or you could get

---

[23] R. Sharma, A. Sharma and C. J. Chen, *Emerging Trends of Nanotechnology towards Picotechnology: Energy and Biomolecules*, http://precedings.nature.com/documents/4525/version/1

[24] F.D. Drake, *Life on a Neutron Star*, « Astronomy », p. 5, December 1973.

[25] Hugo de Garis, *Searching for Phenomena in Physics that May Serve as Bases for a Femtometer Scale Technology*, http://hplusmagazine.com/2011/01/10/searching-phenomena-physics-may-serve-bases-femtometer-scale-technology/

to a femto-source of energy, much more efficient than nuclear power. Or, again, to an artificial intelligence equipped with femtometric level switches, and therefore capable of conceiving new femto-structures much better than we would be able to think of. Which physical phenomena would make this technology possible? And here, De Garis indulges himself: we could use mini-black holes, or the surfaces of neutron stars – or maybe build a neutron star at a femtometric scale – nuclear molecules, and much more – we refer to his article for a full list. At this point, however, De Garis becomes unstoppable. If one can imagine a femtotechnology (10^{-15} m), why not talk about an attotechnology (10^{-18} m), a zeptotechnology (10^{-21} m), or all of the way down to a planck-technology[26] (10^{-35} m)? If these technologies are possible – although we reiterate that, for De Garis himself, this is a mere speculative game – then one could imagine entire civilizations located within elementary particles, which perhaps communicate with each other using the laws of quantum mechanics – which would allow signals to be transmitted literally instantaneously.

Not satisfied with the work of his colleague and friend, Ben Goertzel topped it off, publishing – also in 2011 – an article eloquently titled *"There's Plenty More Room at the Bottom: Beyond Nanotech to Femtotech."*[27] Goertzel's starting point is that which in physics is known as "degenerate matter," a type of matter characterized by a very high density. We will not go into detail; just remember that this type of matter – which is found on, among other things, neutron stars and white dwarfs – has strange properties, and that Goertzel would like to use it to make his femtotechnologies. The only problem, admits the Transhumanist, is that there is no way of knowing whether the degenerate matter can be kept in stable conditions even in a low-gravity environment like the Earth.

The author refers, at this point, to the work of Alexander A. Bolonkin, a Russian physicist who moved to America, and who – in an article entitled *Femtotechnology: Nuclear Matter with Fantastic Properties*[28] – imagines a type of matter that he calls "AB-matter," which should possess the same characteristics of degenerate matter, while being able to remain stable under conditions of gravity similar to those of Earth. Bolonkin then adds – in what seems to us to be a mere intellectual game – that his AB-matter would have extraordinary properties – hardness and

---

[26] The Planck length is, according to contemporary physics, the smallest possible physical distance, below which the concept of dimension loses its meaning.

[27] B. Goertzel, *There's Plenty More Room at the Bottom: Beyond Nanotech to Femtotech*, http://hplusmagazine.com/2011/01/10/theres-plenty-more-room-bottom-beyond-nanotech-femtotech/

[28] A. A. Bolonkin, *Femtotechnology: Nuclear Matter with Fantastic Properties*. 2009. http://nextbigfuture.com/2011/10/femtotechnology-ab-needles-fantastic.html. In addition: ibid. *Femtotechnology: Design of the Strongest AB-Matter for Aerospace*. http://vixra.org/pdf/1111.0064v1.pdf

resistance a million times higher than those of ordinary matter, super transparency, no friction, and so on. The physicist also comes to propose the design of aircraft, ships and vehicles of all kinds made with his AB-matter, thanks to which they would have incredible capabilities – invisibility, the ability to penetrate walls and barriers like ghosts, and protection from atomic explosions and radiation flows.

Also in 2011, Goertzel communicated with Murray Gell-Mann, who – apparently – told him he had never seriously thought about femtotechnology, but that it seemed reasonable to work on this idea. In short, a "maybe." For his part, Ray Kurzweil – hyper-optimist as ever – estimates that femtotechnologies will be something accomplished during the twenty-second century.[29] Waiting for these developments, we prefer to return to our most glaring nano-world.

5.7   The Amazing Bio-medical Nano-machines of Dr. Freitas

Robert A. Freitas Jr., a researcher at the Institute for Molecular Manufacturing, located in Palo Alto and one of the foundations connected to Transhumanism, is undoubtedly one of the pillars of Transhumanist nanotechnology. A graduate in physics and psychology, in 1980 he collaborated on a feasibility analysis of hypothetical self-replicating space factories for NASA. Freitas is still working on his monumental work *Nanomedicine*, a comprehensive discussion of the possible medical applications of Drexlerian nanotechnology. At the moment, Volume I and Volume IIA have come out, both available on-line for free.[30] In 2010, Freitas requested and obtained the first patent related to mechanosynthesis. Beyond that he has written a large number of *papers* on various subjects; in some of these, he analyzed the technical details – or, rather, the constraints that nanotechnologists must respect – of some very specific hypothetical nano-machines with medical and biotechnological purposes. Let's look at these.

We'll start with the clottocytes, that is, mechanical artificial platelets; in practice, nano-robots specializing in closing wounds and stopping bleeding.[31] Platelets or thrombocytes are blood cells of more or less spherical or spheroidal shape, two microns wide and with no nucleus. They live between 5 and 9 days and essentially have the task of stopping the flow of blood from blood vessels. In practice, when we get wounded, the platelets rush to the site of the trauma and become sticky, massing, starting a cascade of specific chemical reactions and sealing the wound.

[29] http://www.edge.org/documents/archive/edge107.html

[30] R. A. Frietas, Jr. *Nanomedicine. Volume I: Basic Capabilities*, Landes Bioscience, Austin 1999. Ibid., *Nanomedicine, Vol. IIA: Biocompatibility*, Landes Bioscience, Austin 2003. Cf. www.nanomedicine.com

[31] R. A. Freitas Jr., *Clottocytes: Artificial Mechanical Platelets*, in: « Foresight Update», n. 41, 30 June 2000, pp. 9–11. cf. http://www.imm.org/Reports/Rep018.html

A healthy adult has, on average, a number of platelets ranging from one hundred and fifty thousand to four hundred and fifty thousand per microliter[32] of blood. According to Freitas's projects, the speed of action of his clottocytes should be between one hundred and a thousand times faster than platelets, managing to close a wound in about a second. In numerical terms, the clottocytes could be ten thousand times more efficient than platelets, so the required volumetric concentration of these nano-machines would be 0.01% of that of the thrombocytes. The clottocytes are also immune to the adverse effects that some drugs – such as aspirin – exert on their biological counterparts; they are also independent of the chemical fluctuations that occur in the bloodstream. Driven by the usual nano-computers, they use oxygen and glucose taken from the blood as an energy source.

Now let's move on to microbiovores. As you can guess from the name, these nano-machines deal with eliminating all kinds of pathogens that are found in the blood.[33] 3.4 micron long and 2 micron wide spheroids, the microbiovores consist of 610 billion accurately arranged atoms in a volume of 12.1 cubic microns. Built along the lines of phagocytes, microbiovores act by swallowing and breaking down – through appropriate enzymes, even artificial ones – pathogens; these biomedical nano-bots are a thousand times faster than their biological counterparts and are eighty times more efficient than macrophages. Actually, rather than a single type of nano-bot, we are dealing with an entire class of nano-machines, of which we can imagine different types. The purpose of the Freitas study is only to show that, on board a microbiovore, there is enough space for all of the systems necessary for it to work.

The respirocytes are the nanotech versions of the red blood cells and – was there any doubt? – they are much better than the originals.[34] Spheres of a micron of amplitude made of diamondoid material, these nano-bots are able to transport and infuse an amount of oxygen in tissues 236 times greater than that of red blood cells. Each respirocyte is composed of eighteen billion atoms and can contain nine billion molecules. The respirocytes can be used as a universal substitute for blood – for emergency transfusions – and, given their completely artificial nature, they do not present risks of viral and bacterial contamination. They can be used to

---

[32] One thousandth of a milliliter.

[33] R. A. Freitas Jr. *Microbivores: Artificial Mechanical Phagocytes using the Digest and Discharge Protocol*, 2001. See http://www.rfreitas.com/Nano/Microbivores.htm, http://www.zyvex.com/Publications/papers/Microbivores.html

[34] R.A. Frietas, Jr., *Exploratory Design in Medical Nanotechnology: A Mechanical Artificial Red Cell*, in: « Artificial Cells, Blood Substitutes, and Immobil. Biotech. », N. 26, 1998, pp. 411–430. http://www.foresight.org/Nanomedicine/Respirocytes.html. In addition: ibid., *Respirocytes: High performance artificial nanotechnology red blood cells*, in: "NanoTechnology Magazine" n. 2, October 1996, pp. 8–13.

treat any type of anemia, as well as a number of fetal pathological conditions, such as asphyxia *in utero* because of the detachment of the placenta, and so on. These imaginary nano-bots could eventually be used in any type of medical condition linked in various ways to breathing, from asthma to the bites of some types of snake. The respiratory system should also make it possible to breathe in oxygen-poor environments or in cases when breathing is impossible – for example, in cases of drowning, strangulation, exposure to nerve gas, overdoses of anesthetics and barbiturates, entrapment in very narrow spaces, suffocation and obstruction of the respiratory tract by food or others. In short, by injecting an adequate quantity of respiratory cells, you could spend long periods under water – or, for that matter, in space, provided that you have an astronaut suit – without breathing and without running the typical risks related to overly rapid decompression during re-emergence. In terms of strengthening, obviously, the respirocytes would provide athletes with incomparable tissue oxygenation, which would allow them to break every record.

However, our favorite nano-machines are the "cromallocytes," lozenge-shaped nano-bots made up of about four trillion atoms. Their task is to act as a vector for the carrying out of gene therapy operations, only with a precision and a degree of control much greater than those possible today. Chromallocytes, in particular, should be used for chromosome replacement: the chromallocyte penetrates the cell walls, navigates to the nucleus, removes the chromatin – that is, the genetic material – and, using a sort of nano-proboscis, replaces it with chromosomes synthesized in a laboratory. A chromallocyte is four microns thick and five microns long. Thanks to these nano-machines, it will be possible to cure a large number of pathologies linked to chromosomes or to defective genes – obviously, a complete therapy will require the simultaneous intervention of many trillions of chromallocytes. In the event that scientific research clearly links aging to the accumulation of genetic errors in the cell nucleus, these devices could represent the dream of every longevist.[35]

Last but not least, it is now the turn of the most complex device proposed so far by Freitas: the vasculoid. Designed together with the nanotechnologist Christopher J. Phoenix, this is a unique macroscopic nanotechnological robotic system capable of carrying out all of the functions of blood. Composed of about five hundred trillion nano-bots, they weigh two kilograms and fit the shape of all of the blood vessels in the human body. Originally proposed by Phoenix as *robo-blood*, it would essentially replace blood completely. This is an apparently far-fetched idea – like all of those that we have treated so far, after all – but we want to refer

---

[35] R. A. Freitas Jr., *The Ideal Gene Vector Delivery: Chromallocytes, Cell Repair Nanorobots for Chromosome Replacement Therapy*, in: « Journal of Evolution and Technology », Vol. 16, n. 1, June 2007, pp. 1–97. http://jetpress.org/volume16/freitas.html

you to the long and interesting *paper* by Freitas and Phoenix for details.[36] Specifically, the vasculoid is a collection of myriads of nano-machines in close contact with each other covering the entire internal surface of all of our blood vessels. The system uses a net of cilia to transport – in special nano-tanks – everything that normally carries the blood, blood cells, stem cells, nutrients, hormones, and so on. The advantages are: exclusion from the bloodstream of pathogens of all kinds and cancer cells in metastases; facilitation of the activity of the lymphocytes, and therefore of our immune defenses; elimination of all cardiovascular diseases; reduced sensitivity to toxic substances of all kinds and to all types of allergens; faster transport of the sub-products of metabolism, with an increase in duration and resistance; direct control of all individual hormonal and neurochemical reactions; partial protection from small and large traumas, such as insect bites, bullets and splinters, and falls from great heights; and extreme resistance to bleeding. Obviously, the two authors also propose some scenarios related to the installation procedures of the vasculoid, which should, however, start with the complete bleeding of the patient – appropriately sedated.

5.8   Nothing Ever Happens in Fog City

Among all of the ruminations of nanotechnologists, the one that we like best is undoubtedly the "Utility Fog," an interesting engineering fantasy conceived by the nanotechnologist John Storrs Hall.[37] It is about a hypothetical set of microscopic robots that can associate and reproduce physical structures of any kind. In other words, they represent the result of the encounter between the classic nano-machines and the self-configurable modular robots – that is, robots currently being developed that are composed of modules that can take different forms. From a visual point of view, our Utility Fog looks like a sort of "magic mist," which Hall initially conceived as a nanotech substitute for car safety belts. The robots in question – renamed "foglets" – are structures with arms extended in all directions, bearing hooks that allow them to mechanically join each other, transmitting and sharing information and energy and acting as if they form a continuous substance with mechanical and optical properties that can be modified at will. More specifically, each foglet consists of an aluminum exoskeleton, and each arm has a motor for extension and retraction and to open and close the "hand," i.e., a hexagonal

---

[36] R. A. Freitas Jr.; CJ Phoenix, *Vasculoid: A Personal Nanomedical Appliance to Replace Human Blood*, 1996, http://www.transhumanist.com/volume11/vasculoid.html

[37] See J. Storrs Hall, *Utility Fog: The Stuff that Dreams Are Made Of*. http://www.kurzweilai. net/utility-fog-the-stuff-that-dreams-are-made-of. Ibid., *What I want to be when I grow up, is a cloud*. http://www.kurzweilai.net/what-i-want-to-be-when-i-grow-up-is-a-cloud. Ibid., *On Certain Aspects of Utility Fog*. http://www.pivot.net/~jpierce/aspects_of_ufog.htm

structure with three fingers. Basically, it is all a game of foglets with arms that extend and retract at very high speed, hooking and unhooking each other. Every single foglet is equipped with computational capabilities of a certain level, and can communicate with its neighbors. In practice, the robots are usually in the air, scattered around us, invisible, with their arms released, allowing the passage of air between them. In the event of a collision, the arms engage and lock into their current position, as if the air around the passenger had solidified instantly. The impact is distributed over the entire surface of the passenger's body. Not only do foglets occupy a small percentage of the space around us, but their mist is so fine that it can even enter our lungs, cleaning them of any kind of toxic substance. We used the term "nano-machine" for convenience, since, in reality, the Utility Fog is made up of micro-machines, as, in Hall's conception, they should be no larger than a hundred micrometers. Each foglet has the shape of a dodecahedron, with twelve arms extended outwards. If we have to give a more technical definition, we can say that the Utility Fog is "a polymorphic active material composed of a conglomerate of hundred-micron large robotic cells, created through nanotechnologies and equipped with electric micro-motors and computational capabilities." The center body of a foglet is spherical in shape, with a diameter of ten microns. The arms have a diameter of five microns and are fifty microns long. Each foglet weighs about twenty micrograms and contains about five quadrillion atoms. Its movements are precise on the order of one micron; the arms have no joints, but they are telescopic. The difference between a foglet and a nano-machine is not only in size; unlike the latter, the former is not able to self-reproduce. Beyond its safety belt function, high-tech fog promises to revolutionize our daily lives. For starters: instead of building any object one piece at a time, would it not be better to do it using the foglets, and then change it to your liking? Thus, for example, a piece of furniture of a certain style could be transformed into a piece of furniture of another style, and so on, depending on the program that we put into the nano-computer network that animates the foglets. Furthermore, since each foglet has an antenna arm that can manipulate optical properties and refraction, a thin film of foglet could be used to create a stable or temporary television screen. A wall covered with foglets could change color every day, the floor decorations could change at will. With regard to the latter, it would never get dirty – goodbye to washing the floors, thanks to the foglet – and it could change the tactile sensations it gives us – sometimes becoming rough, sometimes soft. In short, we could fill our house of the future with a little Utility Fog and use it to make literally any object or piece of furniture that we need emerge from the wall or the floor, through "condensation" of the foglets, essentially, and then make it disappear when we do not need it anymore. Better yet: do we want to visit a friend? Our respective houses could "synchronize," assuming the same appearance; then, a foglet reproduction of our friend appears in our house, and one of ours in his. The "magic mist" around us records our actions and our copy at home reproduces them, and vice versa. And we also should not underestimate the protective potential of the Utility Fog: according to Hall, it is huge, so that a house properly filled with Utility Fog could protect its inhabitants from physical effects like the deflagration of a nuclear weapon within ninety-five percent of the impact area. Beyond that, the foglets can also remove bacteria, pollens, mites, and so on. A house full of this magical fog would prevent any domestic accident; indeed, all chores could be done by the Utility Fog itself, eliminating cuts and accidental falls from stairs, slips, and so on. Even such accidents that tend to happen to children would be prevented, as well as those due to objects that fall on top of them, or as the result of electric shocks and domestic pollution. If, instead, we create a real *Fog House* made of foglets, it could self-repair, change appearance as we please, feed itself with solar energy and isolate us from heat and cold.

If, instead, you had a car with foglets, you could change the model every day; moreover, you could fill the cockpit with foglets programmed to behave – from the point of view of your perception – like air, so that you would not see or feel them, so as to guarantee a total "safety belt" that protects every single centimeter of your body. For Hall, with such a system, you could crash at the speed of a hundred miles an hour without even finding your hair in disorder. Now, imagine a road covered with foglets that talk to each other, coordinating and preventing accidents. In fact, we could do everything with foglets: in this case, our car would no longer be a specific set of nano-robots, but a pattern that moves in a flow/road of foglets, like a wave that transports us, sort of like what happens in video games – in which a car is an image that moves through the fixed pixels of a screen. In fact, the car could also be completely transparent, similar to air, so you and your passengers would seem to literally fly without any support.

Alongside this, Hall's invention could give a new meaning to the concept of "telepresence" and "distance work." For example, our personal foglet cloud could collect data on the movements of our hands and transmit them to another very distant cloud, which would reproduce them in real time. Not only that, our cloud could act as a real "invisible armor," thus guaranteeing us invulnerability: thanks to it, we could fight with an alligator, amplifying our movements and protecting ourselves from the teeth of the animal. An adequate amount of Utility Fog could be used to simulate the physical properties of any macroscopic object, including air and water, so that it is indistinguishable from the original – although only from an optical point of view, of course. So, if we are set in a super-cloud, we can create or unpack objects suddenly, levitate them, levitate ourselves, demonstrate super strength, change our external shape to resemble that of a tiger, a rhino, a dinosaur. To do this, we obviously need an operational program with an archive that contains the patterns of the objects we need. With a Personal Utility Fog, we could literally keep all of our properties and personal belongings "stored" on a device similar to a CD, extractable" and "activated" as needed. Maybe we could even equip ourselves with a remote control device and, if we really want, we could also give it the shape of a classic magic wand. The Utility Fog will give us, in short, similar powers to those of the Krell, the aliens represented in the 1956 film *Forbidden Planet*, who had a technology capable of materializing their desires. Here, too, the problems would not be lacking, and, to avoid them, the nanotechnologists who create the magic fog would have to program it so that it only obeys fully conscious commands, and not those from the subconscious, like the Krell desire machine – as well as not responding to ambiguous or equivocal commands. With foglets, we would be able to physicalize the psychological concept of "personal space," establishing a rule by which all of the foglets within a certain distance from a given person are under their exclusive control, and that personal spaces cannot be merged except by mutual agreement. This could prevent any kind of violent crime, and our utopian Fog City would become a much quieter city than modern metropolises. The sort of theft perpetrated with a physical act – with dexterity or not – would be impossible, while hacking and fraud would still be possible. And, anyway, stealing an object made of foglets would not make any sense.

In addition to forming an extension of the senses and individual muscles and giving us invulnerability, the Utility Fog would therefore be able to serve as an infrastructure for the whole society. Fog City does not have permanent buildings, asphalted roads, cars, buses, trucks. It can look like a park, a forest; one day it can become like ancient Athens and the next day like Gotham City. The foglets could be mass-produced and occupy the earth's entire atmosphere, replacing every physical instrument necessary for human life. By exerting a concerted effort, the foglets could "magically" carry things and people everywhere; "virtual" buildings – composed of foglets – could be built and dismantled in a few moments, when needed, allowing for the replacement of cities and streets with farms and gardens.

If, as good Transhumanists, we decided to transfer our minds into a computer, then we could also make new bodies of foglets capable of operating in any terrestrial environment or in outer space – excluding the Sun and certain other extreme environments. In a virtual environment, we could be anything, any animal, a tree, a puddle, a paint layer on a wall – this would be possible in the real world too, if our bodies were made of the Utility Fog.

Not that these foglets are omnipotent, and Hall also informs us of what they would not be able to do – for example, they would not be able to perform activities that include chemical manipulations of matter. They could simulate a flame, but not produce high temperatures. They could prepare food like we do, but they could not simulate it – eating food made of foglets would be like eating sand.

So, what about the Utility Fog? We stick with what we said at the beginning: that this is an interesting engineering fantasy; to this, we add that it seems less problematic than the Drexlerian assemblers, if only because it operates at micrometric dimensions. Will it ever be realized, thus guaranteeing us super-human abilities? Impossible to say it; but do not worry, because, beyond the Transhumanist circles, in the much wider world of military research, there are those who are already trying to get super-powers. Seriously.

6

The New Flesh Rising

6.1 More than Human

Faster than light, able to fly, practically invulnerable, Superman is the forefather of a long series of characters distinguished not only by a more or less garish costume and a secret identity, but also – and above all – by a large series of powers of all kinds. Apart from some *outsiders*, like Batman and Iron Man, the bulk of superheroes differs from us common mortals for intrinsic capacities caused by exceptional events that enable them to perform actions that a human being, however gifted, would never be able to do. And so, Hulk has literally monstrous strength, and he can jump over the atmosphere and hold his breath underwater for hours; Spider-Man naturally adheres to vertical walls and has a "spider sense" that warns him of danger, while the Human Torch can ignite and reach the temperature of a nova; and these are just some of the many superheroes who have been haunting comics, movies, video games and TV series for decades. In short, we are now accustomed to hearing about superpowers, and there is no reason to be surprised if many expect that, thanks to the science of the near future, these capacities are destined to become real. To the point that there are those who, obviously lacking patience, have decided to get to work on obtaining these powers. This is how the concept – very Transhumanist, but now rapidly spreading outside of this context – of *human enhancement* was born, a fairly neutral term that hides the "human, too human" ambition to become post-human.

What, then, is this human enhancement of which so much is heard in Transhumanist circles? And what do these people still want? Is immortality not enough for them? Evidently not; and, on the other hand, how can we blame them? Eternity takes a long time to pass, and some more or less radical retouching of our nature will certainly be necessary. Thus, human enhancement refers, in essence, to

R. Manzocco, *Transhumanism - Engineering the Human Condition*, Springer Praxis Books, https://doi.org/10.1007/978-3-030-04958-4_6

a long series of procedures – some decidedly futuristic, or perhaps impossible, others on the horizon – aimed at overcoming our current physical and mental limitations in a more or less permanent way. More specifically, such enhancement technologies should be based on genetic engineering, nanotechnology, robotics, artificial intelligence, and neurotechnology, all disciplinary domains that should soon converge on us and make us something *more than human.*

Depending on the social context in which you hang out, the term "human enhancement" can be synonymous with the genetic improvement of humans, of human/machine fusion, of more or less sophisticated prostheses, of the use and abuse of nootropics, or of pure and simple doping. For Nick Bostrom, human enhancement is a very old thing, since it would have already started with alchemists[1] – who wanted to obtain the elixir of long life and the philosopher's stone – which was not only supposed to be able to change lead into gold, but also to change people, restoring innocence and happiness to the weary of the world. According to Julian Savulescu – Oxford University ethics scholar and Bostrom colleague – enhancement is "any change in the biology or psychology of a person who increases normal species-specific functioning above a certain level defined statistically."[2] In other words, you are enhanced if you do something that is impossible for all other human beings.

However you want to define it, the strengthening of our physical and mental abilities is certainly not a novelty for our species, and, indeed, in certain sectors – such as sports, but also military technology – it is a common practice. More generally, this label can be used to define the current reproductive technologies – such as the diagnosis and pre-implantation selection of embryos obtained through *in vitro* fertilization, a practice that has aroused much controversy; physical enhancement and improvement – from plastic surgery to doping, from prosthetics to exoskeletons, from pacemakers to various artificial organs under research; and mental enhancement – from nootropics to more or less intrusive neuro-stimulation systems. Mentioning the ideas of the post-humanist Roberto Marchesini, we could also say that the human is a species characterized by hybridization with the non-human, in which using a computer, wearing glasses or a hearing aid, or even just keeping a schedule of commitments is a form of empowerment, and that humanity has tried to overcome itself from the moment it came into the world.

Without dwelling too much on these classifications, we would like to deal with the latest technologies, which also strikes us as the most intrusive in anatomical and, we would say, ontological terms – that is, those essentially related to both the conservation and alteration of human nature. If you insist on asking us for a definition, we will say – paraphrasing the famous declaration of the American judge

---

[1] P. Moore, *Enhancing Me. The hope and the hype of human enhancement*, John Wiley & Sons, Chichester 2008, p. 15.

[2] Ibid. p. xi.

Potter Stewart – that human enhancement is like pornography: when you see it, you recognize it.

As you can imagine, this is a controversial topic, in particular, because of the fact that it can be linked – and critics of Transhumanism do link it – to eugenics and the serious drift suffered by this practice because of National Socialism. Precisely for this reason, beyond the speculations concerning the nature of human enhancement – can it be done? what will it consist of? when it will arrive? - the Transhumanists have long reflected on the ethical implications of all of this.[3] But what interests us here is the scientific aspect of the matter. In other words, we would like to understand what is already being done in terms of human enhancement, what we can do in the near future and, especially, where Transhumanists would like to go.

6.2   Genetic Doping, Nootropics and Other Amenities

So, let's start by taking a look at the "state of the art" in terms of human enhancement. Let's begin with the "nootropics." Also known as *smart drugs*, this is a very diverse set of substances – drugs, supplements, herbs, and so on – that have as their purpose the strengthening of cognitive skills, memory, concentration and resistance to mental fatigue. Thus baptized for the first time in 1972 by the Romanian scholar Corneliu E. Giurgea,[4] the nootropics – from the Greek *nous*, "mind", and *trepein*, "bend" – have come a long way in recent decades, becoming – at least in certain circles, like those of students and academics in general – of common use. And among the Transhumanists as well; let us not forget that the latter, taken as they are with heterodox ideas, generally try not to miss out on the possibility of self-improvement using the best that avant-garde scientific research can offer. To this is added their intrinsic tendency to experiment with new things, even and especially on themselves. The selection of nootropics – legal or a little less legal – is as varied as their mechanisms of action: in addition to "blockbusters" such as caffeine and nicotine, we have, for example, stimulants such as Adderall and other amphetamines; Modafinil, a drug used for narcolepsy, and which stimulates a concentrated wakefulness even in those who are not affected by this disorder; Ritalin, used for attention deficit and hyperactivation syndrome (ADHD), which should increase focus; piracetam, used in the treatment of Alzheimer's, and much more. In short, drugs for thinking better, more clearly and

---

[3] Savulescu J. and N. Bostrom (eds), *Human Enhancement*, Oxford University Press, New York 2011. Moreover: J. Hughes, *Citizen Cyborg: Why Democratic Societies Must Respond to the Redesigned Human of the Future*. Westview Press, Boulder 2004.

[4] C. Giurgea, *Vers une pharmacologie de l'active integrative du cerveau: Tentative du concept nootrope en psychopharmacologie*, in: «Actual Pharmacol.», N. 25, pp. 115–156. Paris 1972.

maybe faster. Keep in mind, however, one fact: these drugs especially affect people who suffer from certain disorders, and this could indicate the existence of a sort of cognitive "glass ceiling," a biological roof that limits our capacity in an unsurpassable way, and which our desire to surpass anyway could lead us to resort to types of intervention not yet on the horizon; nevertheless, the fact remains that the same neurologists have been talking for several years about the possibility of using specific drugs for dementia to enhance memory.[5] In addition to nootropics, pharmacology has made available to us another class of "potentiating" drugs: antidepressants. Psychiatrist Peter Kramer[6] coined the term "cosmetic psychopharmacology" to refer to the use of antidepressants, and especially of Prozac, to alter personality characteristics, to make shy people more confident, obsessive people more relaxed, alienated people happier and happier with themselves. This is what is called the "cosmetic use of anti-depressants," for patients who want the drug but have not received a diagnosis of clinical depression and do not suffer from other mental problems; moreover, it has long been known that musicians and performers often use propranolol to reduce anxiety.[7] It is true that philosophers and geneticists have already speculated, beginning in the 1980s, about the possibility of using gene therapy to enhance human mental abilities,[8] but this does not mean that the pharmacology has caught up with these ideas; it is, however, possible that the future will provide us with other surprises. This is the case, for example, of U0126, a drug still under study, which seems to be able to selectively eliminate specific memories. A few years ago, the well-known neuroscientist Joseph LeDoux and his team at New York University managed to remove a specific memory – linked to a frightening event – from the brain of a group of rats.[9] It cannot therefore be ruled out that, in the future, pharmacology will allow us to have precise control of our memories, a prospect that is both fascinating and disturbing.

As for the increase in our physical potential, there is classic doping, of which everything has already been said and on which it is not worthwhile dwelling. The real news, if it is news, is so-called "genetic doping," i.e., the ability to enhance the physical skills of professional athletes through gene therapy. There is certainly

---

[5] D. Cardenas, 1993. *Cognition-enhancing drugs*. "Journal of Head Trauma Rehabilitation," Vol. 8 (4), 1993, pp.112–114.

[6] P. D. Kramer, *Listening to Prozac. A Psychiatrist Explores Antidepressant Drugs and the Remaking of the Self*, Penguin Books, New York 1993. In addition: D. J. Rothman, *Shiny, happy people: the problem with "cosmetic psychopharmacology,"* "New Republic," n. 210 (7), 1994, pp. 34–38.

[7] J. Slomka, *Playing with propranolol*, « Hastings Center Report », n. 22 (4), 1992, pp. 13–17.

[8] J. Glover, *What sort of people should there be? Genetic Engineering, Brain Control, and Their Impact on Our Future World*, Penguin Books, New York 1984.

[9] K. Smith, *Wipe out to single memory*. http://www.nature.com/news/2007/070305/full/news070305-17.html

interest in the sports world,[10] although it is less certain that genetic doping is yet possible. For example, a few years ago, there was the case of Ye Shiwen, a 16-year-old Chinese two-time Olympic gold medalist for swimming in London 2012 who was at the center of a controversy concerning genetic doping. The Chinese athlete had indeed improved her personal time enormously over those few days, to the point that the American John Leonard, director of the World Swimming Coaches Association, said that the girl's record was "suspect and incredible," suggesting that the authorities who had submitted her to anti-doping tests had also subjected her to specific genetic tests. And, while a head of the Chinese anti-doping authority, Jiang Zhixue, replied that Leonard's claims were completely unreasonable, Ted Friedmann, chairman of the World Anti-Doping Agency genetics panel, stated that he would not be surprised at all to discover that genetic enhancement was already secretly being used by some of the athletes. Friedmann tried to find ways to identify cases of genetic doping and prevent its spread. "This is a technology mature for abuse," he said. But it must be said that Friedmann's positions on the subject are rather heterodox and far from being accepted in the medical and sports communities. Anyway, what about this possibility? Are "Franken-athletes" really around the corner? Certainly, some forms of gene therapy – for example, against cystic fibrosis – are being tested on humans; the DNA of athletes being altered – for example, to increase their muscles and vigor, or to make their blood capable of transporting more oxygen than normal – is something that could certainly occur in a very near future, and something that is therefore worth discussing now.[11] At the moment, the World Anti-Doping Agency does not have a reliable method for determining whether someone has undergone gene therapy and, going on what we know now, some abnormal characteristics – a blood richer than normal in red cells, or a higher than average level of hormones – could also be "simple" characteristics of an extraordinary body. Anna Baoutina, a researcher at the National Measurement Institute in Sydney, says that the great advantage of genetic doping is that it is very difficult to identify with respect to chemical doping; the doped gene is very similar to the natural genes found in the body. For Patrick Schamasch, medical director of the International Olympic Committee (IOC), viruses used to insert genes into the body leave traces that can be identified; However, it is

---

[10] The case of H. Lee Sweeney, a researcher at the University of Pennsylvania Medical School, who, in the late 1990s, inserted a gene that stimulates IGF1 – a growth factor with anabolic properties, managing to increase the muscle mass of guinea pigs, sometimes by as much as 30% – is eloquent in this regard. Sweeney says he received – not even a week after the publication of the results – phone calls from two coaches from as many high school sports teams, eager to know if there was any way to subject their athletes to this treatment. See http://legacy.jyi.org/volumes/volume11/issue6/features/iyer.php

[11] See A. Miah, *Genetically Modified Athletes. Biomedical Ethics, Gene Doping and Sport*, Routledge, London 2004.

certainly possible that new GMO viruses that leave no trace will soon be developed. Dominic Wells, a gene therapy expert who has studied the possibility of genetically modifying athletes, reassures us, however, that it is not yet possible to use gene therapy for doping, even if, to be honest, Franken-athletes are not so far away.[12] And, in this regard, the latest novelty in terms of genetic doping – and probably the first form of genetic enhancement that athletes have used illegally – is Repoxygen. Developed by Oxford Biomedica, it is a form of gene therapy that pushes the body to produce additional reserves of erythropoietin, a hormone that regulates the production of red blood cells and which can be used by athletes as a doping substance; the technique in question is still in its pre-clinical trial phase – on mice – and has not yet been systematically tested on humans. Apparently, it may not be possible to trace it in the athletes' bodies. And if you believe that there are no people willing to undergo such intrusive measures, it means that you have never heard of so-called "enhancement surgery," which is the practice – for example, among baseball players – of undergoing surgery precisely to improve athletic potential – laser eye surgery to enhance sight, pitchers who have their elbows rebuilt with stronger ligaments taken from other parts of the body, and so on. Less intrusive than genetic manipulation, of course, but nevertheless significant.

And after genetic doping, we move on to prostheses, more or less sophisticated instruments that aim to replace a part of the missing body due to congenital conditions, traumas or pathologies. Within this definition, we can include everything that is artificial and stably attached to our body, from silicon breasts to dentures, from artificial arms to bionic legs, from hearing aids to the famous penile implants in vogue among politicians and elderly entrepreneurs. Of course, at the moment, nothing comes close to the accuracy, reliability and natural appearance of the legs, arm and artificial eye of Steve Austin, the protagonist of the famous TV series of the '80s, *The Six Million Dollar Man*; you can be sure, however, that we will get there.

Currently, prostheses for legs and arms are built with materials that combine lightness and strength, such as plastic or carbon fiber; to this, we must add the growing number of technological findings aimed at increasing the conscious control of the user on the artificial limbs; for example, myoelectric limbs have electrodes that allow them to pick up electrical signals from the patient's muscles, a technique known as electromyography, and use them to operate the prosthesis, e.g., closing or opening the artificial hand. There are also more sophisticated systems: one can, in fact, find robotic arts equipped with biosensors that collect

---

[12] See J. Naish, *Genetically modified athletes: Forget drugs. There are even suggestions some Chinese athletes' genes are altered to make them stronger*, 'Daily Mail', 1 August 2012. http://www.dailymail.co.uk/news/article-2181873/Genetically-modified-athletes-Forget-drugs-There-suggestions-Chinese-athletes-genes-altered-make-stronger.html#ixzz2WNEBwD4F

signals from the user's nervous system, and so on. This with regard to technology; And what about upgrades? Can we say that patients with bionic prostheses are "enhanced" compared to those who have a healthy body? In general, prostheses are grafted in order to compensate for an anatomical and/or functional deficiency of the patient, that is, to allow her/him to lead a more or less normal life. However, our question makes sense, given that, in certain contexts, we have had to seriously question whether a certain prosthesis would grant a competitive advantage to its owner. For example, this is the case, a well-known one, with Oscar Pistorius, the South African athlete equipped with a transtibial lower prosthesis, which, apparently, allows him to save 25% of his energy. At the moment, these are small things, but there is no doubt that bionic prostheses will become more and more efficient, to the point of being able to overcome the performance of our natural limbs – a turning point at which we must necessarily ask, paradoxically, whether it is worthwhile to replace natural limbs with artificial ones in a wholly healthy person.

On our list, we cannot forget artificial organs, which aim to give essential functions back to the organism that the latter is not able to sustain. The need for this technology is also dictated by the chronic scarcity of organs necessary for transplantation, as well as by the desire to prevent organ trafficking. Over the years, various devices have been developed in order to make up for this or that dysfunction. We have, for example, brain pacemakers – stimulators that act on various, more or less "deep" areas of the brain – used for everything from depression to epilepsy, from tremors caused by Parkinson's disease to other neurological problems.

Then, we have cochlear implants, which aim to restore at least partial hearing to those who have lost it; ocular prostheses – typically micro-cameras connected to the retina or the user's optic nerve – which currently have limited functionality; various models of artificial heart, which, for now, are able to support patients only for limited periods of time. As for the liver, something seems to be in motion; HepaLife is developing a bio-artificial liver model based on stem cells – solely for support, as the patient awaits a transplant. In addition, Colin McGucklin and his colleagues from Newcastle University are working on an artificial liver that could be used, in a few years, to repair a partially damaged natural liver, or that could be used outside of the patient's body to perform hepatic dialysis. For its part, MC3, a company in Ann Arbor, is working on development of real artificial lungs, while various research groups are trying to create semi-biological substitutes for the pancreas – for the purpose of treating diabetes. The list of the types of organ that we are trying to replace with artificial counterparts goes on, but we will stop here – persuaded that we have given you an idea.

6.3   Do-It-Yourself Super-Powers

Surely, Transhumanists do not intend to risk health – and, indeed, immortality – with chemical doping or with more or less risky drugs; but neither do they have the intention to stay inactive, waiting patiently for someone else to get to work on human enhancement. And so, they have developed a tendency, more or less widespread, to experiment on themselves, trying this or that solution. And, indeed, the first experiments – amateurish more than anything – with healthy individuals have already started. This is the case with Todd Huffman, a graduate in neuroscience, Transhumanist and collaborator with Alcor; with the help of some "body modification" artists – practitioners of a set of rather extreme body modification practices, including tongue bisection, subcutaneous horn implantation, and the like – Huffman had a magnet implanted in his left ring finger.[13] The magnetic part of the small implant – as big as a grain of rice – is made of neodymium, an element belonging to the group of "rare earths." One of the central concepts of much of Transhumanism is so-called "substrate independence," for which certain functions do not depend on the type of material substrate that hosts or produces them. So, it is conceivable that certain mental and sensory functions are carried out by a non-biological substratum, to which the organic one can also be connected. Thus, Todd's goal was to understand whether adding a whole new kind of sense – tied to a non-organic substrate – could push his brain to alter his structure and end up generating a new, unprecedented sensory component of our vision of the world – at the same time changing the way our intellect works. Essentially, it was a matter of developing a simple and elegant method for creating a whole new sensory experience that did not fall within the five classical senses – a bit like what happens with other animal species, capable of identifying magnetic fields or other non-conventional sensory data sources, for example.[14] The surgery – not so trivial – consisted in putting the implant into contact with a set of hundreds of nerve endings present in the ring finger; in this way, the small magnet was able to take advantage of the highly developed pressure sensors located at the tips of the fingers. No nerve was connected directly to the magnet, but when the magnet was stimulated by an external source, such as an electric motor or a transformer, it began to oscillate, and the "sensors" transmitted everything to the brain. It was not really a new sense organ in the embryonic phase, but a new way of using a classical sense, touch, educating the brain to recognize these signals. A simple beginning, then, or maybe just a provocation. But with a *caveat*: the internal chemical environment of our body is quite corrosive – just not as much as the blood of the

---

[13] See http://www.wired.com/gadgets/mods/news/2006/06/71087

[14] See, for example: H. C. Hughes, *Sensory Exotica: A World beyond Human Experience*, MIT Press, Cambridge 1999.

Alien – and the corrosion of the neodymium carries various toxic substances with it; in other words, do not try this at home.

Huffman's intervention was rather superficial; it was nothing compared to those Kevin Warwick underwent. A researcher and professor of cybernetics at the University of Reading, Warwick has worked on artificial intelligence, "deep brain stimulation" and robotics. The work for which he has become very famous is his "Project Cyborg," started in 1998 with the implant of a radio transmitter on a chip into his upper left arm.[15] The device transmitted an identification code, which was read by some transponders inserted into the doors of his work area; the latter then automatically opened when Warwick passed. Other transponders turned on the lights, or caused a computer to say "Good morning, Professor Warwick!" The self-proclaimed cyborg kept the implant in for 9 days; at the time, Warwick said that he felt a "special connection with the world" – luckily, during that time, Murphy's Law did not manifest itself, thus avoiding leaving him in the dark or closing a door in his face. In 2002, the academic-cyborg proceeded with the insertion of a second implant; this time, it was a small plate to which a hundred and a half millimeter long electrodes were connected. Unlike the first implant – which consisted of a subcutaneous graft – this was connected directly to a nerve, and connected on the other end to a cord that emerged from Warwick's arm through an incision. Thanks to it, the Warwick team was able to collect signals coming from the nerve, or transmit signals it to it. For example, one of the tests carried out by the researcher was based on a form of sonar similar to that of high-tech bumpers. With them, Warwick tried to get a real "ultrasonic vision"; every time his sonar approached something, it sent a small discharge to his nerve, allowing him to navigate through his laboratory with his eyes closed. One of the things the scholar has noted with greater pleasure is the speed with which his nervous system learned to listen to the new signal.

The second set of experiments examined Warwick's nervous system outputs; in practice, the researchers tried to understand which nerve impulses could be collected with the chip and whether they could be used to direct independent robotic prostheses. By moving his hand, Warwick allowed his team to record the nerve impulses associated with it and use them to move a robotic limb connected to a dedicated computer. The experiment was even successfully carried out remotely, via the internet – Warwick was in the US and the robotic arm in Reading. The third experiment also involved the researcher's wife, who underwent the implantation of a simple electrode into the left arm, connected to her husband's implant with a cable, so that, through electrical impulses, the two could communicate using a sort of Morse code. Warwick kept the implant in place for 3 months and, after

---

[15]Among the books published by Warwick on these topics, we mention: *I, Cyborg*, University of Illinois Press, Champaign 2004; and *March of the Machines: The Breakthrough in Artificial Intelligence*. University of Illinois Press, Champaign 2004.

extraction, underwent medical tests to verify that the direct connection of the chip to his nerve had not damaged it – which it turned out had not happened. In analyzing the experiment, the British journalist Pete Moore – who produced a skeptical but curious look at human enhancement – stressed that these implants present a certain risk of infection, as the incision necessary to pass the cables constitute an open access point for bacteria; that, in any case, there is a risk of permanent damage to the nerve; and, finally, that a possible reaction of the immune system cannot be ruled out. Another point that Moore emphasizes is that, in the context of human enhancement, there is a lot of hype, but that the practical results are, for now, very limited; for example, Warwick's proposal to equip firefighters with an "ultrasonic view" useful for moving through environments invaded by smoke clashes with the fact that human eyes have six million cones and one hundred and twenty to 150 million rods – and can therefore easily capture light waves, which are very short – while the sonar – which has to deal with much longer waves – will never provide the same resolution as our natural sight. This is just an example, and also a rather specific one; yet it stresses quite well the gap that separates ambitions from concrete capabilities in this area.[16] To this, we must add that, in the course of the 2000s, a new factor came into play, which could change – even radically – the cards on the table: the United States Army.

6.4   Imagination and Military Power

The stereotype of the military is that of a disciplined person, duty-bound and, above all, devoid of imagination. But those who think thusly must have taken movies like *Dr. Strangelove* too seriously; otherwise, they would have a better appreciation for the experiments that the US Army – often "on the edge" – has organized over the past few decades. Just think, for example, of MK-Ultra, the project through which, on the basis of presumed similar initiatives by the Chinese Communists, the US tried to realize a "brainwashing" procedure between the '50s and '60s, although without success. Or think about the Stargate Project – nothing to do with the homonymous films and TV series – that, between the '70s and the '90s, represented an attempt to study alleged paranormal phenomena – for example, precognition, clairvoyance, trips outside of the body, and so on – from the point of view of possible military applications. Thus: do not say that the high officers of the Pentagon are not people with open minds. Precisely for this reason, we are not astonished to read the amazing enterprises of DARPA – the Defense Advanced Research Projects Agency, the federal agency for military research in

---

[16] P. Moore, Op. Cit. It is interesting, however, that, according to Moore, although the Transhumanists are a relatively small group, the impact of their thinking on the policy-making of some countries – like the United States, but not only them – would be relevant.

short, very secret and well-financed. Founded in 1958, and located in Arlington, Virginia, DARPA[17] has an annual budget of $2.8 billion and only 240 permanent employees. The working methodology of DARPA is substantially indirect, in the sense that its managers conceive and plan this or that project, subdividing it into sub-projects whose realization is then delegated – obviously together with adequate funding – to this or that university research center, to this or that company. As often happens with American things, even this agency was born out of a shock, and, in particular, the technological defeat due to the Soviets, who, in 1957, launched Sputnik 1. To avoid other bad outcomes, and to guarantee the American armed forces a constant technological superiority, Dwight D. Eisenhower decided to create such an agency – initially giving it the name of ARPA – with the mission of pushing the frontiers of scientific and technological research far beyond immediate military needs. And this is how the story of these "apprentice sorcerers" – who have been nothing if not lucky – begins, a story that, during the '70s, gave us, among other things, the first embryonic version of the Internet – at that time called, coincidentally, the Arpanet. This is to say that several of the inventions supported and funded by DARPA have gone to have a significant impact on society. If the original mission of the agency was to *prevent technological surprises* from adversaries, the current one aims instead to *surprise them technologically*. The work of DARPA is therefore interdisciplinary and has evolved over time; if, at the end of the '50s, it contributed to the birth of what would become GPS, during the '60s, the agency worked on computers, behavioral sciences and material sciences. The '70s saw not only the birth of the Arpanet, but also a more general commitment by DARPA to air, sea and land technology, for the development of laser technologies for defense against space missiles, for submarine warfare, and for the application of defense computation. It is difficult to summarize the number and variety of projects on which the "Pentagon wizards" have worked, even during the last few years. Among the most curious, we remember the Battlefield Illusion project, for the creation of illusions on battlefields,[18] the BigDog/Legged Squad Support System, MeshWorm, a worm-shaped robot, capable of slipping into hard-to-reach places, and Proto 2, an artificial arm controlled by thought. And then, there are the remote-controlled robotic insects, different types of exoskeleton, and even the Transformer, an armored flying car. In 2011, DARPA also organized the *100-Year*

---

[17] www.darpa.mil

[18] The aim of the project is to "manage the sensory perception of the adversary, in order to confuse, delay, inhibit or mislead his actions." The goal will be achieved by understanding the way in which "human beings use their brains to process sensory inputs," which will allow them to "develop auditory and visual hallucinations" that "will provide a tactical advantage to our forces." Shachtman, *Darpa's Magic Plan: "Battlefield Illusions" to Mess With Enemy Minds try to create illusions on the front lines*, «Wired», February 14, 2012. http://www.wired.com/dangerroom/2012/02/darpa-magic/

Starship symposium, with the aim of encouraging the public to think seriously about interstellar flight, with the idea of achieving this result within a century[19] – bite the dust, *Star Trek*. And then, of course, there is the program for the super-soldier – or rather programs, as it is actually a very varied set of projects and objectives. Joel Garreau, a professor at Arizona University and author of *Radical Evolution. The Promise and Peril of Enhancing Our Minds*, a book in which he treats the themes in question, has long been involved in the topic. Garreau is keen for people to remember that, however amazing, the agency's projects are not day-dreams, but real attempts to radically change the human being:

> Imagine if the soldiers could only communicate with thought … Imagine if the threat of a biological attack was without consequences. And contem-plate, for a moment, a world where learning is as easy as eating, and replac-ing damaged body parts is as easy as going for a fast-food drive-thru. (…). It is important to remember that we are talking about science in action, not science fiction.[20]

In short, DARPA seems to have taken the famous army slogan *be all you can be* very seriously; indeed, perhaps it wants to get something more: to be able to create human beings that are completely unstoppable. So, let's see how the agency aims to achieve this goal by taking a look at the projects it has chosen to divulge.

Much Better than Iron Man

Exoskeletons are mobile machines consisting of a wearable structure containing a system of engines that give users much greater strength than that of a normal human, as well as the ability to perform very strenuous activities for a prolonged time and practically effortlessly. Normally developed for military purposes – to carry heavy loads in and out of battlefields – exoskeletons can also be used by firefighters in particularly arduous rescue operations or in the medical field – for the rehabilitation of stroke patients or paraplegics. The models on the market are many, for example, the XOS by Sarcos and – in a clear tribute to the almost hom-onymous Marvel superhero – the HULC by Lockheed Martin. If these models have military uses, HAL 5 by Cyberdyne has medical purposes – that is, to help people with walking and mobility problems. With the "Future Warrior" program, DARPA aims to create enhanced exoskeletons that will allow soldiers to lift unbe-lievable weights effortlessly, walk or run for hours, and so on.

---

[19]T. Casey, *Forget the Moon Colony, Newt: DARPA Aims for 100 Year Starship*, «CleanTechnica», January 28, 2012. http://cleantechnica.com/2012/01/28/fmoon-colony-newt-darpa-has- 100-year-starship/

[20]Joel Garreau, *Radical Evolution. The Promise and Peril of Enhancing Our Minds, Our Bodies – and What It Means to Be Human*, Doubleday, New York 2004, pp. 22–23.

Metabolic Domination

Joe Bielitzki manages the "Metabolically Dominant Soldier" program. Very ambitious, the project in question aims to fiddle with the internal machinery of human cells – starting from cellular metabolism – in order to increase resistance to fatigue and physical strength. One of the options taken into consideration by this program is the manipulation of mitochondrial DNA and modification of the number of mitochondria inside cells, so as to increase their efficiency in producing energy – for example, by allowing a soldier to walk practically without limits wearing an eighty-pound backpack. Another goal is to develop technologies that allow soldiers who are seriously injured to go into hibernation; this would allow them to survive for a short time, even without oxygen, waiting for help. Still another objective is to optimize the use of oxygen, allowing soldiers to sprint for 15 min at Olympic levels with a single breath, to a degree that would shame Usain Bolt; it seems, in fact – according to Bielitzski – that human beings do not manage oxygen efficiently, wasting much of what they capture with a single breath. Another front is that of food; in this case, we want to manipulate DNA in order to give the soldiers' bodies the ability to convert fat into energy more efficiently, so that they can go several days without eating – eliminating the need to carry food. If successful, such a project – once widespread outside the military – would sweep away the multi-million dollar business of diets in an instant.

Forbidden to Sleep

A slightly *over the edge program* is "Continuous Assisted Performance," managed by John Carney. Its goal is to eliminate the need for sleep during an operation, while maintaining a high level of physical and mental performance. Precisely for this reason, this program is studying dolphins and whales, animals in which, as is known, the cerebral hemispheres sleep in turns, in order to help them avoid sinking into the sea and drowning. The idea would be to find a way to transfer this capacity to ourselves. In this regard, DARPA is funding the research at 360 degrees; for example, it financed experimentation on a drug – tested on army helicopter pilots – that was able to extinguish the "sleep switch." The compound allowed the volunteers to stay awake for more than 40 hours, maintaining a high level of concentration – one that, incredibly, improved after almost 2 days without sleep. In this area, another technology that DARPA is testing is transcranial magnetic stimulation.

Prohibited from Getting Sick

Carney has also been entrusted with the "Unconventional Pathogen Countermeasures" program, which seeks to develop a strategy capable of blocking any type of pathogen once and for all, making soldiers immune to any transmissible disease. One of the goals is to discover a genetic core common to all or almost all of these life forms, and find a way to block it. One example would be to find an enzyme present only in bacteria and not in us. Another system could be what Carney calls "genomic collapse," something that "glues" itself to the genome of pathogens so tightly that it cannot be read and replicated.

Zero Injuries and Zero Pain

Kurt Henry manages the "Persistence in Combat" program; it, in turn, gave financing to a company, Rinat Neuroscience,[21] in order to develop a "pain vaccine" that could block intense pain in 10 seconds and whose effects would last 30 days; if successful, the treatment in question would revolutionize pain therapy. Another scholar funded by DARPA is Harry Whelan, a neurologist at the Medical College of Wisconsin. Whelan is studying a process known as "photobiomodulation," a technique of cellular stimulation that consists in the use of near-infrared light to treat injuries and wounds, whose healing this procedure seems to accelerate. Henry is also working with a number of researchers – although details currently remain elusive – who have discovered that the cascade of bodily reactions that stop bleeding can be stimulated by signals from the brain. If this were somehow to be consciously controlled, we could train the soldiers to stop bleeding within a few minutes.

Programs of All Kinds

Obviously, this is just the tip of the iceberg, and there are many more projects that have been launched in the past or are currently active. At DARPA, Alan Rudolph – biologist and current director of the International Neuroscience Network Foundation – has worked on the development of a brain/computer interface, whose possible military applications would be numerous: think of the possibility of controlling a distant fighter with one's mind, or using an exoskeleton by guiding it directly with thought, rather than with the stimuli coming from the muscles. It would be like inserting the brain into a different vehicle, removing it from the human body and putting it into a larger, much stronger robot. Rudolph has also worked on the development of a chip capable of simulating brain circuits, and that would therefore be able, in the future, to replace them, and perhaps to enhance them – for example, by increasing the speed of information processing. Among DARPA's past projects, we know something about the "biomotor," a set of technologies aimed at using the resources found in the human body as fuel – and that could be used to operate devices implanted in our body. Then, we have the "Engineered Tissue Constructs" program, which is based on the idea of

---

[21] Subsequently acquired by Pfizer.

reconstructing customized organs and parts of the body on demand, as well as carrying out the construction process inside the body, without the need for a transplant.

The Armor Inside Us

Those that we have presented to you thus far were projects publicly promoted by DARPA in 2002, and described in great detail by Garreau in his book. Over the years, the agency's research programs have evolved and multiplied. And so, in 2007, at DARPAtech – the agency's regular meeting – Michael Callahan launched a new slogan: "to make human beings act more like animals." A slogan obviously associated with a new program: "Inner Armor." The purpose of the project is two-fold.[22] The first objective is to allow soldiers to be able to act in extreme environments – at high altitudes, in very hot places and in the deep sea. For each of these conditions, there are animal species that are able to handle the situation well. And so, for example, the Indian goose can fly for days at heights comparable to those of the Himalayas without a break. Some microorganisms thrive in volcanic chimneys, which emit boiling steam and in which temperatures comparable to those of Venus are recorded. Then, there is the sea lion, an animal able to redirect its blood flow and slow its heartbeat in order to stay underwater for hours. Callahan would essentially like to study and apply these biological tricks to soldiers, obtaining divers able to increase the flow of oxygen towards their main organs by 30 or 40%, imitating the reflection of immersion.[23] It would also not be bad if the divers could do it automatically; how DARPA intends to achieve this result, however, we don't know.

The second goal is to make kill-proof soldiers. Dumps full of chemical and radiological materials teem with microorganisms that easily resist these conditions and, starting from the study of these creatures, Callahan would like to create a set of "synthetic vitamins" that can prevent chemical and radioactive poisoning, thus allowing soldiers to wander in otherwise lethal environments. Not only that; the procedures in practice by the army today provide for the protection of soldiers against a limited number of dangerous pathogens; among the purposes of "Inner Armor" would therefore be the realization of preventive universal immune cells, which protect against *all* types of infectious disease. To this is added another objective, namely, the development of a system for predicting the evolution of pathogens, so as to prevent the emergence of a new pathogen with a vaccine.[24]

---

[22] http://archive.darpa.mil/DARPATech2007/proceedings/dt07-dso-callahan-armor.pdf

[23] The diving reflex is a set of reactions that take place in many mammals – especially marine – at the time of immersion in water, in order to reduce the consumption of oxygen. It includes the reduction of heartbeat and the concentration of blood in some organs, especially the heart and brain.

[24] N. Shachtman, *"Kill Proof," Animal-Esque Soldiers: DARPA Goal*, August 7, 2007. http://www.wired.com/dangerroom/2007/08/darpa-the-penta/

Silent Talk

Do you remember the famous "mental fusion" from Star Trek, with which the Vulcans are able to enter into direct communication with the minds of others? Well, DARPA has set up a more or less similar project. Codename: "Silent Talk." The aim of this program is to explore futuristic mind-reading technologies with appropriate tools that can detect electrical signals in soldiers' brains. The ultimate goal is to develop high-tech helmets capable of transmitting and receiving thought – a sort of telepathic internet, in short – thus allowing entire armies to keep in touch without radio. It is with this in mind that DARPA funded the research of Jack Gallant and his team at the University of California at Berkeley. In 2011, these scholars carried out the following experiment: Gallant and his colleagues made a number of volunteers look at some Hollywood movie trailers and, thanks to a simple scan of their brains via magnetic resonance imaging, they managed to reconstruct the frames of such video clips. And this is only one of the innumerable studies in progress as regards the reading of the mind; other research, for example, aims to "extract" from the brain the words that the latter is listening to.[25]

Better than Lizards

Led by Robert Fitzsimmons, the "Regenesis" program wants to study the mechanisms that allow certain animals – such as lizards – to regrow their limbs – a phenomenon that, although to a much lesser extent, is also present in human beings.[26] The aim is, as usual, to activate this capacity in American soldiers in particular.

These are the projects that could lead us, in the future, to the super soldier and to any possible impact on civilians in terms of human development. In any case, after receiving some negative feedback from the press and the public, DARPA decided to "fly low," renaming its projects in less rebellious and more reassuring terms. And so, "human enhancement" has become "optimization," while "Regenesis" is now called "Restorative Injury Repair." But a question remains: why is a super-secret agency like DARPA so chatty? Is it only a communication strategy that aims to "impress" the public and impress the enemy? Maybe. But there is another possibility, namely, that, with respect to the projects that we have

---

[25] I. Morris, *Hitler would have loved The Singularity:* "Daily Mail", February 6, 2012. http://www.dailymail.co.uk/debate/article-2,096,522/The-singularity-Mind-blowing-benefits-merging-human-brains-computers.html # ixzz2WNEJP4GE

[26] Perhaps not everyone knows this, but children up to the age of eleven are often able to regrow their cut fingers – if the cut only affects the phalanx, excluding the joint. In the 1970s, Cynthia Illingworth, a pediatrician at the Children's Hospital in Sheffield, discovered the phenomenon, a discovery also covered by *Time* magazine in 1975. See http://www.time.com/time/magazine/article/0,9171,913436,00.html

illustrated, media disclosure means that the planning phase is almost over and that the agency's executives are ready to try to convince the public of the need to accommodate the changes that these projects will inevitably bring.[27] But the most important question regards the consequences that such technologies could have on our identities. In other words: is a human who does not eat, does not sleep, does not feel pain and never stops still recognizably human? At DARPA, these philosophical questions obviously do not worry them; the alteration of human nature produced by such research is, from their point of view, only an involuntary consequence.

Do not think, however, that DARPA is the only American institution that seriously takes into consideration the possibility that, over the next few decades, human capabilities will undergo a process of strengthening. Published in 2002, *Converging Technologies for Improving Human Performance*[28] is a report commissioned by the National Science Foundation and the Department of Commerce, edited by Mihail C. Roco – the nanotechnologist of the NNI – and William Sims Bainbridge – a known scholar who specializes in, among other things, the sociology of religion. In its 415 pages, the report collects descriptions and comments on the state of science and technology in four related fields, known collectively as NBIC – nanotechnology, biotechnology, information technology and cognitive science. To be considered is, in particular, their potential to improve health, overcome disability and facilitate human enhancement in the military and industrial sectors. Among the perspectives examined, there is that for which understanding the mind and the brain will allow us to create new species of artificially intelligent systems that will generate economic well-being at unimaginable levels. According to the report's authors, within 50 years, intelligent machines could create the wealth necessary to provide food, clothing, shelter, education, medical care, a clean environment and physical and financial security for the population of the entire world. In short, the engineering of the mind offers us the opportunity to bring humanity into a new golden age. It makes a certain impression to see an official document declare that "it is time to rekindle the spirit of the Renaissance" to achieve "a golden age that will be a turning point for human productivity and quality of life." The authors foresee, among other things, that the direct connection between the human brain and machines will transform work in the factories, control cars, guarantee the military superiority of those who adopt it, and allow for new sports, art forms and ways of interaction among the people; wearable sensors will enhance each person's awareness of their health conditions, their

---

[27] M. Snyder, *US Super Soldiers Of The Future Will Be Genetically Modified Transhumans Capable Of Superhuman Feats*, « The American Dream», August 31, 2012.

[28] M. C. Roco and W. S. Bainbridge (ed), *Converging Technologies for Improving Human Performance*. Springer, New York 2004. http://www.wtec.org/ConvergingTechnologies/Report/NBIC_report.pdf

environment, pollutants, potential risks and information of interest; the human body will be more durable, healthy, energetic, easy to repair and resistant to many types of stress, biological threat, and the aging process; new technologies will compensate for many physical and mental disabilities and will completely eradicate the handicaps that have plagued the lives of millions of people; anywhere in the world, individuals will have instant access to the information they need; engineers, artists, architects and designers will experience enormously expanded creative skills, thanks to the fact that we will gain a greater understanding of the nature and functioning of human creativity; ordinary people, but also politicians, will have an enormously improved awareness of the cognitive, biological and social forces that control their lives, and this will allow for better adaptation, greater creativity and more efficient decision-making processes; formal education will be revolutionized by an increased understanding of the physical world from the nanometric to the cosmic levels.

Not bad, the prospects opened up by DARPA and the National Science Foundation; but not extreme enough to meet the expectations of the Transhumanists. Who, as usual, have thought of doing things by themselves, elaborating a long list of enhancements that they would love to possess.

6.5   The Flesh of the Future

Obviously, the *upgrades* of human nature that the Transhumanists would like to see carried out go far beyond military applications and, if put into practice, would substantially change the functional organization and chemical composition of our body. An old film by David Cronenberg, *Videodrome*, comes to mind at this point, in which the protagonist goes through a series of progressive mutations that lead him to transform himself into the "New Flesh" – a term taken from the Gospel, in reality, but that Cronenberg re-reads in a much more disturbing light – a new state of being that is also characterized by a deeper understanding of reality.

First of all, in order to evaluate the ethical aspects of human development as they intend, the Transhumanists have also launched a research project supported by the European Union – named *ENHANCE* – which aims to deepen all of the philosophical issues related to cognitive enhancement, life extension, enhancement of mood and an increase in physical performance.[29] Funded by the European Commission in the context of the Sixth Framework Program (2002–2006) and destined to last 24 months, *ENHANCE – Enhancing Human Capacities: Ethics, Regulation and European Policy*, was started in October 2005.[30] It is interesting to

---

[29] http://ieet.org/index.php/IEET/more/the_eus_enhance_project

[30] This is the announcement: http://www.unisr.it/view.asp?id=6124

note that, at the European level – more than official, therefore – it is recognized that many of the medical biotechnologies *currently* under development could be used, in theory, to enhance the physical and mental capacities of humans. The project announcement reaffirms the way in which these technologies – together with nanotechnologies – could be applied, in practice, to make human beings think better, feel happier, improve their sporting abilities or extend their life spans. There is also talk of the possibility that these enhancements may modify our self-perception, transforming our society into a "post-human society." The *ENHANCE* project, in short, was launched to investigate the degree of development within disciplines such as biology, bio-gerontology and neuroscience, within the frame-work of four areas, namely, cognition, mood, physical performance and lon-gevism. The project was coordinated by the medical ethical scientist Ruud ter Meulen and the institute to which the Center for Ethics in Medicine of the University of Bristol belongs, but *ENHANCE* also saw the involvement of various institutes and characters related to Transhumanism, like Julian Savulescu from the Center for Practical Ethics at the University of Oxford, Nick Bostrom from the Future of Humanity Institute and the Swedish neuroscientist and Transhumanist Anders Sandberg.

It goes without saying that the enhancement discussed in the course of *ENHANCE* is, from the point of view of the Transhumanists, only a beginning; many other wonders await us in the near future. To begin with, according to authors like George Dvorsky, we would be moving towards a real "telepathic civilization."[31] Dvorsky's reflection starts from the work of Chuck Jorgensen and his team at NASA's Ames Research Center; these scholars have developed a sys-tem that collects and converts the nerve signals that arrive at the vocal cords into words emitted by a vocal synthesizer, this in order to help those who have lost the ability to speak to communicate, as well as for astronauts or people who work in noisy environments.[32] The system operates in the reverse way as cochlear implants, which capture acoustic signals and convert them into nerve signals that the brain can decode. By combining these two technologies with a radio transmitter and a computerized acoustic and neural conversion device, one could completely bypass the world of sounds and enter the age of telepathy – promptly renamed by Dvorsky "techlepathy." For the Transhumanist thinker, the human species is destined to become a telepathic one – a change that will irrevocably alter all relationships and interactions between people. In short, as we have already noted in this chapter, the technology that interfaces directly with our brains is expanding more and more, and techlepathy is perhaps very close to becoming reality. Jorgensen himself

---

[31] G. Dvorsky, *Evolving towards telepathy. Demand for more powerful communications tech-nology points to our future as a "techlepathic" species*, "Sentient Developments", May 12, 2004. http://www.sentientdevelopments.com/2006/03/evolving-towards-telepathy.html

[32] See http://www.nasatech.com/NEWS/May04/who_0504.html

confirmed this possibility to Dvorsky. And, in fact, Jorgensen's final goal is to develop a computerized system – based on special equations that allow for the deciphering of neural signals – that is completely non-invasive[33] and registers both incoming and outgoing neural signals at the same time.

For Peter Passaro – a researcher at Georgia Tech's neural engineering laboratory who deals with the mapping and decoding of nervous activity – the technology to create an implantable cell phone already exists; it's just about rolling up our sleeves and making it happen. According to him, the next step would be to connect directly to the centers of language, thus allowing us to directly transmit verbal thought. Warwick also agrees with Dvorsky; according to him, however, not only will it be possible to transmit thoughts or words, but – after a while – all kinds of signals produced by our brain and body, such as our emotional states, our position, and so on.

Dvorsky predicts that the advent of the first telepathic civilization will take place in three phases. First, we will develop and use subvocal devices; then, we will arrive at the one-way transmission of thought; finally, there will be two-way communication of thoughts, emotions, and so on – perhaps through a future version of the internet. Thanks to techlepathy, human cooperation and teamwork – whether it's the work of a rescue team or a rock band in concert – will be greatly enhanced. In this regard, artists will certainly be able to create unimaginable artistic expressions, directly transferring emotional and conceptual experiences in a telepathic way; even intimate and loving relationships could reach a degree of intimacy never dreamed of, even by romantic poets. And it is also possible that the famous "six degrees of separation" are destined to undergo a drastic downsizing.[34]

If it is true – we add – that "we are all alone," that we are all locked up in our heads and separated from others by the wall of language, then we can see human history as a progressive path of the mutual approach of individuals, for we have

---

[33] In fact, in order to function or perform remarkably, a technology doesn't necessarily need to be grafted into the human body. In this regard, Transhumanist Alexander Chislenko coined the term "Fyborg," in 1995, short for *functional cyborgs*, to indicate people empowered by electronic or mechanical devices not inserted into the body. We see examples of cyborgs every day, in those who use mobile phones, i-Pads, earphones or simple glasses. See http://www.lucifer.com/~sasha/articles/techuman.html

[34] The "six degrees of separation" are a sociological theory in which every human being is connected to every other human being through a chain of personal relationships that includes, at most, six people. And so, if we take, for example, an inhabitant of Rome and one of Ulan Bator, the first will certainly know a second person, who will know a third person, who will know a fourth, and so on, until we reach the inhabitant of the Mongolian capital, for a maximum of five intermediaries. Launched for the first time in 1929 by the Hungarian writer Frigyes Karinthy in the tale *Chains*, this theory was then studied by several sociologists, including Stanley Milgram. Dvorsky suggests the possibility that techlepathy would be able to bring people who live far away to such a point as to dramatically reduce the average number of intermediate degrees.

learned – through increasingly articulated systems of analysis and interpersonal communication, through literature, philosophy, and psychology – to better understand that which goes on in the heads of others, their subjective experiences and their inner lives. Perhaps with techlepathy, we will experience, for the first time, the experience of others in a direct way – as is known, the subjectivity of others is a very relevant philosophical theme, which, with techlepathy, could perhaps reach a turning point.

The future does not only have telepathy in store for us, of course; the Transhumanist Michael Anissimov had fun in this regard, drawing up a list of the upgrades we could see in the coming decades.[35] To begin with, in 30 or 40 years, we could have artificial nanotech antibodies faster and more efficient than natural antibodies, and made of polymers or diamonds, capable of communicating with each other with acoustic impulses; thanks to them, we could be able to walk in rooms contaminated by dangerous viruses in T-shirts and shorts.

And then, there is super-sight. We already have very advanced microscopes at our disposal that weigh very little; we are also studying the sight of birds of prey, some of which have such a sharp view that they can see a hare a mile away; then, we have compact devices that can explore the whole electromagnetic spectrum, from x-rays to radio waves, and everything in between. There are already artificial retinas that can provide low resolution vision to the blind. In the future – reasons Anassimov – by combining all of these techniques and this knowledge, we will have a super-sight that covers the whole spectrum and that can also see details at a microscopic level. In this context, the biggest challenge is not to create a superior artificial eye, but to reshape the visual cortex so that it can process the data and deliver it to the rest of the brain without it feeling overwhelmed.

In early 2006, Ray H. Baughman and his colleagues at the University of Texas in Dallas developed artificial muscles in synthetic polymers a hundred times stronger than ours and powered by alcohol and hydrogen. If – fiddling a bit with bionics – we could replace our biological muscles with polymeric ones, we could lift heavy objects weighing even twenty or thirty tons. At the moment, it is only science fiction, of course, but, in life, you never know; and then, who would not want to be as strong as the Incredible Hulk? Transhumanists would obviously love to adopt such muscular fibers, including for the additional advantage that they entail, that is, the fact that, thanks to their resistance, they could shield the human body from bullets and other forms of physical aggression.

The Beauty Corner. It is logical to imagine that the technologies currently used for aesthetic purposes – such as liposuction and plastic surgery – would end up being perfected to such an extent that they become much more precise, easier to perform, less invasive and less expensive. The aesthetic interventions of the future

[35] Michael Anissimov, *Top Ten Cybernetic Upgrades Everyone Will Want*, http://www.acceleratingfuture.com/michael/blog/2007/01/ten-transhumanist-upgrades-everyone-will-want/

will therefore render extreme beauty a common thing; does this imply that we human beings will no longer appreciate it? Far from it, Anassimov reassures us: our brain is programmed to appreciate beauty anyway; an attractive person is attractive, with or without other attractive people around – we are not very sure about this, but, for now, we will suspend judgment.

Then, combining the mind-brain interfaces of the future with the Utility Fog, we will also have a form of telekinesis – in practice, driving our personal "cloud" nanotech, we can move objects of all kinds just by thinking about it. But Anassimov goes further: by merging with nanotechnologies, according to him, we could become self-poietic – that is, literally capable of creating ourselves – and allopoietic – that is, capable of creating other things – in a completely new way. We could insert nanotech manufacturing units inside ourselves, which will allow us to produce nano-robots and, using raw materials found in the environment, build all of the objects we want from our body and our will. We could also reconstruct parts of ourselves that have been damaged or amputated, literally infusing life into the inanimate matter that we will absorb from time to time; in short, an autopoietic and allopoietic *molecular manufacturing*.

And again: would you like to fly? Not with a plane or a helicopter, but just fly like birds do? In recent years, several types of portable jet wings have been developed – to be placed on the back, such as the one built by the former Swiss military pilot Yves Rossy and used to fly over the Alps. With nanotechnologies – perhaps with fullerenes – in the next few years, the resistance and lightness of these devices will increase dramatically, while the costs will collapse; so, we could achieve robot wings much lighter than us, so that they can bend and be worn under clothes. In short, we will eventually see personal flight – not in the sense of personal flying vehicles, but just people flying. Goodbye forever to airplane flights?

Natasha Vita-More also has her own proposal for enhancement: *smart skin*, a bio-synthetic fabric that maintains and enhances the characteristics of natural skin.[36] The outer layer would be composed of cellular-type structures assembled with nanotechnology. Smart skin would be designed to repair, rebuild and replace itself; it would contain nano-bots distributed through the epidermis and able to communicate with the brain, obtaining from it indications on the conformation and the superficial tone to be set. Smart skin would continuously transmit intensified sensory data to the brain; as regards the relationship with others, it can communicate the emotions we experience, reproduce symbols, images, colors and the like. Finally, it would be able to tolerate high doses of environmental toxins and shield the body from radiation. In short, we understand that, rather than taking off our own skin and replacing it with an enhanced version, Vita-More aims to nanotechnologize what we already have. Also, since it seems that our skin is inhabited

---

[36] N. Vita-More, *Nano-Bio-Info-Cogno Skin*. http://ieet.org/index.php/IEET/more/vita-more20120318

by almost two hundred species of bacteria, for the Transhumanist, injecting enhancing nano-bots into it would not seem so intrusive.

Finally, we cannot forget about – as usual, *last, but not least* – Ray Kurzweil's proposals. For the inventor and futurist, in the coming decades, the physical and mental systems of our body will undergo a radical updating; we will use nano-robots to upgrade and eventually replace our organs.[37] By 2030, we will have finished retro-engineering the human brain and non-biological intelligence will merge with our biological brains – optimistic as ever, our Kurzweil. And, as is known, sex has been separated from reproduction since the Sexual Revolution – that is, one can have sex without reproducing and one can reproduce without having sex. For Kurzweil, the same can happen with food and nutrition. In reality, there are already rough procedures for carrying out this separation: think of the drugs that block carbohydrates, which prevent the absorption of complex carbohydrates, or the various substitutes for sugar. Meanwhile, our favorite Transhumanist tells us, the development of more sophisticated drugs, which block caloric absorption at the cellular level, is underway. But we will not stop here and, as soon as the right technologies are available, we will proceed to re-engineer our digestive systems, in order to disconnect the sensory aspects from their original purpose, i.e., provide nourishment – thus updating a system that nature has calibrated on scarcity.

For Kurzweil, the human body 2.0 will arrive in a progressive way, a piece – or organ – at a time. In an intermediate phase, the nano-bots in the digestive tract and bloodstream will intelligently and accurately extract the nutrients we need, will request any other necessary nutrients through our own local wireless network and eliminate all food that is not needed. Already, now, there are machines that can navigate in the bloodstream; in fact, dozens of projects are in progress for the creation of "biological microelectromechanical systems" (bioMEMS), micrometric robots destined to immerse themselves in our cardio-circulatory system, with various diagnostic and therapeutic functions – a quick tour on Google will give you an idea of how vast and fruitful this research field is.

In the 2020s, we will be able to directly introduce the nutrients that we need into our bodies with nano-bots; a possible scenario is that of the "nutritive dress," which is a garment – like a belt or a t-shirt – containing nano-bots loaded with nutrients that enter and exit the body through the skin. And so – guarantees Kurzweil – while we will taste every kind of menu, a completely separate process will put all of the nutrients we will need into our blood. You can also imagine nano-bot eliminators that will decompose the waste substances, greatly easing the load to be eliminated through the intestine. "One might comment" – says Kurzweil – "that we do obtain some pleasure from the elimination function, but I

---

[37] Ray Kurzweil, *Human Body Version 2.0*. http://lifeboat.com/ex/human.body.version.2.0

suspect that most people would be happy to do without it." After a while, we will no longer need nourishing clothes; the metabolic nano-bot reserves will be available everywhere, embedded in the surrounding environment. Thanks to the nano-bots, we will be able to keep ample reserves of everything in our bodies, thus being able to go without eating for very long periods. Once the nano-botic process is optimized, we can do without the digestive system completely. And, apparently, Kurzweil is a big fan of Freitas; for the supreme *baby-boomer*, with the Freitasian robotic blood, we will eliminate the need for a heart, while, with the respirocytes, we can do without the lungs – thus managing to go to places where there is no air to breathe. But it does not end there. Specific nano-bots will do the work of the kidneys, while other nano-bots will supply us with bio-identical hormones. However, do not worry, all of these processes of change will not be immediate, but will go through several implementation phases, with continuous improvements in design.

So, let's see. We have eliminated the heart, lungs, red and white blood cells, platelets, pancreas, thyroid and all of the organs that produce hormones, kidneys, bladder, liver, the lower part of the esophagus, stomach and intestines. There remains the skeleton, the skin, the sexual organs, the mouth, the upper part of the esophagus and the brain. The skeleton is stable, but can be infused with nano-bots, which will make it even more solid, resistant and able to repair itself. Even the skin can be improved with nano-engineered materials that will protect us even more from the physical and thermal effects of the environment, while increasing our capacity for intimate and hedonic communication. Same for the mouth and the upper part of the esophagus. We finally arrive at the nervous system. We began our journey with computers that filled entire rooms, then we found them on the desk and, a little later, in our pockets. Soon, we will put them directly into our bodies and our brains and, in the end, we will be more robotic than biological. By 2030, we will have billions of nano-bots inside the capillaries of our brain, which will communicate with each other as well as with our neurons. One of the possibilities offered by this enhanced brain will be the ability to immerse ourselves in a total virtual reality, replacing the normal signals of our sense organs with those generated by nano-machines. We could also have a great variety of virtual environments, in which we will be able to change appearance and personality, becoming other people; we could also project our sensory experiences, as well as their neurological correlates, on the web, so that other people can experience what we experience, and vice versa – a bit like in the curious movie *Being John Malkovich*. We will use nano-bots to expand our minds, greatly increasing the number of neural interconnections and adding virtual connections through communication with the nano-bots. This will expand all of our mental capacities, from memory, to thought, to pattern recognition.

And let's not forget Kurzweil's proposal regarding the *upgrade* of the cell nucleus, which is the replacement of the latter with a special nanotech pair, that is, a nano-computer and a nano-bot. The nano-computer will contain all of the information from the DNA and, in connection with a specific nano-bot, will be able to carry out the synthesis of the proteins. So, we could defeat all biological pathogens – except prions – and we would eliminate all DNA transcription errors – one of the major sources of aging, the inventor admits.[38]

Although Kurzweil and colleagues' projects for our bodies are extreme, those for the mind – as we have already begun to glimpse – are even more so. It is therefore time to take a closer look at the human brain, to understand how Transhumanists intend to reshape it.

---

[38] See Ray Kurzweil, *The Singularity Is Near. When Humans Transcend Biology*, Penguin, New York 2005, pp. 232–233.

7

Colonizing the Mind

7.1 Behind Our Eyes

With a metaphor destined to become very famous, in 1942, the famed neurophysiologist Charles S. Sherrington defined the brain as "the enchanted loom."[1] As well as being a scientist, Sherrington was also a poet, and he never missed the opportunity to combine his two passions, coloring brain activity with poetry. And it is truly enchanted, that bizarre kilogram and a half of matter that lies behind our eyes, which, for some unknown reason, is capable of thinking. Not that we have not tried to understand the brain; indeed, especially in the last two decades, thanks in part to magnetic resonance – which has given us the opportunity to look inside of this biological puzzle – our knowledge on the subject has grown considerably. But the mystery remains, as we see now.

First of all, let's ask ourselves: what exactly is inside of our skull? In fact, there are various ways of dissecting – metaphorically – and organizing brain matter, so much so that, over the years, neurological research has offered us different models and metaphors. Despite all of the limits of the case, we confess without shame that our sympathies go to the theories of the American neurologist Paul MacLean, and, in particular, to his theory of the triune brain.[2] Proposed for the first time in the '60s, this model starts with the idea that the human brain is the result of a stratification process started well before the birth of our species. Basically, our brain would be composed, in a sense, by *three* superimposed brains – and not necessarily ones strictly coordinated with each other, indeed, ones that are often

---

[1] C. S. Sherrington, *Man on his nature*, Cambridge University Press, Cambridge 1942, p. 178.
[2] P. D. MacLean, *The Triune Brain in Evolution: Role in Paleocerebral Functions*, Springer, New York 1990.

© Springer Nature Switzerland AG 2019
R. Manzocco, *Transhumanism - Engineering the Human Condition*,
Springer Praxis Books, https://doi.org/10.1007/978-3-030-04958-4_7

autonomous – the reptilian, the mammalian, and the more specifically human; as a real *bricoleur*, nature has produced this system little by little, through trials and errors, opting opportunistically from time to time for the solution that seemed the most suitable at the moment. The reptilian brain dates back three hundred million years and represents the most primitive aspects of our nervous system, regulating the basic aspects of our physiology – breathing, heartbeat, and body temperature – and instinctive reactions – sex, aggression, nutrition and flight. It includes the bulb of the brain; the Varolio bridge, which has various functions, including, apparently, an important role in dream production; the midbrain, which, among other things, processes the incoming sensory information; and the cerebellum, linked to motor control. The mammalian brain, which emerged about a hundred million years ago – and which, as we can guess from the name, we share with other mammals – is instead able to produce more articulated emotional responses and acquire complex memories. The key structures of this section are the amygdala, home to intense emotions such as fear and anger; the hippocampus, linked to long-term memory and spatial navigation; the hypothalamus, which deals with endocrine activity and many somatic functions; and the cingulate gyrus, responsible for the cortical control of the heartbeat and blood pressure and linked to voluntary attention. Finally, there is the neo-cortex, particularly developed in *Homo sapiens* – a species that, according to the most recent discoveries of paleoanthropology, would have appeared about two hundred thousand years ago. In the neo-cortex lie the so-called "higher faculties": language, abstract thought, and self-awareness. In particular, it is branched into right and left hemispheres, in turn divided into lobes: the occipital lobe, located in the back of the head and responsible for processing visual stimuli; the temporal lobe, responsible for vision, auditory processes and recognition of the objects we see; the parietal lobe, which integrates the information coming from the senses; and the frontal lobe, which has a central role in the decision-making processes. This is only a model, and we have also presented it in a simplistic way; we decided to espouse it as such not only for the sympathy that inspires us, but also for the fact that it seems to be particularly dear to some Transhumanists.

But it is not only the brain; among our interests, there is also a much more elusive object: the mind. What is it about? Since, as is well known, philosophers have not reached a satisfactory agreement on even the definition of their favorite activity – that is, philosophy – it is easy to imagine how the concept of mind is equally problematic. We can define the mind, preliminarily, as the set of cognitive faculties that allow us to be conscious, to remember, to perceive, to think. In short, it would be the minimum "equipment" necessary to be what we are, to distinguish ourselves from inanimate things and, to a lesser degree, from animals, because it seems that the latter also have a mind in their own way, some more than others, depending on the species. We often speak of "higher" faculties of the mind, such as judgment, the ability to think rationally, and so on – though, over the years, emotionality has also been redeemed in various ways. The mental faculties therefore include thought, imagination and awareness, including self-consciousness. And then, there are the "contents" of the mind, "objects" that reside precisely there: thoughts, memories, emotions, intuitions, "images." This last term in quotes represents an enormous philosophical problem: are these real images, like pictures and photographs, only more blurred? Or is it just a manner of speaking? And if the latter is true, then what are "mental images"? As you can see, it is a philosophical quagmire; better to stay out of it. Equally intricate are the problems of free will – are we really "free," responsible for our choices and capable of self-determination, or are we just "puppets" in the hands of our genes, our brain chemistry, or the culture we belong to? – and the status of our mental contents – our ideas, our concepts, do they exist independently of our minds, and are they, in fact, "somewhere," or are they just signs or symbols we need, and that will disappear with us?

All of these problems – and many others – represent the heart of a modern discipline known as the "philosophy of mind." Not that philosophers did not take note of these things before; and not that other philosophers, far from the philosophy of the mind, do not deal with them. Transhumanists seem to us, however, to be linked mostly to the so-called analytical or Anglo-Saxon philosophy, which is just like the philosophy of mind, and therefore we will focus, above all, on this last discipline, which deals with the nature of the mind, mental states, mental properties and the relationship of the mental with the physical; in practice, the mind-brain relationship, central to contemporary philosophical thought, and central to one of the concepts most dear to Transhumanists, the "mind-uploading," i.e., the ability to "load the mind" into a different support from the biological one. Here, prepare now for a nice series of "-isms," which abound in the philosophy of mind.

The first -ism on the list is "dualism," the idea that the mind and the brain are two completely different substances; if Plato can be considered the spiritual father of this idea – for the fact that he distinguished between the material world, in becoming, and the world of ideas, eternal and immaterial – Descartes is the main modern representative of it. For the latter, the *res extensa* – in short, the body – was extended, that is, occupied space, and mechanically determined, that is, subject to the rigid laws of the cause-effect relationship; the *res cogitans* – that is, the mind – was immaterial, that is, it did not occupy any space, and had the ability to think and freely determine itself. Already, we see here, in a nutshell, the philosophical issues of this conception: if mind and body are two such different substances, to the point of interpenetrating and "ignoring" each other, how is it possible that the two communicate, allowing, for example, my mind – that is, me – to control my body? Descartes tried to get by with his famous pineal gland – which was supposed to act as a "place of contact" between res cogitans and res extensa – but without much success. The greatest modern representatives of the dualism of substances can be considered Karl Popper and John Eccles, for whom the mind "interacts" in some way with the body; Popper hypothesized the existence of three "worlds," the world of material entities, that of cultural entities – mathematical ideas and more – and that of the mind, in constant interactive relationship with each other.

Monism, on the other hand, is a set of positions that, in different measures, *identify* mind and brain. For functionalism – among the many names that subscribe to it, we can mention Hilary Putnam and Jerry Fodor – mental states are *functions* of the mind and/or brain; where do they come from, then? One possibility is that mental functions or states *emerge* from the brain; in practice, the natural world would be composed of *layers*, one superimposed on the other, and each one having properties not simply reducible to the lower layer, but somehow emerging from it through an ontological "leap." And so, the organic world would emerge from the inorganic one, the more complex life forms from the less complex ones, and thought – or the mind – from non-sentient life. Functionalism represents a form of "dualism of properties," that is, it starts from the existence of a unique substance – matter – but the existence of different types of property, that is, physical and mental, is admitted. There are also linguistic approaches to the mind-body problem, such as the one embodied by Ludwig Wittgenstein – in the first phase of his thought – and by his followers, who consider this question completely illusory and destined to be dissolved during a process of linguistic "purification" that gives up on terms – like the metaphysical ones, typical of so much philosophy of the mind – apparently meaningful but actually not. The behaviorist approach is also interesting, started by John Watson and then relaunched by Burrhus F. Skinner (although, here, we leave the territory of the philosophy of mind and enter that of psychology). Behaviorism is, as is known, hostile to the concept of introspection and in favor of an analysis of the mind in terms of observable and measurable behaviors. Also curious – if only because of its terminological origin, inspired by the American rock band Question Mark & the Mysterians – is mysterianism, embodied by, among others, Colin McGinn, for whom human beings are "cognitively closed" with regard to their own minds. According to him, we would be deprived of the procedural skills for the formation of the concepts necessary to fully grasp the causal chains that underlie our minds – and, in an almost crypto-Transhumanist way, McGinn suggests that, in the future, we might be able to change our biology, and actually become able to understand ourselves.[3]

Particularly important, from the point of view of our discourse on Transhumanism, is cognitive science and the computational approach to the human mind, in turn associated with studies on artificial intelligence. Cognitivism is a movement within psychology, born around the 1960s and based on the idea

---

[3] C. McGinn, *The Mysterious Flame. Conscious Minds in a Material World*, Basic Books, New York 1999.

that the mind is comparable to a computer program: in practice, the brain would be the hardware and the mind the software. Here, too, we have greatly simplified things: cognitivism is much more than that and, over the decades, has undergone a very complex evolution, sometimes intersecting with studies on artificial intelligence. Within the latter sector, the well-known philosopher of the mind John Searle made a distinction destined to become classic, that is, between weak AI and strong AI. The proponents of the first aim, in simple terms, to adequately simulate the mental states present in human beings in machines, while the supporters of the second want to create an artificial mind – obviously contained in a special computer – that is actually endowed with human awareness and intelligence. The ultimate goal is to develop an artificial mind capable of overcoming the Turing test – named after the famous computer science pioneer Alan Turing. The scientist's argument goes something like this: we arrange a human being in a room and, through a special terminal, we make her/him converse with someone else who is in the next room, without the first seeing or hearing the other party in any way. But there is a clause: the latter could be either a human being or a computer. If the first subject is not able to determine whether he/she is talking to a human or a machine, that is, if the latter is so sophisticated as to be able to deceive the subject of the experiment, then it can be said that it is as intelligent as a member of our species, and the test is passed.

We must not forget that, alongside analytical philosophy and the cognitive approach, there is also so-called "continental philosophy," which, proceeding from Nietzsche, then passes through Edmund Husserl, Martin Heidegger, Michel Foucault, Jacques Derrida, and many others again; even these thinkers have something to say about humans and the knowledge that concerns them. For example, Heidegger, although not actually an existentialist, elaborated, in *Being and Time*, an exceptional existential analytic, which takes humans into consideration as a totality "thrown into the world" and projected towards the future and towards possibility. And again: during the Twentieth Century, we have witnessed the advent of the human sciences, which, in the form of sociology, linguistics, psychology, cultural anthropology, and so on, have elaborated such a harvest of knowledge, ideas and theories related to human nature so as to prevent even the most motivated academic from mastering them. And think about the fact that we have not even considered Buddhist and Hindu theories of the mind – far more sophisticated than commonly believed.

Those that we have presented to you are just a few examples, just to make you understand that human knowledge is very articulated and often contradictory; the access routes to that itinerant mystery that is humanity are numerous, growing and probably destined to develop ever more vertiginously over the centuries and millennia to come.

Transhumanists, however, have quite clear ideas. Or, rather, some of them have them and, as we shall see, tend to opt for a relatively reductionist conception, which mixes cognitive science, neuroscience and strong artificial intelligence.

In our opinion, this is a limiting approach, but, all in all, a commonsensical one: you have to start somewhere if you want to reconstruct human nature from top to bottom, and, in fact, it seems that the neurological-cognitive-computational approach lends itself better than others to the development of more or less structured proposals on how our precious enchanted loom should be reshaped through technology. But before tackling the Transhumanist ideas about it, it's better to see what we can do now.

7.2   Neurotechnologies: The State of the Art

It is difficult even to begin to summarize the number, extent and variety of research programs that, in different degrees and according to different ways, involve the human brain and the technologies that refer to it. As usual, we begin with a definition: with the term "neurotechnology," we define all of those devices and procedures capable of interacting directly with the human and animal nervous systems, as well as all of the research related to this field. Neurotechnologies have been around for about fifty years, but have only developed markedly over the past two decades. This field of research seems destined to undergo a strong acceleration as a consequence of the international project "Decade of the Mind," launched in 2007 with the aim of stimulating strong public investments in the study of the mind and the rooting of the latter in brain activity. One scientific field strongly linked to neurotechnology is so-called neuro-engineering, a discipline that aims to use engineering techniques to study, repair or – hopefully – strengthen the brain. This framework brings together computational neuroscience, neurology, electrical and electronic engineering, robotics, cybernetics, nanotechnologies and computer engineering. Currently, neuroengineering research is focused on understanding the way in which the motor and sensory systems of our brains encode and transmit information, and this in order to learn how to manipulate them, so as to be able to connect the brain to neuro-prostheses or brain-machine interfaces. It is a fairly recent field of research, so much so that the first dedicated scientific journal, the *Journal of Neural Engineering*, was born only in 2004 – simultaneously with the *Journal of NeuroEngineering and Rehabilitation*.

Currently, neurotechnology can be roughly divided into two branches, i.e., imaging and stimulation. In the first case, we start by mentioning functional magnetic resonance, which is based on the detection of the influx of blood and oxygen in this or that area of the brain, a fact that corresponds to its greater activation – which allows us to see the brain "at work," understanding which area does what. Computer tomography is based on X-rays, while positron emission tomography is based on the activation of specific biological markers – that is, it monitors the influx of substances used by the brain, such as glucose. Finally, we have magnetoencephalography, which is based on the mapping of the electrical activity of the brain.

In the context of stimulation,[4] we first have transcranial magnetic stimulation, which, as you can understand from the name, aims to interfere with the electrical activity of this or that area of the brain using specific magnetic fields. This technique is used to treat a large number of conditions, from migraines to depression, from obsessive-compulsive disorder to Parkinson's disease. Brain pacemakers, medical devices implantable in the brain, whose function is to send electrical signals to the nervous tissue, are much more invasive. Depending on the area in which they are implanted, we speak of either cortical stimulation or deep brain stimulation. It is also possible to implant them near the vertebral column or cranial nerves. The aims of such an intervention are mainly medical – preventive or curative. Among the diseases treated with these devices, we have epilepsy – for the purpose of blocking epileptic attacks, Parkinson's disease – to reduce symptoms such as tremors, limb stiffness and more, and chronic depression. Deep brain stimulation is generally based on three components: the external pulse generator – a battery-powered neurostimulator embedded in a titanium structure, an insulated polyurethane cable with some platinum-iridium alloy electrodes inserted into the brain, and an additional isolated cable that connects the two. Generally, the pulse generator is placed subcutaneously, in the clavicle or sometimes in the abdomen.

We now enter the field of neural prostheses, devices capable of supporting or replacing missing brain functions by stimulating the nervous system and recording its activity. They typically involve electrodes that measure nerve activation and report the tasks to be performed to prosthetic devices. The technologies already available or in more or less advanced development are different; among them, we mention cochlear implants, which restore hearing to the deaf. Visual prosthetics are still in an elementary development phase. In addition, we are studying motor

---

[4] Electrical stimulation of the brain is as old as neurology itself. If, in 1870, Eduard Hitzig and Gustav Fritsch showed that electrically stimulating dogs' brains produced bodily movements, in 1874, Robert Bartholow did the same with regard to human beings. Well-known are the experiments conducted between the '50s and the '70s by the famous Spanish physiologist José Delgado, who managed to achieve a certain control of human and animal behavior through electrical stimulation. The scholar invented the "stimoceiver," a radio-controlled microchip implanted in the brain and capable of transmitting electrical impulses for the purpose of modifying basic behaviors and sensations, such as aggression and the perception of pleasure. Later, Delgado wrote a rather controversial book, *Physical Control of the Mind: Toward a Psychocivilized Society*. In it, the author claimed that, after having tamed and civilized the nature that surrounded him, man had to civilize his inner self – obviously through procedures similar to those he had studied. See J. Delgado, *Physical Control of the Mind: Toward a Psychocivilized Society,* Harper and Row, New York 1969.

prostheses, which aim to stimulate the muscular system in place of the brain and spinal cord. Then, we have artificial limbs, which we talked about before.

Cochlear implants are composed of several sub-systems. The sound is picked up by a microphone and transmitted to an external processor – usually placed behind the ear – which translates it into digital data; the latter are then translated into radio signals transmitted via an antenna to a device located inside of the patient, which sends them – in the form of electrical impulses – to the cochlea. Visual prostheses are based on electrical stimulation of the retina, and include three approaches: the epiretinal (electrodes placed above the retina), the subretinal (under the retina) and the extraocular transretinal (just outside of the eye).

Now, we turn to brain implants, devices connected directly to the patient's brain. One of the main objectives of these tools – at the research level – is to circumvent areas of the brain damaged by strokes or head injuries. The brain implants work by stimulating, blocking or recording – all through electrical impulses – the signals produced by single neurons or groups of neurons. One of their aims is so-called "sensorial substitution," which consists in the development of substitutes for sight and hearing – an additional approach to that of visual and cochlear implants. The progress achieved in recent years has been considerable, particularly that related to the direct control – that is, through thought – of artificial devices, such as robotic arms, and so on. Currently, brain implants are made up of a most diverse number of materials, from tungsten to silicon, from a platinum-iridium alloy to stainless steel; soon, however, more futuristic materials will come into play, such as carbon nano-tubes.

So, we arrive at the BMI – brain-machine interface – that is, all of those technologies that aim to build a direct interface between the brain and one or more external devices, in order to help or enhance the former. The idea of BMI is actually old news, in the sense that the first systematic researches related to it started as early as the '70s at the University of California, under the aegis of – guess who – DARPA. The main difference between the neuro-prostheses mentioned above and BMI is the fact that the former connects the nervous system to an instrument, while the latter connects the brain to a computer. Neuro-prostheses can be connected to any part of the nervous system – including, therefore, the peripheral one – while a BMI connects to the central nervous system. However, these are largely overlapping fields of research and experimentation, so much so that studies on neuro-prostheses are often closely related to those on BMI, as the use of PCs with special programs is inevitable.

There are various laboratories around the world that have successfully tested BMI on animals; for example, there is the well-known case of the monkey who, in 2008 – at the University of Pittsburgh Medical Center – managed to move a robotic arm with its thought.[5]

---

[5]M. Baum, *Monkey Uses Brain Power to Feed Itself With Robotic Arm*, «Pitt Chronicle», September 6, 2008. http://www.chronicle.pitt.edu/story/science-technology-monkey-uses-brain-power -feed-itself-robotic-arm

Among the main scholars of BMI worth mentioning is the Brazilian Miguel Nicolelis, who works at Duke University in Durham; the approach that he promotes is the use of many electrodes scattered along a large area of the brain, in order to reduce the variability of the response produced by a single electrode. Other important research groups – which have developed both BMI and algorithms able to decode neuronal signals – are those of John Donoghue of Brown University, Andrew Schwartz of the University of Pittsburgh and Richard Andersen of Caltech.

One distinction that we can make between the various BMI devices is that relative to the degree of invasiveness: and so, we have invasive, partially invasive and non-invasive BMIs. A typical example of an invasive BMI is the famous BrainGate by Cyberkinetics, a US company linked to DARPA[6]; BrainGate is a chip – equipped with ninety-six electrodes – that was implanted, for nine months in 2005, into the brain cortex of Matt Nagle, a tetraplegic patient who, thanks to the device, was able to control a mechanical arm for the first time. The man also managed to control a cursor on the screen of a PC, to change the channels of a TV and to turn lights on and off, all through thought. Partially invasive BMIs are based on the implantation of the related devices into the skull of the patient, but not directly into the brain; they use various techniques to measure brain electrical activity, such as electrocorticography – a procedure similar to electroencephalography. Finally, we have non-invasive BMI, which read the brain activity through non-invasive techniques, such as electroencephalography or magnetic resonance.

Parallel to neurotechnology, we must mention a topic of great interest to Transhumanists, namely, brain simulation – in other words, the attempt to reproduce the overall brain activity of increasingly complex animals on a computer, all the way up to humans. Several animal brains have been mapped and, at least in part, simulated using appropriate software. For example, the brain of *Caenorhabditis elegans* – a simple nematode worm – that consists of only 302 interconnected neurons with about five thousand synapses, was mapped in 1985[7] and partially simulated in 1993.[8] Despite only having to deal with a simple nervous system like that of C. elegans, we still have not yet been able to understand how it produces

---

[6] http://www.cyberkinetics.com/

[7] M. Chalfi, J. E. Sulston, J. G. White, E. Southgate, J. N. Thomson, S. Brenner, (1985). *The neural circuit for touch sensitivity in Caenorhabditis elegans*, "The Journal of Neuroscience", 5 (4), pp. 956–996, 1985. http://www.jneurosci.org/content/5/4/956.full.pdf

[8] E. Niebur and P. Erdos, *Theory of the locomotion of nematodes: Control of the somatic motor neurons by interneurons*, «Mathematical Biosciences», 118 (1), pp. 51–82, 1993. http://www.ncbi.nlm.nih.gov/pubmed/8260760

the relatively complex behavior of this organism.[9] The fruit fly brain has also been carefully studied and a simplified simulation has been obtained.[10]

In this field, the role of "guru" perhaps goes to Henry Markram, a South African-Israeli scholar and head of the "Blue Brain Project" at the École Polytechnique Fédérale de Lausanne.[11] Launched in 2005, the project in question represents an attempt to create a synthetic brain through the retro-engineering of the mammalian brain, down to the molecular level; the final goal is to study the architectural and functional principles of the brain, with the hope of shedding some light on the nature of consciousness.

The Blue Brain Project is based on an IBM Blue Gene super-computer with the NEURON software installed – developed by Michael Hines, a Yale University scholar – not only capable of simulating a neural network, but also one based on realistic models of neurons. The initial aim of the project – achieved in 2006 – was to simulate the neocortical column of a rat, i.e., the smallest functional unit of the neocortex. Now, these scientists are pursuing – under the heading of the "Human Brain Project" – a twofold objective, that is, to carry out the simulation at the molecular level – to study the effects of gene expression on our nervous system – and to simplify the procedure of simulation of the cortical column, so that we can then simulate several interconnected columns – in order to arrive at the simulation of an entire human neocortex, which is made up of about a million cortical columns.[12]

What we have given you is just a smattering, and the field of neurotechnology and neuro-engineering represents much more than that; inviting you to go deeper into this fascinating area on your own, let's move on to the proposals put forward by Transhumanists, and let's talk about the possibility of transferring one's mind into a different, non-biological substratum. Here, we are finally at *mind-uploading*.

7.3   Escaping from the Body

Yet another *evergreen* of Transhumanism, mind-uploading is also known as *mind transfer* or *whole brain emulation*. It is basically the copying or transfer of the individual conscious mind from the brain that hosts it into a substratum of a

---

[9] R. Mailler, J. Avery, J. Graves, N. Willy, *A Biologically Accurate 3D Model of the Locomotion of Caenorhabditis Elegans*, 2010 International Conference on Biosciences. pp. 84–90, March 7–13, 2010. http://www.personal.utulsa.edu/~roger-mailler/publications/BIOSYSCOM2010.pdf

[10] P. Arena, L. Patane, PS Terms, *An insect brain computational model inspired by Drosophila melanogaster: Simulation results*, The 2010 International Joint Conference on Neural Networks (IJCNN). http://ieeexplore.ieee.org/xpl/login.jsp?tp=&arnumber=5596513&url=http%3A%2F%2Fieeexplore.ieee.org%2Fxpls%2Fabs_all.jsp%3Farnumber%3D5596513

[11] http://bluebrain.epfl.ch/

[12] http://www.humanbrainproject.eu/

different, artificial type – but still capable of supporting all of its functions. In practice, the brain – dead or alive – will be scanned and mapped in a very detailed way; the data collected in this way will have to be loaded onto a fairly powerful computer – and, *ça va sans dire*, one not yet existing – which will then have to reproduce all of the cognitive and behavioral aspects of the "uploaded" subject. The mind, thus freed from the flesh, can then occupy a robotic body or another biological body synthesized for the occasion, or it may decide to live on the Web or in a virtual reality of its liking – completely indistinguishable from the real one.

Put down like this, it seems like a mere question of procedure, that is, it is only a matter of waiting for the development of sufficiently powerful computers and sufficiently advanced scanning techniques, *et voila*, digital immortality is served. In reality, there is no lack of philosophical problems, as we shall see. Of course, whole-brain emulation is considered by many non-Transhumanist scholars to be the natural outlet of computational neuroscience and strong AI studies.

To begin with, let's take a look at the most strictly philosophical and scientific of problems, starting from the main characteristic of the human brain: its enormous complexity. It represents about eighty-six billion neurons, according to the most recent estimates,[13] connected through "hooks" called axons and dendrites, which form the interneuronial synapses, crossed by signals modulated by specific chemical substances, the neurotransmitters. And all of this incessant electrochemical activity produces or "secretes" the mind. The idea that it is possible to "load" the latter onto a different substratum is based on the idea that it has no "spiritual" quality or, rather, that it is not a substance separated from the body, immaterial and belonging to God. Another assumption is obviously that, one day, machines will be able to think – an idea supported by some eminent neuroscientists and *computer science* scholars, such as Marvin Minsky,[14] Douglas Hofstadter,[15] Christof Koch and Giulio Tononi.[16] This assumption is necessary for the fact that our uploaded mind then needs a computational substratum that will allow it to continue to "live" – that is, "run" on a computer powerful enough to at *least* reproduce human capabilities and characteristics. Underneath it all, then, is the need to have sufficient computational capacity, which, at the moment, is not available – although

---

[13] J. Randerson, *How do many neurons make a human brain? Billions fewer than we thought*, «The Guardian», February 28, 2012. http://www.theguardian.com/science/blog/2012/feb/28/how-many-neurons-human-brain

[14] M. Minsky, *Conscious Machines*, in *Machinery of Consciousness*, Proceedings, National Research Council of Canada, 75th Anniversary Symposium on Science in Society, June 1991. http://kuoi.org/~kamikaze/doc/minsky.html

[15] http://spectrum.ieee.org/computing/hardware/tech-luminaries-address-singularity

[16] C. Koch and G. Tononi, Can machines be conscious?, « IEEE Spectrum», Vol. 45 n. 6, pp. 55–59, 2008. http://ieeexplore.ieee.org/xpl/articleDetails.jsp?reload=true&arnumber=4531463

some futurologists, like Kurzweil, believe that it will be available within a few decades. However, it is currently difficult to know how much computing power is needed for such an enterprise – but we can imagine that it must be high, given the astronomical number of interconnections that will need to be mapped.

The central problem, however, remains that of the identity of the uploaded mind: the skeptics – and I confess that I am among them – believe that, even if the reproduced mind contained *exactly* the same memories, the same emotions and the same psychological traits as the original, it would not identify with the latter, but would remain what it is, i.e., a copy – a fact that would be even more evident if the copying could be done in such a way as to avoid dissecting the original brain, i.e., with the subject still alive. It is also necessary to establish how much of a person's memory must necessarily be preserved so that one can speak of the "same" person; for example, if there is no memory of childhood, does it still count?

But if a person is definable as a pure and simple collection of information, then copying it onto a computer would be a form of immortality, so-called "digital immortality." The first to make this proposal seems to have been, in 1971, a bio-gerontologist at the University of Washington, George M. Martin.[17] So, it seems that even mind-uploading is nothing more than an attempt to deceive death, procuring immortality. A secular immortality, which would allow us to escape forever from the slavery of the flesh, through our "avatar" – similar to that of Second Life, but with more details.

In reality, however, the question is far from resolved; the Transhumanists who are in favor of mind-uploading generally deal with the question by saying that personal identities are *fuzzy*, that they have blurred edges. For Bruce F. Katz, an artificial intelligence scholar and professor at Drexel University in Philadelphia, the Self is actually illusory,[18] and thus one can imagine copying certain psychological characteristics into a new medium and considering the copy to be in perfect continuity with the original.[19] Anders Sandberg removes the problem by saying that, "if we can accept growing old, we can probably accept being transferred into a computer."[20] In doing so, the Transhumanist thinker makes a real conceptual leap, which bypasses the famous problem of the "Theseus ship." This is a paradox that arises from the following question. Let's say we have an object

---

[17] G. M. Martin, *Brief proposal on immortality: an interim solution*, in: "Perspectives in Biology and Medicine", n. 14 (2), 1971.

[18] This affirmation could, however, be objected to by asking why, since the Self is illusory, one must go through so much trouble to preserve or copy it.

[19] See Bruce F. Katz, *Neuroengineering the Future. Virtual Minds and the Creation of Immortality*, Infinity Science Press, Hingham 2008.

[20] P. Moore, *Enhancing Me. The hope and the hype of human enhancement*, John Wiley & Sons, Chichester 2008, p. 61.

composed of several parts – like a ship, in fact, but possibly a living being too – and that we replace the latter a little bit at a time. In the end, when we have replaced *all* of the parts, are we dealing with the *same* starting object? Plutarch speaks of the paradox in question in his *Life of Theseus* and, as you can imagine, he has generated endless philosophical debates on the nature of identity and continuity – already existing even before Plutarch wrote about this topic, to tell the truth. In short, it is the usual philosophical dilemma related to identity and change, that is – if we want to put it down in more philosophical terms – to the relationship between Being and Becoming. Which, applied to the theme that interests us, consists of the question: if we replace the brain of a person with artificial parts a little at a time, or if we "copy" the mind to a different medium, do we end up with the same person or with a different person? It is needless to continue to discuss this question – believe me, we will not find our way out of it – but let's just keep in mind that the hypotheses on the subject are many, and all seem to carry more or less good reasons in their favor.

Philosophical questions aside, how do we achieve mind-uploading? One could try a very detailed scanning of all of the inter-neuronal connections of the person's brain that we want to upload – provided that our starting hypothesis, namely, that the human personality is "contained" in the cerebral micro-structure, is correct. However, if the subject in question is not already dead, then it would be murder, as the brain tissue would be minutely destroyed. This kind of destructive scanning should already be minimally possible; for example, three years ago, a Stanford team managed – thanks to a special high-resolution photographic system – to map a small portion of the murine brain, identifying the synapses in detail.[21] The road ahead is still long and, at the moment, we are missing too many things, such as neuronal epigenetics – that is, the activation patterns of genes expressed in neurons. Moreover, even if we could take a snapshot of the brain, we would capture a single moment in time, and not its dynamic nature, which is made of a continuous flow of connections, more like a film than a photograph. Additionally, knowing about a connection is not enough: we must also know what it does, how every single neurotransmitter goes through it, and so on; we essentially need a super-precise chemical map of the brain.

Another possibility would be to freeze the brain and then scan and analyze the organ in question in detail one piece at a time – in this case, the patient should already be dead and, better still, subjected to cryonic suspension. This way, you could work very calmly, using, for example, a scanning electron microscope.

If you are dealing with a brain that is still alive, you could also try to create a functional three-dimensional map of the brain, using, for example, one of the

---

[21] *New imaging method at Stanford reveals stunning details of brain connections*, "Medical Daily", November 17, 2010, http://www.medicaldaily.com/new-imaging-method-developed-stanford-reveals-stunning-details-brain- connections-234704

already-mentioned brain imaging systems – but this is something that we cannot yet do, since the resolution of current imaging systems is not yet sufficient, lacking the necessary computational capacity. As we said, neural simulation is a field of research existing outside of Transhumanism that has led us to map and, at least partially, to simulate the nervous system of different animal species.

According to Sandberg, we would need a sort of hybrid scanner: "it cannot be a traditional optical microscope or an electron microscope. (...) It will be a combination of different techniques; it will be a new type of microscope that has not yet been built. The components seem to be already in circulation, more or less."[22] According to Peter Peters, a leading researcher in the field of electron microscopy, at the moment, one can scan only 0.2 cubic millimeters of brain tissue a day, while a complete scan would require one hundred and ninety million days. For the future, Sandberg proposes, rather than the use of a single microscope, "a kind of uploading assembly line, with hundreds or even thousands of microscopes that perform an automatic scan at the same time."[23] A basic problem remains, namely, the fact that it is so often assumed that the brain functions as a computer – and we know how computers work, because we have built them – while it is probably, in fact, an incredibly more complex system, and especially that it dynamically self-modifies, so that knowing the state of the brain's network at a given moment is just the beginning. Even Sandberg admits that, currently, there is no technology capable of mind-uploading, let alone doing so without injuring the brain. In the end, the thinker calls into question the usual nanotechnologies, and, in particular, proposes flooding the brain with nano-machines and letting each of them hook onto a particular neuron and collect the information we need. Aware, then, of philosophical problems, Sandberg and his colleague Nick Bostrom prefer to talk about whole-brain emulation. For his part, Katz circumvents philosophical problems by embracing one of the hypotheses described above in relation to the Theseus ship, and proposing the "progressive transfer of the mind": the brain is connected to neural prostheses, which essentially become a part of it; then, other prostheses are connected to them, and so on; finally, we proceed to the gradual "extinction" of the original brain, a gradualism that should guarantee ontological continuity between the original and the uploaded consciousness. Neither is this version of mind-uploading without problems, though. How gradual does this gradual process have to be in order to preserve the individual identity? Is it just a matter of speed? Or is there a specific ontological difference between the first and the second types of mind-uploading?

At this point, we ask ourselves: apart from immortality, why should we transfer our mind into a non-biological support? In reality, say the proponents of

---

[22] P. Moore, *Enhancing Me. The hope and the hype of human enhancement*, John Wiley & Sons, Chichester 2008, p. 56.

[23] Ibid. p. 57.

mind-uploading, the advantages would be many. To begin with, the speed: an uploaded mind could think much faster than one contained in a brain, even if it is not necessarily smarter. While, in fact, our neurons are "messaging" through electrochemical signals at a maximum speed of one hundred and fifty meters per second, an artificial mind, going at the speed of light – three hundred thousand kilometers, or three hundred million meters, per second – would be two million times faster. Furthermore, while neurons produce, at most, a thousand "action potentials" – in practice, activation peaks – per second, in this case, the chips are also two million times faster, and growing. In practice, for such a mind, a subjective year would last a matter of seconds.

Secondly, we could travel in space at a ridiculous cost – "loading" a whole community of virtual people onto a spaceship, allowing them to lead their lives in a simulated environment as they wait to arrive at their destination – or we could even transmit the minds of the astronauts from one place to another via wireless.

One can even imagine the possibility of "dividing" into multiple copies of oneself, which could then live different lives, only to reunite at the end, recomposing the initial self – obviously enriched beyond measure by new experiences; one can also imagine splitting into multiple selves, travelling through the Galaxy following different paths and then meeting at the far end of it, celebrating a "reunification party."[24] More generally, we could alter and amplify our emotions, have a virtual body capable of changing sex and appearance as we please and immerse ourselves in hyper-realistic virtual worlds. Thus, we could take a "psychic holiday from ourselves," freely modifying our more or less rooted psychological categories to become another mind/person partially or completely different from who we are. Finally, there is the theme of multitasking: human beings work in a serial way, that is, they take a task and they divide it into lower units of activity, which they then execute one at a time. An uploaded mind could instead be programmed to perform several tasks simultaneously, in parallel.

Let's now take a look at the proponents of mind-uploading; the first on the list is obviously Marvin Minsky. Then, we have Joe Strout, an activist who, in 1993, created the first website dedicated to the topic, baptized the Mind-Uploading Home Page[25]; he saw in this possibility an alternative to the resurrection of the body by scientific means. Then, we obviously have the two "giants" of so much of European Transhumanism, Anders Sandberg and Nick Bostrom, authors of a very detailed report on the simulation of the human brain: *Whole Brain Emulation: A Roadmap.*[26] We also mention Kurzweil, who will be dealt with later on, and for

---

[24] This is an idea proposed by Transhumanist Keith Henson, who called it the "Far Edge Party." Cf. https://lifeboat.com/ex/bios.h.keith.henson

[25] http://www.ibiblio.org/jstrout/uploading/

[26] A. Sandberg, N. Bostrom, *Whole Brain Emulation: A Road Map.* Technical Report # 2008 3, Future of Humanity Institute, Oxford University, Oxford 2008. http://www.fhi.ox.ac.uk/brain-emulation-roadmap-report.pdf

whom artificial intelligence will make possible the mind-uploading and retro-engineering of the human brain.[27] And then there is Ian Pearson – the head of British Telecom's futurology unit, who asserts that human beings will be able to transfer their consciousness into computers, thus achieving virtual immortality, by 2050.[28] Finally, we get to know Randal A. Koene, guru of mind-uploading and Dutch scholar of computational neuroscience, neuro-engineering and information theory. Among the ranks of the Transhumanists, and director – between 2008 and 2010 – of the neuro-engineering department of the Spanish high tech company Tecnalia, Koene founded the Society of Neural Prosthetics and Whole Brain Emulation Science,[29] an association expressly dedicated to mind-uploading. Subsequently renamed carboncopies.org,[30] the organization – in fact, a network – monitors the technical progress that is to lead us from the creation of partial neural prostheses to the prosthetic simulation of brain functions as a whole, and, in the immediate future, aims to build a network of researchers interested in this scope. The long-term goal is obviously the creation of an SIM, i.e., a Substrate-Independent Mind.

These, then, are the ideas and characters related to mind-uploading. Meanwhile, the debate is fervent; proof of this is the June 2012 edition of the *International Journal of Machine Consciousness*, dedicated entirely to the theme of mind-uploading. And so, for Michael Hauskeller, philosopher at the University of Exeter, there is no reason to believe that a simulation of the human brain, however accurate, will end up producing consciousness, as it is not at all obvious – from a logical-metaphysical viewpoint – that the simulated mind could be considered the continuation of the original self.[31] According to Randal A. Koene, to achieve mind-uploading, there would not even be a need for a full understanding of the human brain, but only the ability to replicate the functional behavior of its basic computational elements, that is, synaptic responses, and so on, a goal already achievable, according to him, with today's technologies.[32]

---

[27] R. Kurzweil, *The Singularity is Near*, Viking Books, New York 2005, pp. 198–203.

[28] *Brain downloads 'possible by 2050'*, 'CNN.com International,' May 23, 2005, http://edition.cnn.com/2005/TECH/05/23/brain.download/

[29] http://www.minduploading.org/

[30] http://www.carboncopies.org/

[31] Michael Hauskeller, *My Brain, My Mind, and I: Some philosophical assumptions of mind-uploading*, "International Journal of Machine Consciousness", Vol. 4, n. 1, 2012, pp. 187–200. http://www.worldscientific.com/doi/pdf/10.1142/S1793843012400100

[32] R. A. Koene, *Experimental research in the whole brain emulation: the need for innovative in vivo measurement techniques*, "International Journal of Machine Consciousness", Vol. 4, n. 1, 2012, pp. 35–65. "International Journal of Machine Consciousness", Vol. 4, n. 1, 2012, pp. 187–200. http://www.worldscientific.com/doi/pdf/10.1142/S1793843012400033

According to Harvard scholar Kenneth J. Hayworth, mind-uploading is certainly possible, but far beyond our current capabilities.[33] For Patrick D. Hopkins, philosopher at Millsaps College, the "transfer of the mind" is not possible, or, rather, it would not be a real transfer, and this conception would betray a basic dualistic prejudice – that is, in treating the mind and the mental schemes as "objects" that can be transferred, Transhumanists would refer – contrary to their "official" positions, which are of a materialistic type – to concepts that are no more clear than those of the "soul" or "ghosts."[34] Finally, there are also those who, like the Finnish scholars Kaj Sotala and Harri Valpola, enjoy imagining scenarios in which human minds – uploaded or connected through neural prostheses – end up merging permanently, dissolving the boundaries of personal identity.[35]

7.4   Unlimited Intelligence

Implicit in the idea of interfacing brains and computers or transferring the mind into a digital medium is the possibility of amplifying our intellectual abilities; to extend them far beyond the limits granted by the nearly four billion years of evolution that have preceded us.

The idea of increasing human intellectual abilities – known as *intelligence amplification, machine augmented intelligence* and *cognitive augmentation* – is, in fact, not so recent, having been proposed in the '50s and '60s by some of the pioneers of cybernetics and computer science. In other words, it is the use of computer technology to increase human intelligence.

The first to introduce the concept of *intelligence amplification* was the British scholar William Ross Ashby, who – in his 1956 work *Introduction to Cybernetics* – spoke precisely of "amplifying intelligence." "Problem solving" is, for Ashby, an appropriate term, in the sense that almost all logical problems – such as those that can be found in puzzle magazines – can be reduced to a common form, which consists in the request – addressed to the "problem solver" – to choose a certain element within a given set. And if intelligence can be reduced to "appropriate

---

[33] K. J. Hayworth, *Electron Imaging Technology for Whole Brain Neural Circuit Mapping*, "International Journal of Machine Consciousness", Vol. 4, n. 1, 2012, pp. 87–108. http://www.worldscientific.com/doi/pdf/10.1142/S1793843012400057

[34] P. D. Hopkins, *Why Uploading Will Not Work, or The Ghosts Haunting Transhumanism*, «International Journal of Machine Consciousness», Vol. 4, n. 1, 2012, pp. 229–243. http://www.worldscientific.com/doi/pdf/10.1142/S1793843012400136

[35] K. Sotala and H. Valpola, *Coalescing Minds: Brain Uploading-Related Group Mind Scenarios*, "International Journal of Machine Consciousness", Vol. 4, n. 1, 2012, pp. 293–312. http://www.worldscientific.com/doi/pdf/10.1142/S1793843012400173

choice," then we can imagine that this capacity can be expanded beyond current limits.

But the first to propose an enhancement of the human intellect through interaction with machines was Joseph Licklider, American computer scientist and psychologist, who, in a famous technical article from 1960, *Man-Computer Symbiosis*, hypothesized that associating human beings and computers would allow the various aspects of the two members of the couple to compensate each other. Licklider imagines a future in which the human brain and the computer will enter into a very close symbiotic relationship, and that this unparalleled pairing will be able to think like humans have never been able to do and process data much better than a computer.[36]

After Licklider, it is the turn of the American engineer and inventor Douglas Engelbart, who was the first to share the vision in which machines would allow us to think better. In particular, for Engelbart, technology allows us to manipulate information, and this will result in the development of even more advanced technologies. The scholar's goal is to develop technologies that can directly manipulate information and create systems for facilitating intellectual work in individuals and groups. His thought was expressed in a 1962 report, *Augmenting Human Intellect: A Conceptual Framework*, in which the author claims, among other things, that "enhancing the human intellect" means.

...increasing the ability to approach complex problems, to gain comprehension to suit the particular needs and to derive solutions to problems. Increased ability in this respect: better-understanding, better comprehension, the possibility of gaining a degree of comprehension in a situation that was already too complex, speedier solutions, better solutions, and the possibility of finding solutions to problems that before seem insoluble.[37]

Engelbart's "complex situations" include professional problems pertaining to diplomats, managers, social scientists, biologists, physicists, lawyers and planners, regardless of how these problems arise. These ideas then led to the birth of the Augmented Human Intellect Research Center – at the Stanford Research Institute International, in Menlo Park – and led to or contributed to the emergence of physical and conceptual tools such as hypertexts.

And so, we finally arrive at *augmented cognition* – codename AugCog – a field of contemporary research that lies among computer science, neuroscience and

---

[36] See J. C. R. Licklider, *Man-Computer Symbiosis*, "IRE Transactions on Human Factors in Electronics," Vol. HFE-1, pp. 4–11, March 1960. http://groups.csail.mit.edu/medg/people/psz/Licklider.html

[37] D. Engelbart, *Augmenting Human Intellect: A Conceptual Framework*, Summary Report AFOSR-3233, Stanford Research Institute, Menlo Park, CA, October 1962, our translation. http://www.dougengelbart.org/pubs/augment-3906.html

cognitive psychology. Behind AugCog lurk the usual suspects, the guys at DARPA, who, in this case, aim to develop systems for measuring the processing of information by the human being and the cognitive state of their user. Another goal of augmented cognition is to create closed circuit systems that modulate the flow of information that the user receives according to the cognitive abilities of the latter. Then, there is the enhancement of attention, a very useful skill in situations when you have to make decisions very quickly – like on a battlefield. In short, it seems that the agency wants to develop specific high-tech tools, like the CogPit, an "intelligent" cockpit that manages to "get in tune" with the pilot, for example, by monitoring her/his brainwaves, filtering out useless information, and so on.[38] All of this ideally leads us to a further device – at the moment, purely imaginary – dear to theorists of man-machine interaction, writers of science fiction and Transhumanists: the exocortex.

The device in question is an artificial and external information processing system that should increase the cognitive processes of the brain. It should be composed of external processors, programs and hard disks capable of systematically interacting with the human cerebral cortex – in practice, through a direct connection with it – to the point that the exocortex should be perceived by the user as an extension of one's mind. The term "exocortex" was coined in 1998 by the Canadian cognitive psychologist and computer scientist Ben Houston, to indicate precisely the close brain/computer coupling theorized by Licklider and Engelbart.[39] Prior to this formal definition, science fiction writers such as William Gibson and Vernor Vinge – respectively, in *Neuromancer* and *The Peace War*, both released in 1984 – imagined similar devices. But the first to spread this concept in the world of science fiction was a writer very dear to Transhumanists, the Brit Charles Stross, who, in *Accelerando* (2005), writes of human beings surrounded by an exocortex composed of distributed agents and personality threads supported by clouds of Utility Fog.

The exocortex then takes us by the hand and leads us to another typically Transhumanist notion, that of "super-intelligence" – the term by which Transhumanists refer both to a hypothetical artificial mind superior to ours and to the intellectual abilities shared by the latter and by our post-human descendants. Bostrom defines super-intelligence as a mind much smarter than the best human brains in virtually every field, including scientific creativity, general wisdom and social skills.[40] There are two things that are worth noting. First of all, that

---

[38] The ideas of DARPA in this regard are numerous, as can be seen in the special documentary produced by them, *The Future of Augmented Cognition*. http://ieet.org/index.php/IEET/more/augcog2007

[39] http://exocortex.com/

[40] N. Bostrom, *How long before superintelligence?*, Linguistic and Philosophical Investigations 5 (1): 11–30, 2006. Our translation. http://www.nickbostrom.com/superintelligence.html

Transhumanists do not have particular prejudices towards the means that should be used to achieve this super-intelligence: we can, in fact, arrive at it through machines – creating artificial minds capable of self-enhancement – or through biotechnologies – that is, generating human beings who are increasingly more intelligent; the latter is, among other things, the hypothesis put forward by David Pearce, who speaks of "bio-Singularity" in this regard.[41] One can also get there by using both means, or by merging machines and human beings into a superior synthesis. Secondly – in the face of the critics who think Transhumanism is philosophically naive – Transhumanists distinguish between "weak" and "strong" super-intelligence. The former is at the level of human intelligence, but it is faster; it can also be seen as a way to quantitatively overcome the computational limits of *Homo sapiens*, as Bruce Katz illustrates.[42] Our first mental limit is undoubtedly that of the so-called *working memory*, which some confuse with short-term memory, that is, the ability to remember, more or less in detail, events that have happened a few moments earlier. Working memory is actually the ability to handle a certain number of mental objects at the same time, a function governed by the famous rule of "magic number seven, plus or minus two." Basically, our immediate awareness can manage between five and nine mental objects at the same time, depending on the circumstances – be it images, numbers or anything else. The Princeton psychologist George A. Miller discovered this limitation in the 1950s.[43] This ability is fundamental to our cognitive processes, because the more we manage to manage, the more detailed our arguments and our deductions will be. It is not clear whether or not this ability can be improved with exercise, but it could certainly be upgraded with our future mind-computer interfaces. Then there is the question of long-term memory – of particular importance for all of the aspirants to immortality, since they will have to find a way to efficiently store their millennial memories. Katz cites the work – somewhat dated, to tell the truth – by the American psychologist Thomas Landauer, for whom our long-term memory has a truly laughable capacity, about one hundred megabytes, in practice, the capacity of a

---

[41] D. Pearce, *The Biointelligence Explosion. How to self-improve organic robots will modify their own source code and bootstrap our way to full-spectrum superintelligence*, 2012. http://www.biointelligence-explosion.com/

[42] Bruce F. Katz, *Neuroengineering the Future. Virtual Minds and the Creation of Immortality*, Infinity Science Press, Hingham 2008.

[43] George A. Miller, *The Magical Number Seven, Plus or Minus Two: Some Limits on Our Capacity for Processing Information*, «Psychological Review», Vol. 101, No. 2, 1956, pp. 343–352. Recent studies seem to have shown that this "magic number" is less than seven; most likely, it is no more than four. http://www.psych.utoronto.ca/users/peterson/psy430s2001/Miller%20GA%20Magical%20Seven%20Psych%20Review%201955.pdf

hard disk from the '80s – a period to which the scholar's article goes back.[44] We do not guarantee the reliability of this research – one hundred megabytes seems rather small to us – but our long-term memory certainly has its limits, which the neurotechnologies of the future may find ways to overcome. But it does not end there; in this regard, Katz lists our limits in terms of the speed and efficiency of the processing of information, the limits of application of rational thought (the human capacity for irrationality is well known), the capriciousness and randomness of the creative process, and the impossibility of focusing on more than one problem at a time – in the face of those who preach, perhaps with the aim of putting pressure on employees, the virtues of multitasking.

On the other hand, the "strong" super-intelligence represents an intellectual level that is situated above the human level from a *qualitative* point of view. We are therefore in the presence of a mind that goes beyond our own, and which we are not constitutively able to understand.[45] In this regard, Michal Anassimov tells us that "the true super-intelligence is something radically different – a person capable of seeing the obvious solution that the entire human race has missed, conceiving and implementing advanced plans or concepts to which the greatest geniuses would never think, understand and rewrite their cognitive processes at the most fundamental level, and so on".[46]

Transhumanists, however, tend to address issues at 360 degrees, and this also applies to the question of intelligence. In particular, a topic that some of them have dealt with – and that normally goes beyond the interests of artificial intelligence theorists and AugCog experts – is that of wisdom. Various definitions can be given of this virtue; we like to consider it as the ability – acquired through experience – to navigate smoothly in the reality of every day, to move while reducing the frictions that life entails and to better manage the situations and relationships in which we find ourselves, regardless of the hand of cards dealt to us by the chance. Certainly, immortal human beings would have at least a thousand years or even millennia of life in front of them, so we can imagine that humans biologically similar to us, but with a thousand years of experience behind them, would have a unique perspective on the world and a style of management of reality infinitely more far-sighted than that of any "wise man" who ever existed. In spite of this, it is possible to imagine further "help" to human wisdom, in the form of the proposal put forward by Natasha Vita-More, namely, that of a new version – or, if we want,

---

[44] T. K. Landauer, *How much do people remember? Some estimates of the quantity of learned information*, «Cognitive Science», 10, pages 477–493, 1986. http://csjarchive.cogsci.rpi.edu/1986v10/i04/p0477p0493/MAIN.PDF

[45] http://www.transvision2007.com/page.php?id=260

[46] Michael Anissimov, *Top Ten Cybernetic Upgrades Everyone Will Want*, http://www.acceleratingfuture.com/michael/blog/2007/01/ten-transhumanist-upgrades-everyone-will-want/

an *upgrade* – of classically understood wisdom. Her idea is to combine two future technologies, AGI – *artificial general intelligence* – and nanotechnological macrosensing. The first consists of a field of research *in progress* that aims to develop an artificial intelligence capable of learning, thus acquiring new knowledge – unlike traditional artificial intelligence, which has knowledge already embedded in its programming and is aimed at immediate problem solving. AGI is a concept similar to that of the Seed AI theorized by Yudkowsky, that is, an artificial intelligence programmed for self-change and self-understanding, triggering a virtuous circle of self-improvement.[47] The macrosensing nanotech is an idea by Freitas, which consists in disseminating throughout the human body – and, in particular, the nervous system, although not just that – nano-sensors able to monitor the somatic and extrasomatic states, i.e., the psycho-physiological conditions of the subject and the data coming from the sense organs, in order to optimize – more than our natural sensorial apparatus would – the management of the data that come to us from the external and internal worlds. Associating these two technologies and then combining them with the human brain would offer the latter a permanent guide to life, a sort of "personal oracle," in short – almost like the demon Socrates spoke about – ready to give us advice on request, and with a degree of intimacy and connection with our ego that is variable at will. In short, a super-super-ego, without the repressive and psychopathogenic function of the Freudian Super-ego.[48]

Now it's time to mention another core project of the Transhumanist enterprise, a plan differs markedly from the non-biological or anti-biological proposals that we have seen so far. We are talking about the plan elaborated by one of the fathers of the World Transhumanist Association, David Pearce. And we have to keep in mind the following apostille: in sharp contrast with the other Transhumanists, the British philosopher proposes a restructuring of the human mind based not on neuro-engineering, but on biotechnologies.

---

[47] E. Yudkowsky, *What is Seed AI*, 2000, http://www.singinst.org/seedAI/seedAI.html

[48] N. Vita-More, *Wisdom [Meta-Knowledge] through AGI/Neural Macrosensing*. http://www.natasha.cc/consciousnessreframed.htm

8

The Bright Day of the Soul

8.1 The Hedonistic Imperative

Just to be blunt, what is the point of living forever if you cannot live happily ever after? In other words, do we want our ordinary lives, with all of their messes and hassles, to go on indefinitely? To keep working like dogs, to be bored during a rainy Sunday afternoon, to suffer for love or for the lack of it, to feel lonely, and so on? In short, why would someone repeat the banality of life forever and ever?

Do not worry, folks: Transhumanism has all of the solutions for your existential issues and your potential ennui. Let me introduce you to a brilliant independent British philosopher, David Pearce, author, in 1995, of a quite visionary manifesto, *The Hedonistic Imperative*.[1]

The goal, ambitious indeed, is to "eradicate suffering in all sentient life." Reprogramming our nervous system, of course, and discarding our Darwinian past, courtesy of the usual tools recommended by Transhumanists. That is, biotechnology and molecular nanotechnology. And this is not just about our bodies, but the whole planetary ecosystem, each and every living creature, so that no other species has to suffer in the blissed and blessed world that the post-humans will engineer for everybody.

Pain and sorrow exist because they fulfill an evolutionary need, of course. After all, we need to feel pain when we touch something extremely hot, otherwise we would not withdraw our hand and, therefore, we would heavily damage our bodies. But what if we could find a way to replace our old, gene-driven

---

[1] https://www.hedweb.com/

© Springer Nature Switzerland AG 2019
R. Manzocco, *Transhumanism - Engineering the Human Condition*,
Springer Praxis Books, https://doi.org/10.1007/978-3-030-04958-4_8

motivational system with another one based exclusively on gradients of bliss, or wellbeing, rather than the classic, universal – across the Animal Kingdom – pain-pleasure axis? Even better: we could recalibrate this axis in such a way that we could experience degrees of pleasure, happiness, serenity and motivation impossible to imagine for our evolutionarily limited brains.

So, it seems that Pearce is a man with a plan, and a very bold one, after all – he admits that his project is sketchy, though, and that his is just a manifesto, in need of further elaboration and validation from the neuroscientific community.

The language that he uses is clearly reductionist: so, for example, covering the topic of motivation, he uses terms like "dopamine-overdrive," in order to indicate the possibility of actually enhancing traits like curiosity, and an exploratory and goal-oriented behavior. After all, as much as we like to enjoy a deep and serene "serotoninergic" state – serotonin is a neurotransmitter connected to the regulation of mood, among other things – we don't want to be "blissed-out" and get stuck in a sub-optimal condition that is not really compatible with our survival and evolution. What we do want is to manipulate our hedonic treadmill, raise our hedonic set-point as much as possible – millions or billions of times, if this even makes sense – and enjoy a combination of bliss and strong, unbeatable motivation to explore, create, and just live.

The hedonic treadmill, also known as hedonic adaptation, is the tendency of human beings to quickly return to a more or less stable level of happiness in spite of major positive or negative life events. The term was coined by Philip Brickman and Donald Campbell in their 1971 essay *Hedonic Relativism and Planning the Good Society*.[2] During the 1990s, the concept was further developed by the British psychologist Michael Eysenck into the "hedonic treadmill theory," which compares the human pursuit of happiness to an athlete on a treadmill – in other words, you have to keep walking in order to remain in the same place. Connected to the idea of hedonic adaptation is the concept of a "happiness set point," which states that human beings, on average, maintain a constant level of happiness throughout their lifespan. Psychologically speaking, the purpose of hedonic adaptation is to keep the individual sensitized to their surrounding environment, that is, to avoid complacency, keep individual motivation going and foster acceptance of the situations that cannot be changed. In the study *Lottery Winners and Accident Victims: Is Happiness Relative?*, Brickman, Dan Coates, and Ronnie Janoff-Bulman studied lottery winners and paraplegics, showing that, after an initial, often dramatically positive or negative change, the happiness level of the subjects returned to their usual average levels.[3] In 2006, Ed Diener, Richard Lucas and Christie Scollon

[2] P. Brickman; D. Campbell, *Hedonic relativism and planning the good society*, in: M. H. Apley, ed., *Adaptation Level Theory: A Symposium*, Academic Press, New York 1971, pp. 287–302.
[3] P. Brickman; D. Coates; R. Janoff-Bulman, *Lottery winners and accident victims: Is happiness relative?*, «Journal of Personality and Social Psychology». 36 (8): 917–927, 1978.

concluded that individuals do have different hedonic set points, that is, people are not hedonically neutral, but have different individual average levels of happiness – and that those levels are at least partially heritable and genetically determined.[4] A statement backed by a 1996 study by David Lykken and Auke Tellegen, based on thousands of sets of twins who were monitored for 10 years, which concluded that almost 50% of individual happiness is influenced by genes.[5] There is even some evidence that our hedonic set points could be raised through chemical means, in particular, through a specific compound, NSI-189, currently under experimentation by an American biotech company, Neuralstem Inc.

So, going back to the *Hedonistic Imperative*, we want gradients of bliss, and not a perpetual uniform state of wellbeing, because, otherwise, we would be completely insensitive to the stimuli coming from the outside world, perpetually trapped in our inner paradises. Just as 200 years ago, before medical science developed strong, potent pain-killers, everybody believed that physical pain was an inescapable, even necessary feature of human existence, so, nowadays, we tend to think that mental pain and a less-than-optimal mood are just part and parcel of what it means to be human. The idea that, in the near future, we will be able to alter our mental states and live super-happily ever after seems bizarre indeed. Post-humans will have a different opinion, of course; after leaving behind their primordial Darwinian mind, they will enjoy states of "magical joy" that are bio-technologically intensified, multiplied, and enormously differentiated.

Pearce's buzzwords are "serotonergic serenity" combined with the goal-seeking energy of a "raw dopaminergic high." Euphoric peak experiences[6] will be channeled, controlled and genetically diversified. Another concept to keep in mind is that of "intentional object" – philosophically speaking, "intentionality" is the "object-directedness" or "aboutness" of our thought. So, intentional objects are those objects – situations, things, persons, even concepts – that we think about, worry about, and are happy about. Of course, from a reductionist and biological perspective, we connect our emotional lives to these objects for evolutionary reasons, because they serve some purpose related to the survival of our genes. But what if we could re-engineer these "peripheral routes" to our personal fulfilment,

---

[4] E. Diener; R. Lucas; C. Scollon, *Beyond the hedonic treadmill: Revising the adaptation theory of well-being*, «American Psychologist», 61 (4): 305–314, 2006.

[5] D. Lykken; A. Tellegen, *Happiness Is a Stochastic Phenomenon*, «Psychological Science», 7 (3): 186–189, 1996.

[6] The concept of "peak experience" was developed by the American psychologist Abraham Maslow; by it, we mean, in short, all of those more or less rare moments in which the subject feels "in tune with the world," or "fully master of himself," but, at the same time, "fused with the Whole." It is something like a quasi-mystical experience with a strong existential connotation, for which the subject experiences a strong "fullness of meaning" in relation to his/her life and his/her place in the world. See A. Maslow, *Religion, Values and Peak Experiences*, Viking, New York 1964.

and increase the number and variety of our intentional objects – the things we are happy about – by rational design? After all being in a "hyper-dopaminergic" state would definitely increase the number and variety of the activities and goals we pursue, and our post-human descendants might enjoy modes of existence and interests that are, for us, unimaginable.

The targets of Pearce's ambitious plan are the brain's pleasure centers, located in the mesolimbic dopamine system, the headquarters of the brain's reward circuitry, extended from the ventral tegmentum to the nucleus accumbens, and projected to the limbic system and the orbitofrontal cortex. The combination of hyper-dopaminergic and hyper-serotoninergic states, carefully engineered, in order to avoid mania and psychosis, would offer us sights, music, mystical states and intuitions more majestic than we can understand. From an informational perspective, the absolute location of our experiences of the pain-pleasure axis doesn't count. What counts is the difference in gradients, which means that the post-human gradients of bliss would be as informative as the current painful raw sensations we experience. What matters is the function, not the raw sensation associated with it. Pearce does point out that his project is sketchy, that, from a neuroscientific viewpoint, it might be simplistic or plainly wrong. It is just a manifesto, though, open to any kind of improvement and update we want to add. The plan entails the modification of the meso(cortico-)limbic dopamine system, increasing the number of its neurons. The dendrites and the axons of these neurons innervate the higher cortical regions of the brain, in what Pearce calls the "encephalisation of emotion." Enriching our emotional and hedonic lives implies the multiplication of the mesolimbic dopamine cells and the reduction of their feedback inhibition mechanisms, paying attention, at the same time, to the possible side effects of this process of hyper-stimulation, notoriously connected to mental illnesses such as schizophrenia. The philosopher privileges a "twin-track approach," consisting in boosting both the dopaminergic and serotoninergic systems. Of course, this approach looks, and is, crude, and the details need to be worked out by the scientists of the future, in what Pearce calls "hedonic engineering" or "Paradise engineering." So, it seems that a major transition is waiting for us – and for the whole Animal Kingdom. Those lucky enough to live in the middle of it, whose lives will experience both sides of this evolutionary transition – the hellish "human, too human" side and the heavenly post-human one – will probably feel as if they have just woken up from a strange, blurred and nightmarish dream.

Our post-human descendants will empower the dopaminergic and serotoninergic sides of our sensorium, and, at the same time, ban the molecular substratum of any and every nasty experience – molecules like nociceptin, substance P, and bradykinin, to name a few.

But we want to make sure that the post-humans are not just blissed-out, but rather super-happy about external (or internal) intentional objects. In other words, after dismissing the tyranny of the genes, we want to re-define the encephalisation of emotions, not to de-encephalise them. We don't want a uniform, perpetual orgasm, but a good – or great – mood "about" something. A "something" that we will get to choose, thus satisfying our "second-order" desires – desires about our desires. In other words, by re-designing the axonal and dendritic arborisation of the neo-cortex – which allows us to rationalize our emotions, to connect them to this or that intentional object – we will bootstrap ourselves into fulfilling our second-order desires and decide who or what we want to be, and be happy about things that, right now, we cannot actually understand.

For example, "an unprecedentedly vivid sense of reality, a perpetually enriched feeling of meaningfulness and significance, a sense of heightened authenticity, and never-ending raw-edged excitement – or intense serenity and spiritual peace. (…) A musician might then hear, and have the chance to play, music more exhilarating and numinously beautiful than his or her ancestors ever dreamed of; the celestial music of the spheres heard by privileged medieval mystics will be as a child's toy tin-whistle in comparison. (…) The sensualist will discover that what had previously passed for passionate sex had been merely a mildly agreeable piece of foreplay. Erotic pleasure of an intoxicating intensity that mortal flesh has never known will thereafter be enjoyable with a whole gamut of friends and lovers. (…) A painter or connoisseur of the visual arts will be able to behold the secular equivalent of the beatific vision in a million different guises, each of indescribable glory." A whole new plane of being will open up to us, and, of course, our language will have to evolve accordingly. But Pearce is not done yet: "Still in a personal vein, fragile self-esteem and shaky self-images will be beautified and recrystallized afresh. For the first time in their lives, in many cases, human beings will be able wholeheartedly to love both themselves and their own bodily self-images. (…) Love will take on new aspects and incarnations too. For instance, we will be able, not just to love everyone, but to be perpetually in love with everyone, as well; and perhaps we'll be far more worth loving than the corrupted minds our genes program today. (…) Another aspect of post-Transition love may be found even more surprising. Individual personal relationships may at last be bonded truly securely, should we so desire. Throughout the ages, dreadful pain has been caused by the soul-destroying cruelties of traditional modes of love. We acknowledge, in the main, that we hurt the most those we love. (…) After the Transition, on the other hand, one will be able to love somebody more passionately than ever before. In the post-Darwinian era, one will be safe in the knowledge that one will never hurt them, nor be hurt by them in turn. True love really can last forever, though responsible couples should take precautions. If one desires a particular relationship to remain uniquely and enduringly special, then the mutually coordinated design of each other's neural weight spaces can ensure that a distinctively hill-topped plateau in the new hedonic landscape structurally guarantees that each other's presence is always uniquely fulfilling."

But, wait a minute, if we are not going to be blissed out, if we are to go through different gradients of bliss, wouldn't a state of diminished pleasure be comparable to a state of pain? Pearce couldn't disagree more: if we are faced with two alternatives, like picking the lesser of two evils, this doesn't make the lesser evil somehow pleasurable. It is still perceived as painful. Likewise, the lesser of two delights will still be delightful. And let's not forget the possibility of developing and designing brand-new emotions, so alien to our Darwinian mind-set that the post-human mind would probably look opaque to us. Expanding our somato-sensory cortex, allowing it to interpenetrate the rest of our nervous system, will allow the Self to become "One" with our own neural micro-cosmos, acquiring a currently unimaginable direct knowledge of ourselves.

Our psychedelic future will also include brand new types of experience: "Heaven knows what further incommensurable modes of what-it's-like-ness ("qualia") will be disclosed when much more far-reaching changes in the architecture of excitable cells are engineered." And the basic "trinity" that rules our stream of consciousness – the cognitive, affective and volitional aspects, that is, "thinking," "feeling" and "willing" – might be widened, and new, totally different forms of consciousness might be added to the family. In other words, the post-human minds of the third and fourth millennia will probably be alien to us, cognitively so distant from ours as to be closed to our comprehension. "Systematic experimental manipulation of consciousness via psychoactive agents will complement the third-person perspective of physical science. Exploration will be most prudently conducted by ecstatics, native-born or otherwise, rather than by gene-disordered Darwinian minds. (...) Will there come, eventually, a post-personal era in which discrete, gene-generated superminds choose progressively to coalesce; or will the fragmented island universes left over from the depths of the Darwinian past continue in semi-autonomous isolation indefinitely? If consciousness is ontologically fundamental to the cosmos (...), do superstrings vibrating at energies orders of magnitude higher than ours support modes and intensities of experience correspondingly greater than those of the current low-energy regime? Or do they really lack what-it's-like-ness altogether?"

Another one of Pearce's interesting slogans is "breaking the Tyranny of the Intentional Object": our emotions are encephalized, which means that we are genetically programmed to be happy about something that serves the purposes of our genes. Another name for this is Peripheralism, that is, the idea that we must look for satisfaction at the periphery of our being, into the external world and the Objects – things, situations and so forth – that it offers. The tricky aspect of all of this is that, no matter how much we engineer our environment – in other words, no matter how perfectly our desires are satisfied – the hedonic treadmill will step in and make us bored – we could call it "the tragedy of raising expectations." Accessing the core of our brain will allow the hedonic engineers of the future to hijack our hedonic treadmill, raise our hedonic set point and, more importantly, increase our list of Intentional Objects, the things that we will be happy about, in unpredictable and surely wonderful ways.

And what about the informational role of pain, of our too human "nociception"? Nowadays, we can build robots that can deal with the environment and are able to avoid dangerous stimuli without the "raw" feeling of pain. Which means that, in the near future, we might either biotechnologically reprogram our peripheral nerves to act quickly without the pain or – using robotics or nanotechnology – offload the informational role of pain to artificial prostheses or any device with which we decide to merge: "Buddhists focus on relieving suffering via the extinction of desire; yet it's worth noting this extinction is technically optional, and might arguably lead to a stagnant society. Instead it's possible both to abolish suffering and continue to have all manner of desires." And, of course, in a few generations, the neurophysiological substratum of boredom – which does have an adaptive value, as it pushes us to seek novelty – will be erased and substituted with something else that is less obnoxious. So, becoming bored in the classical sense will be physiologically impossible.

And let's not forget about wire-heading – stimulating the pleasure centers of the brain intercranially does not produce tolerance. Pearce quotes classic research on mice to prove his point, which is that it is possible that pleasure and happiness do not cause habit and then boredom, as happens with Peripheral Intentional Objects. The point here is not to find a way to be blissed out – evolutionarily speaking, a very bad strategy, as blissed out people would be unlikely to reproduce. Rather, the idea is to stress the fact that, with careful hedonic engineering, the post-humans will be able to enjoy perpetual paradisiac happiness without ever getting bored with it: "Recall wireheading is direct stimulation of the pleasure centres of the brain via implanted electrodes. Intracranial self-stimulation shows no physiological or subjective tolerance i.e. it's just as rewarding after two days as it is after two minutes. Uniform, indiscriminate bliss in the guise of wireheading or its equivalents would effectively bring the human experiment to an end, at least if it were adopted globally. Direct neurostimulation of the reward centres destroys informational sensitivity to environmental stimuli. So assuming we want to be smart – and become smarter – we have a choice. Intelligent agents can have a motivational structure based on gradients of ill-being, characteristic of some lifelong depressives today. Or intelligent agents can have our current typical mixture of pleasures and pains. Or alternatively, we could have an informational economy of mind based entirely on (adaptive) gradients of cerebral bliss – which I'm going to argue for."[7]

---

[7] https://www.abolitionist.com/wireheading.html

But there is more: "[T]here is no biological reason why each moment of one's existence couldn't have the impact of a breathtaking revelation. As the phenomena of *déjà vu*, and its rarer cousin *jamais vu*, strikingly attest, a sense of familiarity or novelty is dissociable from the previous presence or absence of any particular type of Intentional Object with which such feelings might more normally be associated. So the kind of thrill one might first have got witnessing, say, the Creation can in principle become a property of every second of one's life."

As to the alleged creative virtues of suffering, Pearce underlines that things like hopeless despair and deep suffering are just that, and creativity can actually be magnified and enhanced through the appropriate engineering of our motivational system: "Personal growth is more likely to unfold if one's appetite for life gets steadily keener. This will occur if one's experiences get progressively richer and more rewarding. Odysseys of self-exploration across the hedonic landscape can offer scope for ever-deepening self-discovery and idealized self-reinvention. (...) It is worth distinguishing between the destiny of the humanities and the sciences after heaven has been biologically implemented. For a start, the exquisite aesthetic experiences on offer to our genetically enriched descendants may inspire an unprecedented flowering rather than a withering of the fine arts. Our current enjoyment of, say, Van Gogh's 'Sunflowers' or Leonardo's 'The Last Supper' will seem distracting tickles in comparison. Those who would deny that beauty is in the eye of the beholder might, or might not, be impressed by the disposition of paint on canvass which inspires these rhapsodies. (...) Tomorrow's technologies of fine-grained emotional control may enable early post-humans, for instance, to amplify their most treasured second-order desires for, say, cultural excellence, intellectual acumen and moral integrity while banishing the baser carnal passions. It's conceivable (...) that our distant descendants will enjoy some kind of ceaseless rapture – perhaps contemplating unimaginably sublime beauty or love or elegant mathematical equations. Or, less portentously, hilariously funny jokes." This is a possibility, of course, but not a likely one; after all, the hedonic engineers of the far future will want to avoid stagnation, so they will probably try to counterbalance these ecstatic states with more dynamic ones. And "'timeless' bliss doesn't have to feel static. Mastery of the neurochemistry of time perception may allow each here-and-now to have a vast temporal depth, a rich internal dynamics, and subjectively to last an eternity."

8.2 Reprogramming Predators

"And the wolf shall dwell with the lamb and the leopard shall lie down with the kid, and the calf and the young lion and the fatling together and a little child shall lead them," says *Isaiah 11:6*. Sadly for them, this looks like nothing more than a utopian fantasy, and a pretty naïve one at that. We know how things work in the natural world: animals kill and devour each other, starve, die of thirst, and so on. Even if we implemented – as Pearce suggests – a global form of veganism, the Animal Kingdom would still be plagued by physical suffering. But things don't have to go that way: our post-human descendants might be able to re-engineer the whole ecosystem, phasing out or reprogramming the predators and controlling the reproductive rate of their prey with a form of depot-contraception – a long-acting contraception that would allow our descendants to manage the ecosystem without the need for predators. So, Pearce's Abolitionist Project is even more ambitious than we thought: redesign the ecosystem of the whole planet, erasing – through genetic manipulation – suffering from the entire natural world. Animals should not be eaten, says Pearce, they should be looked after, as we do with our babies. Nanotechnologies and the Internet of Things will become so developed that they will allow us to control and manage every single cubic meter of the planet. Nanotech, in particular, will reach every remote corner of the ocean, erasing suffering up to the last unknown mollusk. So, there you have the plan: immune contraception for the prey, nanotechnology to reach every corner of the ocean, and the reprogramming of Nature's beloved "psychopaths," the predators.[8]

What our post-human descendants will do is to build a cruelty-free world. Global veganism – that is, forcing human beings to become vegans – doesn't look feasible. But biotechnology can offer us a nice alternative: artificially cultured, cruelty-free meat. Lab meat, basically, which is already under research by a few companies.[9] And, as soon this artificial meat becomes easy to produce, cheaper than regular meat and maybe more palatable and delicious, this would bring to an end the current meat industry.

Pearce strongly criticizes the contemporary "conservation biology," which, rather than reducing the suffering of free-living animals, perpetuates it. As we have seen, the British philosopher's plan entails something more than perpetual bliss for all post-humans, as it aims for the complete abolition of suffering throughout the animal world. For obvious reasons – there are not that many spots on the higher levels of the food chain – carnivore species are small in number, so reprogramming them should not be such a big problem, as long as we develop adequate technologies. There are two options on the table: the first one is to phase out the predators, basically, to sterilize them and then let them go extinct; the second one is to manipulate their brains, in order to keep their killer instincts under control – at the same time feeding them with artificial meat. As a possible strategy, Pearce mentions the so-called "ratbots,"[10] rats whose behavior is controlled through elec-

---

[8] https://www.hedweb.com/abolitionist-project/reprogramming-predators.html

[9] See, for example, http://www.new-harvest.org/

[10] https://www.wireheading.com/roborats/index.html

trodes implanted into their brains, an indication of what we could do in order to keep the carnivores under control. Plus, the herbivores could be re-programmed to no longer be afraid of the ex-carnivores. So, the wolf might be able to lie down with the lamb after all. The second option would be preferable in the case of "iconic" animals, species that have a symbolic value for us, like lions, tigers, bears, wolves, and so on. Of course, the non-carnivore animals should be kept under control as well, using immuno-contraception in order to prevent their numbers from running out of control. So, in the end, the post-humans of the far future will turn the whole planet into a cruelty-free zoo, looking after all of the other sentient life-forms. And, of course, all the non-human animals could be enhanced, and undergo the same hedonic engineering and intelligence amplification that we will.

As stressed by another Transhumanist, Micah Redding, environmentalism mandates Transhumanism; after all, natural catastrophes have already hit our planet in the past, and might do so again in the future; and the only species able to develop visionary technologies that might save not just humanity, but every other lifeform as well, is *Homo sapiens*. Otherwise, sooner or later – either because of some asteroid or because of the death of our Sun – life on Earth will be gone.[11]

8.3   A Nightmarish Brave New World?

David Pearce's opinion on Huxley's *Brave New World* couldn't be more negative: "*Brave New World* (1932) is one of the most bewitching and insidious works of literature ever written."[12]

Tough but completely understandable words, though; after all, Huxley's novel is brought up every time someone wants to criticize Transhumanist ideals and projects. It has become the symbol of every human attempt to use science to increase happiness – through technological means, of course. And, in using it as such, critics forget that *Brave New World* is not a work of futurology, it is not an attempt to predict the future outcome of our civilization's technological development. It is a work of great literature, for sure, but of the satirical kind; its purpose is to criticize both the communist utopia and the consumeristic Fordist society.

The world depicted by Huxley is not the techno-utopia dreamed of by Pearce; quite the opposite, actually: it is a future society that is completely blissed-out and trapped in a sub-optimal condition. There is no creativity, no hyper-dopaminergic motivation, in that society. In other words, it actually represents the opposite of what Transhumanists are trying to achieve; in another book, *Brave New World*

---

[11] M. Redding, *Environmentalism Mandates Transhumanism*, http://micahredding.com/blog/environmentalism-mandates-transhumanism

[12] https://lifeboat.com/ex/brave.new.world

Revisited – published in 1958 – Huxley himself describes that society as a nightmare.

And, to tell the truth, from a purely narrative viewpoint, the society of *Brave New World* lacks the fervid imagination that characterizes many works of science fiction, capable as they are of opening up possibilities for our future evolution. No matter how far-fetched the Transhumanist vision might be, nobody in that movement is planning to get rid of freedom through behavioral conditioning or abolishing love and passion, or to substitute the latter with shallow entertainment, as happens in Huxley's novel, which, in turn, idealizes the diseased, brutish and painful life that humans lead in the Reservation from which John Savage, the novel's protagonist, comes.

Happiness in *Brave New World* is pursued through a typical peripheralist strategy, mass-consuming classically shallow sex, goods, and drugs designed to numb, not to enrich, the mental lives of its users. Kids are conditioned to accept human mortality – the exact opposite of what Transhumanists are preaching – and the biological nirvana produced by soma has nothing to do with the existentially and artistically meaningful experiences that the post-humans might enjoy, courtesy of David Pearce's Paradise engineering. The brave new worlders are just manipulated dupes, deprived of any philosophically, existentially or aesthetically significant experience, happy victims of a society that, unlike the utopia dreamed of by the Transhumanists, doesn't entail evolution and growth. Later on, Huxley himself developed his own drug-based utopia, *Island*, published in 1962, which depicts a society modeled around, among other things, the author's experiences with the consumption of mescaline and LSD. And let's not forget that, no matter how vivid the experiences with these drugs can be, they are nothing compared with the alleged wonders of Pearce's Paradise engineering.

So, *Brave New World*'s society is not developed, but rather further constricted; it offers fewer options than our society, or the post-human society, not more. It is a benevolent tyranny, in which human beings are systematically produced by machines and genetically programmed to belong to a limited set of castes – the alpha, the beta, the gamma, the delta and the epsilon – hierarchically organized and manipulated in order to be completely satisfied with the *status quo*. Love is forbidden, shallow sexuality is compulsory, pleasure is omni-pervasive. Great art does not exist anymore, and it is seen as a danger to the social order. John Savage, the man from the Reservation – conceived in the classic and forbidden way by two brave new worlders – is in love with Lenina, a beta, but these feelings cannot really be fulfilled. In Pearce's utopia, John and Lenina would have the chance to love each other and live happily ever after – actually "happily forever after," as mortality would be abolished.

8.4   The Bio-Intelligence Explosion

Let me now introduce a concept on which we are going to put a lot of focus later on: the Technological Singularity, the idea that technological progress is not just moving forward, it is actually accelerating, and that we are approaching an essential discontinuity in our history. Soon, technology will produce a greater-than-human intelligence, and this intelligence will mark the end of the human era as we know it. There are at least three main scenarios concerning the Technological Singularity. In the first one, advocated by the Singularity Institute, the creation of recursively self-improving software-based minds – that is, artificial minds able to modify and improve themselves – will culminate in a super-fast intelligence explosion. Basically, the super-intelligence of the near future will be non-human and non-biological, maybe even hostile to our species. The second scenario, promoted by Ray Kurzweil, envisions a future in which men and machines will merge via brain implants, the distinction between biological and artificial will be blurred, and humans will be able to upload their minds onto a non-biological substratum. The third scenario is the one advocated by David Pearce, the Bio-Intelligence Explosion: humans of the near future will rewrite their own genetic code, recursively re-editing their own minds, bootstrapping their own way toward a full-blown super-intelligence, super-sentience and super-happiness based on information-sensitive gradients of bliss.[13] In other words, we will use biotechnology to self-hack and self-improve ourselves, becoming, biologically, post-humans. Pearce envisions user-friendly interfaces – like a "Windows" of bio-hacking – able to change and customize our own genetic code, producing a cognitive and emotional enhancement and an indescribable state of bliss. This idea of an imminent Bio-Intelligence Explosion opens up further scenarios; let us look at a few of them.

One possibility – not really promoted by Pearce; just contemplated – is similar to a universal wire-heading, a society in which every individual sets their own hedonic set-point to the maximum level and lives in a state of perpetual physical and psychical pleasure. Not really a good scenario, as individuals of this society will lack any incentive to reproduce and perpetuate that same society. Extinction would follow. The second possibility – changing our hedonic set-point and creating a new motivational system based on gradients of bliss – is the option chosen by Pearce, and it open ups further options, that is, the exploration of new mind-spaces.

First of all, will our post-human descendants feel tempted to examine the Darwinian state of mind of their human ancestors? Exploring the higher levels of paradisiac consciousness might be more rewarding, but we cannot discount, in

---

[13] https://www.biointelligence-explosion.com/

principle, the possibility that our descendants will be willing to explore the emotional "basement" – horrible, for them – in which we currently live. As an alternative, they might only be able to know what it is like to be a normal human through analogy. Anyway, another possibility is that they will learn to tinker with their states of consciousness in completely novel ways. Let us not forget that the states of consciousness we normally live in are numerically limited: waking life, dream, sleep, and, for a few, lucid dreaming. Nothing will prohibit our post-human descendants from building brand-new, for us, unimaginable, states of consciousness, occupying a wider experiential state-space. So, the post-humans will be super-sentient too, able to shift between a great multitude of radically different modes of consciousness, exploring psychedelia in a way that we will never be able to do. Or they might pursue a lifetime of introspection and tranquil meditation, achieving new and unimaginable levels of self-knowledge. Or they might explore the universe, after regulating their level of "adventurousness" accordingly. Or they might learn to apprehend time in a different manner, altering their perception of the time flow, making every second of their lives as long as they want. Super-sentience also means ultra-high-intensity experiences: ants have, for example, a certain – quite dim – level of consciousness, humans have a higher level – they are more "awake" – so post-humans might have even more of this, to the point that, compared with them, we would look like sleep-walkers. Of course, this means that the post-humans will have to invent millions of new terms to indicate brand new, super-intense sensations, concepts, and states of consciousness. Another option is "reversible mind-melding": just as the Hogan sisters[14] share a thalamic bridge, so might we, engineering a form of deeper mind-melding, create a way of sharing the content of our minds, maybe making possible a form of cross-species mind-melding – so that we might finally be able to know what it is like to be a bat,[15] after all. And let's not forget about the possibility of meeting, one day, a momentous "unknown unknown": "If you read a text and the author's last words are 'and then I woke up', everything you've read must be interpreted in a new light – semantic holism with a vengeance. By the year 3000, some earth-shattering revelation may have changed everything – some fundamental background assumption of earlier centuries has been overturned that might not have been explicitly represented in our conceptual scheme. If it exists, then I've no inkling what this 'unknown unknown' might be, unless it lies hidden in the untapped subjective properties of matter and energy. (...) By the lights of the fourth millennium, what I'm writing, and what you're reading, may be stultified by something that humans don't know and can't express."

---

[14] cf. S. Dominus, *Could Conjoined Twins Share A Mind?*, «The New York Times», May 25, 2011. http://www.nytimes.com/2011/05/29/magazine/could-conjoined-twins-share-a-mind.html
[15] T. Nagel, *What is it like to be a bat?*, in: «The Philosophical Review», 83 (4): 435–450, 1974.

8.5   Saving the Multiverse, One Timeline After the Other

In another paper, *Quantum Ethics? Suffering in the Multiverse*,[16] Pearce introduces an even more ambitious idea – Transhumanists do not lack for imagination, as you can easily ascertain. Why limit ourselves to saving the world, when we can save the universe and deliver it from evil? Pearce gathers a multitude of cosmological theories, in order to show us that there is more suffering in the universe than meets the eye. First of all, consider the "block-universe" theory, that is, the idea that the past, in spite of being, well, past, is as real as the present, and so is the suffering that it contains. Secondly, it is likely – according to several cosmological theories, anyway – that there is more than one universe. An infinite number, actually; while many of them are deprived of any lifeforms, others are populated by creatures able to feel pain. And many are the theories that admit the existence of infinite universes: post-Everett quantum mechanics suggests the existence of an infinitely branching multiverse, composed of every possible combination and variation of alternate realities. But things get even more complicated: "[S]ome speculative work in contemporary theoretical physics suggests that even the multiverse of Everettian quantum mechanics doesn't remotely exhaust the totality of suffering. (...) Suffering may exist in other post-inflationary domains[17] far beyond our light cone; and in countless other "pocket universes" on variants of Linde's eternal chaotic inflation[18] scenario; and in myriad parent and child universes on Smolin's cosmological natural selection hypothesis; and among a few googols of the other 10^{500+} different vacua of string theory[19]; and even in innumerable hypothetical 'Boltzmann brains',[20] vacuum fluctuations in the (very) distant future of 'our' Multiverse. These possibilities are not mutually exclusive. Nor are they exhaustive. Thus some theorists believe we live in a cyclic universe, for instance; and that the Big Bang is really the Big Bounce. (...) Faced with this fathomless immensity of suffering, a compassionate mind may become morally shell-shocked, numbed by the sheer enormity of it all. Googolplexes of Holocausts are too mind-wrenching to contemplate. We might conclude that the amount of suffering in Reality must be infinite – and hence any bid to minimize such infinite suffering would still leave an infinite amount behind. A sense of moral urgency risks succumbing to a hopeless fatalism. (...) Thankfully such moral defeatism is premature. For it is not at all clear that physically realized infinity is a cognitively

---

[16] https://www.hedweb.com/population-ethics/quantum-ethics.html
[17] https://en.wikipedia.org/wiki/Cosmic_inflation
[18] https://en.wikipedia.org/wiki/Chaotic_inflation
[19] https://en.wikipedia.org/wiki/M-theory
[20] https://en.wikipedia.org/wiki/Boltzman_brain

meaningful notion. Infinities that crop up in the equations of theoretical physics have always hitherto turned out to be vicious; and yield meaningless results."

And this idea of erasing suffering from the whole multiverse has such a Buddhist flavor that one might ask if Transhumanists like David Pearce have anything to do with Buddhism. They do, actually, at least partly. At this point, let me introduce the work of another prominent Transhumanist, who is also a former Buddhist monk: James Hughes.

8.6   If you Meet the Buddha on the Road, Engineer Him

James Hughes is Executive Director of the very, *very* Transhumanist Institute for Ethics and Emerging Technologies, as well as a bioethicist and sociologist; he has published a book called *Citizen Cyborg* and is working on a second book devoted to Buddhism and Transhumanism, tentatively titled *Cyborg Buddha*. And he is not the only one in the Transhumanist movement sympathetic toward Buddhist philosophy: for example, another member of the IEET is Michael LaTorra, an American Zen priest who runs the Trans-Spirit list, which promotes discussion of neurotheology, neuroethics, techno-spirituality and altered states of consciousness.

In "Technologies of Self-Perfection,"[21] Hughes asks an interesting question: What would the Buddha do with nanotechnology and psychopharmaceuticals? The idea promoted is quite simple, that is, to promote and enhance virtues and traits of personality cherished by Buddhism using regular and future Transhumanist technologies, instead of – or together with – years and years of meditation and Buddhist practice.

In his writings, Hughes builds his approach on different sets or descriptions of virtues and character traits. For example, he uses the work of the renowned psychologist Martin Seligman,[22] who developed a model of six character strengths, each one provided with a subsidiary series of virtues:

> *Wisdom and Knowledge*: creativity, curiosity, judgment, love of learning, perspective
> *Courage*: bravery, perseverance, honesty, zest
> *Humanity*: love, kindness, social intelligence
> *Justice*: teamwork, fairness, leadership
> *Temperance*: forgiveness, humility, prudence, self-regulation

---

[21] J. Hughes, *Technologies of Self-Perfection*, https://ieet.org/index.php/IEET2/more/hughes20040922

[22] J. Hughes, *Enhancing Virtues: Building the Virtues Control Panel*, https://ieet.org/index.php/IEET2/more/hughes20140728

Transcendence: appreciation of beauty and excellence, gratitude, hope, humor, spirituality

Hughes also mentions the Pali canon's *Buddhavamsa*, which offers a different list of virtues – *paramitas*, or "perfections" – of the Buddhas[23]:

Dana – generosity
Sila – proper conduct
Nekkhamma – renunciation
Prajna – transcendental wisdom, insight
Virya – energy, diligence, vigor, effort
Kshanti – patience, tolerance, forbearance, acceptance, endurance
Sacca – truthfulness, honesty
Adhitthana – determination, resolution
Metta – loving kindness
Upekkha – equanimity, serenity

For his book *Cyborg Buddha*, he focuses on the manipulation of four basic clusters of capacities, resulted from the overlapping of psychological science, classic Greek virtues and Buddhist virtues:

Self-control: *sophrosyne*, restraint, conscientiousness, temperance, *sila*
Caring: *humanitas*, agreeableness, compassion, fairness, empathy, *metta*, *karuna*, *mudita*
Intelligence: *phronesis*, *sophia*, open-mindedness, curiosity, love of learning, prudence, *prajna*
Positivity: *eudaemonia*, (lack of) neuroticism, emotional self-regulation, positivity, bravery, humor, *sukha*
Mindfulness, the effective exercise of attention in executive functions and decision-making
Social Intelligence, the application of intelligence to other minds and relationships
Fairness, the application of intelligence to the most effective means for helping others
Transcendence, the discovery of happiness and fulfillment beyond ordinary pleasures

But what about the Transhumanist part of this Transhumanism/Buddhism crossover? Well, Hughes' basic proposal is to use present and future technologies to enhance all of these virtues and put them under our conscious control, no matter which list we want to endorse. So, for example, the sociologist defends the use of

---

[23] J. Hughes, *Using Neurotechnologies to Develop Virtues – A Buddhist Approach to Cognitive Enhancement (Part 1)*, https://ieet.php/IEET2/more/hughes20121016

pharmaceutical means, first, drugs like Modafinil – which can safely enhance focus and vigilance –, and, then, futuristic drugs that will be developed in the near future for exactly this purpose, as soon as we manage to get rid of all of the taboos against a chemical means to moral perfection.

Hughes stresses that, according to a multitude of research, many – or all – of the traits he mentions might have a genetic component, which we could tweak using gene therapy or another futuristic form of genetic manipulation.

And let's not forget about all of the "brain machines," technologies already available: the list includes Neurofeedback, that is, devices like Muse, which allow users to influence their own brain waves; Transcranial magnetic stimulation (TMS), which involves temporarily depolarizing neurons in the brain with electro-magnetic coils; Transcranial direct current stimulation (tDCS), based on the passage of low levels of current directly through the scalp; Deep Brain Stimulation (DBS), Brain Machine Interfaces (BCI) and Memory Prosthetics – basically, an artificial hippocampus –, all brain implants that have already been developed and that can be used to influence our behavior and our cognitive skills.

But, of course, the real game changer will be nanotechnology, in the Transhumanist sense, that is, nanoscopic machines able to swim through the body, reach the brain and influence its inner workings in ways that will make all contemporary technologies look crude in comparison. On a more Transhumanist note, Hughes thinks – and you should agree, if you are a Buddhist – that neurotechnologies and nanotechnologies will show, ultimately, that the Self does not really exist, it is just an illusion, although a very persistent one. Which invites me to ask whether longing for physical immortality, as Transhumanists usually do, does not clash with this Buddhist doctrine and the related idea of non-attachment. Anyway, Hughes' idea is, among other things, to build a kind of "virtue control panel," a device or piece of software at first, and a brain implant afterwards, that works as an artificial super-ego able to gently nudge our behavior in the right direction, allowing us to strictly adhere to a moral code that we have freely chosen; a sort of inner AI guardian angel that can increase and enhance our control over our appetites and unwelcome desires. So, we might end up installing devices in our brains that constantly monitor and control our behavior, an "augmented cognition" under the control of our rationality and our second order desires.

Touching upon the theme of super-intelligence and super-wisdom, however, inevitably ends up dealing with issues much broader than that of pure and simple intellectual enhancement, thus ending up on a path that leads us straight into the heart of Technological Singularity.

9

All Hail the Technological Singularity

9.1 In Thirty Years

The hyper-technological avalanche that will swamp us, transfiguring us, already has a date of arrival, 2045, and also a name: the Singularity.

This term – which is now part of the daily jargon of a large part of Transhumanism – means an alleged process of accelerating scientific and techno-logical progress that, according to its supporters, would affect human society as a whole, and, in particular – for obvious reasons – that of the West. The speed with which new technologies are developed will increase, and this process will culmi-nate – within a few decades – in the birth of an artificial intelligence *superior* to that of humans. After this event, human and post-human events are destined to take a turn that is, for us, incomprehensible; in other words, we ordinary human beings are not structurally able to foresee, let alone understand, what will happen after the advent of the Technological Singularity.

What are we likely to face in the very near future, before this quasi-mystical "Technological Singularity"? In his manifesto, *Technology vs. Humanity. The coming clash between man and machine*,[1] the Swiss futurist Gerd Leonhard has drafted a list that illustrates very eloquently the huge changes that we are going to live through, which he calls "Megashifts": exponential, combinatorial and recur-sive – that is, self-amplifying – technological changes. Transformations that will exert a joint-action on us, and will *flood* us, bringing about the first steps into the modification of our nature. From an existential viewpoint, we can say that these Megashifts will somehow unhinge the ontological structure of human existence,

---

[1] G. Leonhard, *Technology vs. Humanity. The coming clash between man and machine*, Fast Future Publishing, London 2016.

maybe amplifying its limits, pushing them farther and farther, maybe just changing them in unexpected – positive or negative – ways. We will adjust, I should add, "healing" from this future shock, and becoming something else.

These Megashifts will envelop us gradually, and then suddenly, stresses Leonhard; he also wants to warn us against the risk that we might lose our humanity, becoming less people and more machines.

But things don't have to go that way; the post-human doesn't have to destroy or diminish the human, but can instead generate a higher synthesis, leading us – or our descendants – to a higher level of existence, one that is only partially predictable. Anyway, here are Leonhard's Ten Megashifts:

1. *Digitization.* Everything that can be converted into digital, will be. And we really do mean everything.
2. *Mobilization and mediazation.* Computing will become invisible and completely ingrained into our everyday items and lives, making every tool, piece of technology, item, book, movie or song mobile, enveloped in a layer of augmented reality, safely preserved in the cloud. Our vitals will be constantly monitored at a distance. We will record and store *all* of our daily experiences in our external e-memory.[2] And, of course, forget about the need to learn a new language – not a good innovation, in our opinion, as studying languages is so good for our perspective on the world –, thanks to the coming automatic translators.
3. *Screenification and interface revolutions.* No more reading boring instructions, everything having been dumbed down and made as increasingly visual as possible and decreasingly abstract as possible, courtesy of touchscreens, augmented glasses and lenses, virtual reality and ubiquitous holograms.
4. *Disintermediation.* Business as usual will no longer mean interacting with human intermediaries, a trend that we are already seeing with e-banking, online shopping, renting places for vacations, getting car rides, self-publishing, and so on.
5. *Transformation.* Living in the near future will mean going from being physically separated from our computers and devices to being physically connected to them 24/7. Technology will move from the outside to the inside, first of our bodies, and then of our brains.
6. *Intelligization.* Any item may be made intelligent, thanks to the increasingly sophisticated AI available to us.
7. *Automation.* So many processes have the potential to be made automatic (substituting humans with machines). The first implications that come to mind: technological unemployment – of course –, and abdication – that is, less phys-

---

[2] G. Bell, J. Gemmell, *Total Recall: How the E-Memory Revolution Will Change Everything*, Dutton, Boston 2009.

ical face-to-face contacts among human beings because of more and more sophisticated communication technologies.

8. *Virtualization.* The rule of thumb will be: if you can create a virtual, non-physical version of something – your workstation, for instance – you will. Plus, you will be able simply to buy the digital schematics of the items you need, and 3D-print them out. The Meatspace – aka, real life – will be deeply interlaced with the Virtual Space.

9. *Anticipation.* Computers will be able to anticipate our decisions in every matter; algorithms will predict crime; monitoring our behavior, our intelligent digital assistants will become our "external brains," digital copies of us that will know us even better than we know ourselves; the Internet of Things – spread everywhere, in our homes, on the streets, in every corner of our environment – will be globally connected and able to gather information and track everything, taking care of most of our decisions and, in the end, increasing human de-skilling. Will our "second neocortex" – our external electronic brains, more or less merged with us – diminish us, or will we form a new, previously unseen, emergent system of man-plus-machine? Our money is on the second hypothesis.

10. *Robotization.* Robots of every kind will be everywhere, according to Leonhard, and we would like to add that they will be more and more human-like.

So, there it is, your everyday life of the near future. But all of this would be *before* the Technological Singularity. After that – if it happens at all –, it will be a whole new game.

For Kurzweil – today, the best-known supporter of this idea – the Singularity represents a moment at which the pace of technological change will be so rapid, its impact so strong, that technology will seem to expand at an infinite speed; the world will see new progress and radical changes from moment to moment.

Already, in 1951, Alan Turing was talking about the possibility that, one day, machines might overcome human intellectual abilities.[3] The first trace of the advent of a radical breaking point dates back to the mid-'50s; in particular, the Polish-American mathematician Stanislaw Ulam tells of a conversation he had with John von Neumann, in which he spoke to him about

> …the accelerating progress of technology and changes in the mode of human life, which gives the appearance of approaching some essential singularity in the history of the race beyond which human affairs, as we know them, could not continue.[4]

---

[3] A. M. Turing, *Intelligent Machinery, A Heretical Theory*, 1951, reprinted in «Philosophia Mathematica», 1996, Vol. 4, n. 3, pp. 256–260. http://philmat.oxfordjournals.org/content/4/3/256.full.pdf

[4] S. Ulam, *Tribute to John von Neumann*, "Bulletin of the American Mathematical Society", Vol. 64, n. 3, part 2, May 1958, pp. 1–49.

Von Neumann does not seem to refer specifically to the birth of intelligent machines, but to technological progress in general, as we have always known him to do.

In 1965, the British mathematician Irving John Good spoke for the first time of an *intelligence explosion*; essentially, according to him, if a machine were to be able to overcome the human intellect by even a little bit, it could improve its design in ways unforeseeable by its creators, and this would give rise to a recursive, that is, self-sustaining, process, thus producing even more powerful artificial minds. The latter would then have even better self-design capabilities, and this would lead to even faster artificial intelligence, leading to the birth of an intelligence far beyond ours:

> Let an ultra- intelligent machine be defined as a machine that can far surpass all the intellectual activities of any man, however clever. Since the design of machines is one of these intellectual activities, an ultra- intelligent machine could design even better machines; there would then unquestionably be an "intelligence explosion," and the intelligence of man would be left far behind. Thus, the first ultra-intelligent machine is the last invention that man need ever make, provided that the machine is docile enough to tell us how to keep it under control.[5]

In 1983, it was the turn of Vernor Vinge, former mathematics professor at San Diego State University, *computer scientist* and well-known science fiction author, considered by Transhumanists to be a true icon of their movement. Vinge took up the idea of Good and disseminated it, starting with an article published in January 1983 in the journal *Omni*, in which he uses, for the first time, the term "Singularity" in relation to artificial intelligence:

> We will soon create intelligences greater than our own. When this happens, human history will have reached a kind of singularity, an intellectual transition as impenetrable as the knotted space-time at the center of a black hole, and the world will pass far beyond our understanding. This singularity, I believe, already haunts a number of science-fiction writers. It makes realistic extrapolation to an interstellar future impossible. To write a story set more than a century hence, one needs a nuclear war in between ... so that the world remains intelligible.[6]

Ten years later, during the symposium "VISION-21," sponsored by the NASA Lewis Research Center and the Ohio Aerospace Institute and held on March 30 and 31, 1993, in Westlake, Vinge presented an article destined to become a

---

[5] I. J. Good, *Speculations Concerning the First Ultraintelligent Machine*, in: FL Alt and M. Rubinoff (ed.), *Advances in Computers*, Academic Press, Waltham 1965, Vol. 6, pp. 31–88.

[6] V. Vinge, *First Word*, «Omni», January 1983, p. 10.

milestone in the Transhumanist movement: *The Coming Technological Singularity: How to Survive in the Post-Human Era.*[7] The incipit is decidedly fulminating: "Within thirty years, we will have the technological means to create superhuman intelligence. Shortly after, the human era will be ended." *Will be ended*, which sounds very disturbing. Is this progress avoidable? asks the author. And, if it is not, can events be guided so as to guarantee survival for our species?

According to Vinge, the change made by the Singularity is comparable – in terms of radicalness – to the appearance of humans on Earth. This ontological revolution could take place in at least four ways, namely: the creation of computers that are equipped with self-awareness and an intelligence superior to ours; the "awakening" of a large network of computers in symbiosis with its users; the creation of a "hybrid" human-computer intellectually superior to us; the neurological development of humanity through biotechnologies. Just to throw some dates into the fray, Vinge says he would be surprised if the Singularity occurred before 2005 and after 2030.

And what would be the consequences of this Singularity? Progress would become much faster, as a result of the creation of machines that are even smarter than their predecessors, and in an even shorter period of time. Technological events that we think can only happen in a million years could happen within the Twenty-first century– in this regard, Vinge quotes the science fiction writer Greg Bear, who, in the 1985 novel *Blood Music*,[8] develops a scenario in which radical changes of an evolutionary type take place within a few hours. At the arrival of the Singularity, our old thought patterns will have to be discarded and a new reality will emerge; despite our ability to hypothesize about this now, the results of this ontological turnaround will certainly be surprising, even for those who systematically deal with these issues. Some symptoms will indicate the advent of this revolution; among them, we have the appearance of machines capable of replacing man in increasingly sophisticated works and the ever-faster circulation of ideas, including the most radical ones.

To the question "Can the Singularity be avoided?," Vinge responds with a "depends." On what? On the fact that we do not know whether or not artificial intelligence is achievable, which is not a settled issue at all. For example, the well-known philosopher John Searle, in *Minds, Brains, and Programs*, denies that this is possible[9]; more generally, the thinker maintains that consciousness depends on the substratum, and would therefore be a consequence of biology as we know it. The same applies to Roger Penrose, physicist and promoter of a conception of the

---

[7] http://www-rohan.sdsu.edu/faculty/vinge/misc/singularity.html

[8] G. Bear, *Blood Music*, Arbor House, Westminster 1985.

[9] J. Searle, *Minds, Brains, and Programs*, "The Behavioral and Brain Sciences," Vol. 3, Cambridge University Press, Cambridge 1980.

mind linked to quantum mechanics.[10] Obviously, these are only opinions, but, as we have already seen above, the debate on the nature of the mind and its technical reproducibility is far from concluded. So: if creating an artificial intelligence is not possible, we will never see the Singularity; otherwise, that is, if we *can* create artificial minds superior to ours, then, sooner or later, the Singularity *must* arrive.

Given the risks associated with this event – think no further than *The Terminator*, in which the Skynet global computer network awakens and destroys humanity – one might also wonder if it may not turn out to be the case that the nations of the Earth prohibit research in this area. But this is a pious illusion: the only result that such restrictions would achieve would be to push the search into hiding, or to allow nations that do not ratify such measures to beat others over time. Even the idea of making such artificial minds in a "quarantine" situation – that is, putting them in a system that prevents them from escaping or interfering with the outside world – seems problematic. For a mind superior to ours, finding a way out of a human trap, however cunning, would only be a matter of time. So, confinement would be – at least according to Vinge – decidedly ineffective. Another possibility would be to include some rules in the programming of these ultra-intelligent machines – similar to the Asimovian "Three Laws of Robotics" – that make them intrinsically benevolent. Here, too, there are some problems, and, in fact, we will tell you right away that the debate on the subject is still ongoing.

A further possibility advanced by Vinge – which, among other things, is our favorite – would be to amplify natural human intelligence, through genetic engineering and man-machine symbiosis. Or, rather, the idea would be that of a multiple approach, which considers both the development of artificial intelligence and that of intelligence amplification.

In this regard, Stephen Hawking commented – during an interview given to the German magazine *Focus* – that, in a few decades, the intelligence of computers will surpass that of humans, also suggesting that we develop, as soon as possible, systems to connect brains and computers, so that artificial brains contribute to human intelligence instead of opposing it.[11] Hawking is not the only one to have pointed out the possible dangers related to the birth of a super-human artificial intelligence. Dangers of various kinds exist: it could actually evolve in unpredictable ways, even malicious ones. Or it could compete with humanity for material and energy resources, and win. Alternatively, it could become indifferent to us – and therefore unintentionally dangerous – or it could take care of us too much, depriving us of our autonomy.

In April 2000, Bill Joy – an American computer scientist and co-founder of Sun Microsystems – published a controversial article in *Wired* entitled *Why the Future*

---

[10] R. Penrose, *The Emperor's New Mind*, Oxford University Press, Oxford 1989.

[11] http://www.zdnet.com/news/stephen-hawking-humans-will-fall-behind-ai/116616

Doesn't Need Us, in which he stated that the most powerful technologies of the Twenty-first century, that is, robotics, genetic engineering and nanotechnology, are likely to lead our species to extinction. The dangers of these technologies lie, above all, in the fact that they could be easily used with malicious intent by individuals or small groups. But it doesn't end there: continuing down the road to the development of increasingly intelligent machines, in the space of 30 years, we could have intelligent robots that are like us or even better than us, which could easily do without human beings, causing our disappearance; according to the scholar, biology teaches us that living species hardly ever survive encounters with competing species that prove superior. For Joy, human beings tend to overestimate their ability to design, and this could only lead to involuntary consequences with a lethal outcome – for the designers themselves. Later, after the article was published, Joy challenged scientists to refuse to work on technologies that could harm our species.

Finally, let's mention the theory proposed by the American microbiologist Joan Slonczewski, that is, the so-called "mitochondrial singularity"; according to the scientist, the Singularity would be only the culmination of a gradual process, begun centuries ago, consisting in the fact that human beings have delegated a growing number of intellectual processes to machines, to the point that, in the future, we could assume a "vestigial" type of role towards the technology – as simple "energy suppliers," a bit like what happened to the mitochondria inside the cells.[12]

Speaking of dates, Kurzweil predicts that the Singularity will arrive around 2045, while, according to the British Transhumanist Stuart Armstrong, there is an 80% probability that it will happen within a century.

As for Kurzweil, the inventor enters the debate in question with *The Age of Spiritual Machines*,[13] a book in which he speaks of artificial intelligence and technological advances that we will witness during the Twenty-first century. According to to him, artificial intelligence will be created by the exponential growth of computers' computational capacity, plus the development of an automated system of knowledge acquisition and some types of algorithm, such as genetic algorithms – in simple terms, complex self-corrective calculation procedures. The machines will seem, at some point, endowed with free will, and intelligence will expand beyond the Earth, throughout the universe.

One of the concepts that he is most insistent upon is his "law of accelerated returns." From the Big Bang, says Kurzweil, the frequency of universal events – in the cosmological sense – has slowed down; on the other hand, biological evolution has achieved increasing degrees of order at an ever-increasing speed, up until the

---

[12] http://ultraphyte.com/2013/03/25/mitochondrial-singularity/
[13] R. Kurzweil, *The Age of Spiritual Machines*, Penguin Books, New York 1999.

birth of technology. Not only that: every time a certain technology approaches an insurmountable wall, new technologies are born that overcome it, and this would be the most intimate essence of human history. In short, thanks to a system of positive feedback, biological evolution accelerates, leading us to technological evolution, which, in turn, accelerates, arriving at the computation, which then reaches the Singularity. The law of accelerated returns therefore indicates that, as the order grows, the speed – of development, of evolution, of technological change – increases.

For Kurzweil, the mind is not a set of atoms, but a grouping of patterns, which can manifest itself in different media at different times – and can therefore also be reproduced on a substratum other than the biological one. According to him, spiritual experience consists of a sensation of transcending daily physical and mortal limits, perceiving a deeper reality; it follows that, in the Twenty-first century, machines will also develop a spiritual dimension – and will therefore no longer be definable as machines in the classical sense of the word.

Kurzweil then makes extensive use of the so-called "Moore's Law." Gordon Moore is an American entrepreneur and co-founder of Intel – which, we will recall, produces chips – whose reputation is linked to the law that bears his name – disclosed for the first time in an article from April 19, 1965, in *Electronics Magazine*.[14] On that occasion, Moore declared that, given a fixed price, the number of active elements in a computer was destined to double every 2 years, adding that this rule would remain valid for a decade – while it was still valid. From this, it follows that the power – processing speed, memory, and so on – of computers and digital devices in general would continue to double about every 2 years, thus following an exponential trend. It must be said that the success of this rule also depended on the fact that it was used as a guideline – in practice, as a business target – by the entire semiconductor industry. In the 1988 book *Mind Children: The Future of Robot and Human Intelligence*,[15] Hans Moravec took Moore's Law and generalized it, bringing it to support the possibility that, in the near future, robots would evolve into a new artificial species – from the 1930s to the 40s of this century. Ten years later, in *Robot: Mere Machine to Transcendent Mind*,[16] Moravec further generalized his theories, arriving at positions similar to those of Vinge and associates. And so, we come to Kurzweil, according to whom Moore's Law will "hold up" until 2020, after which we will switch to technologies other than the

---

[14] G. Moore, *Cramming more components on integrated circuits*, « Electronics Magazine», April 19, 1965, p. 4. http://www.cs.utexas.edu/~fussell/courses/cs352h/papers/moore.pdf

[15] H. Moravec, *Mind Children: The Future of Robot and Human Intelligence*, Harvard University Press, Cambridge 1988.

[16] H. Moravec, *Robot: Mere Machine to Transcendent Mind*, Oxford University Press, Oxford 1999.

current ones, such as DNA-based chips, those based on nanotubes and, finally, quantum computation.[17]

Kurzweil brings up the example of Deep Blue, the IBM computer that beat Kasparov in 1997, to show that artificial intelligence is being born. In commenting on Kurzweil's book,[18] John Searle replies that machines can only manipulate symbols, without understanding their meaning. It is the famous example of the Chinese room, which takes place more or less like this. Let's put a man who does not speak Chinese in a room, and let him receive messages written in that language; then, we give him a complete and articulated set of rules, which explains how to manipulate the symbols in question. The man will then produce the appropriate answers, and this without having really understood the meaning of the messages. The person who writes the messages, and who speaks Chinese, might believe that those inside the room really understand that language, but, in reality, they don't. For the philosopher, computation is the manipulation of symbols according to the rules, without *understanding* the meaning. The exponential increase in the computational capacity promised by Kurzweil does not move the problem one bit, and the only way to build a conscious machine is to understand consciousness, which we are very far from doing.

The philosopher Colin McGinn also has something to say on the theories of Kurzweil; in particular, he underlines that the fact that a machine exhibits an external human behavior does not indicate that it has an inner experience similar to ours, and if it does not have it, uploading a mind onto a PC would be to dissolve it.[19]

Continuing with his reasoning, Kurzweil goes towards a vision of very wide breath, claiming that, once impregnated with intelligence, the universe will be able to decide its own destiny, perhaps avoiding the Big Crunch or perpetual dilution: it will be up to the intelligence to decide. For him, intelligence is therefore the greatest force in the universe; as the "computational density" of the universe increases, intelligence will begin to rival the "great celestial forces."

In *The Singularity Is Near: When Humans Transcend Biology*,[20] Kurzweil increases the stakes. The inventor emphasizes that Moore's Law applies not only

---

[17] Quantum computation is a field of research that aims to create computers able to perform calculations using quantum phenomena. The sub-atomic world is characterized by rules very far from our common sensibility; far from being simple "dots" with defined characteristics, particles have properties that are bizarre to us, like that of being in multiple places at the same time. We are simplifying, of course; what matters is that these strange properties could be exploited to build computers that are enormously faster than the current ones.

[18] J. Searle, *I Married a Computer*, "The New York Review of Books," April 8, 1999. http://www.nybooks.com/articles/archives/1999/apr/08/i-married-a-computer/? pagination = false

[19] C. McGinn, Colin, *Hello, HAL*, "The New York Times," January 3, 1999. http://www.nytimes.com/1999/01/03/books/hello-hal.html?pagewanted=all&src=pm

[20] R. Kurzweil, *The Singularity Is Near: When Humans Transcend Biology*, Penguin, New York, 2005.

to integrated circuits, but also to other contiguous sectors – such as transistors, relays and electro-mechanical computers – and even to more distant sectors, such as the material sciences – read: nanotechnologies – and medical technologies. Moreover, although Moore's Law, as we know it now, is destined, according to him, to run out around 2020, Kurzweil has confidence in the emergence of a new paradigm – perhaps based on nanotubes – that will allow for the further continuation of the exponential growth of computational capacity.

Be that as it may, Kurzweil has no problem admitting that the growth of computational capacity cannot by itself create artificial intelligence; the best way to achieve this will be through the retro-engineering of the human brain – in practice, you will have to study and imitate our central nervous system. To do this, we will use the classic brain imaging technologies, which are also destined to undergo an exponential growth of their own resolution – to then be replaced, in the '20s of the Twenty-first century, by nano-bots that scan the brain from inside.

According to Kurzweil – and this is also the central theme of his latest book, *How to Create a Mind: The Secret of Human Thought Revealed*[21] – the essence of an adult human being consists of about 300 million "pattern detectors"; for the futurist, it would have a hierarchical structure that allows for a growing abstraction from one vertical cortical column to another. At the lower levels, the neocortex may seem mechanical, since it makes simple decisions, but, at the highest levels of the hierarchy, it is able to handle concepts such as poetry, sense of humor, sensuality, and so on. The quantitative increase of these hierarchical levels is what has allowed the transition from the intellect of primates to that of human beings, thus leading to the birth of language, the arts and the whole culture. In practice, a quantitative increase would have led to a qualitative leap. So, Kurzweil asks, why not pursue another "jump," bringing the number of our "pattern detectors" from 300 million to 1 billion?

In any case, it will be genetic engineering, nanotechnology and robotics, or, rather, the convergence of these three disciplines, that will push us into the Singularity. Human life will be transformed irreversibly; we will transcend the limits of our bodies and our biological brains. Kurzweil then stresses that the machines of the future will be human – to the point that, let us say, it is right to ask whether it makes sense to talk about machines, or whether it would rather be better to avoid this type of terminology, in favor of something new, something to be invented.

And so, we come to the fateful date, 2045. At some point, after this date, the exponential growth above will reach an insurmountable limit, i.e., the process of miniaturization and increased speed that characterizes the computer world will stop, having reached the limits imposed by the laws of nature; then, in order to

---

[21] R. Kurzweil, *How to Create a Mind: The Secret of Human Thought Revealed*, Viking Press, New York 2012.

further increase its computational capacity, computers will have to increase their size, and this will cause the artificial intelligences to move "outwards." To begin with, the Earth itself will be transformed into an enormous computational substratum, after which the post-human intelligences will begin to expand into solar and interstellar space, in search of matter and energy to be optimized for their own developmental needs. This will lead to a progressive conversion of all cosmic matter into "thinking" matter, up to a real "awakening" of the whole universe. The ride of Singularity will continue with the acquisition of an increasing degree of complexity – moving closer to our classical monotheistic conception of God, but never reaching it. In the event that other universes exist, it is certainly possible that our "intelligent universe" decides to expand further towards them.

Kurzweil builds a scheme of universal history that, at a glance, is very reminiscent of the emergentist conceptions defended by the "complexity theorists," such as Ilya Prigogine – for whom the different levels of complexity from which reality is composed would emerge from one another, through special ontological "leaps." The Kurzweilian scheme identifies six epochs in the evolutionary history of the universe, all read in the light of information theory. In practice, the First Epoch – that of physics and chemistry – would be characterized by the fact that information, understood as "order," would be preserved only in atomic structures; therefore, a limited quantity of order. Then, there would be the Second Epoch – that of biology – in which information would be stored in the structure of DNA. The Third Age is that of brains, in which neural patterns manage to conserve an even greater amount of order. The Fourth Era is that of technology, in which we witness the birth of hardware and software. The Fifth Age, that of the Singularity, will witness the fusion between human and machine, while the Sixth Epoch will see the saturation of the universe by intelligent matter.

Kurzweil believes that advanced extraterrestrial civilizations do not actually exist. According to the law of accelerated returns, when a civilization develops a primitive form of technology, within a few centuries, it reaches the Singularity; the latter then leads the intelligence to rapidly expand into the cosmos, and to fiddle with it. Since, looking at the sky, we do not notice signs of the intelligent manipulation of matter – square galaxies have not been discovered, for the moment – from this, we can deduce that ETs do not exist. Which is better for us, since our destiny will be to saturate the universe, absorbing all matter and energy and optimizing them for computation.

An interesting program, but what do critics and official science say? Among the criticisms that have rained down on Kurzweil's book, the most common is that of the "fallacy of exponential growth"; in this case, the accusation consists in having taken a process that has, at the moment, proved to be exponential and generalized it, extending it to technological development *tout court*. In short, the inventor would have transformed a simple localized tendency into a true law that would govern human reality in a stringent way. There are also several authors who, in various ways, have challenged the idea that progress would accelerate; we will mention a couple of them. According to Bob Seidensticker – popularizer and insider from the world of high-tech industry – the idea of acceleration is the result of a simplistic reading of technological progress – for further details, we refer you to his interesting book, *Future Hype. The Myths of Technological Change.*[22] Instead, according to Theodore Modis – physicist and futurologist – and Jonathan Huebner – physicist and patent expert – the rate of technological innovation is not only not growing, it is slowing down. In short, these are questions of the history of science and technology, and those who deal with these issues know how difficult it is, if not impossible, to develop reliable and well-argued interpretations of events that are still ongoing. In other words, the discussion is open.

It must be said that the scientific community – or at least a good part of it – has generally accepted Kurzweil's theories with a certain degree of skepticism. In 2008, the Canadian psychologist Steven Pinker stated that

> There is not the slightest reason to believe in a coming singularity. The fact that you can visualize a future in your imagination is not evidence that it is likely or even possible. Look at domed cities, jet-pack commuting, underwater cities, mile-high buildings, and nuclear-powered automobiles--all staples of futuristic fantasies when I was a child that have never arrived. Sheer processing power is not a pixie dust that magically solves all your problems.[23]

Neither is the well-known American cognitive psychologist Douglas R. Hofstadter very gentle toward the Singularity. For him, it is a confusing idea, while the books of Kurzweil and Moravec are "a very bizarre mixture of ideas that are solid with ideas that are crazy." The mixture is so homogeneous that it is very difficult to separate good ideas from nonsense; these are intelligent people, not stupid ones, Hofstadter emphasizes. The foolishness, for him, concerns the upload of the mind onto computers, digital immortality, the fusion of personalities in cyberspace and the notion of an acceleration of progress. The scholar admits, however, that he does not have an easy way to distinguish that which is correct from that which is wrong, and it is certainly possible that some of the things that these people say might come true, even if we don't get to know "when."[24]

For David J. Linden, a neuroscientist at Johns Hopkins University, Kurzweil confuses data collection with understanding the brain; the first can also follow an exponential progress – that is, the tools for data collection will be able to be refined

---

[22] B. Seidensticker, *FutureHype. The Myths of Technological Change*, Berrett-Koehler Publishers, San Francisco 2006.

[23] http://spectrum.ieee.org/computing/hardware/tech-luminaries-address-singularity

[24] G. Ross, *An interview with Douglas R. Hofstadter*, "American Scientist Online", January 2007. http://www.americanscientist.org/bookshelf/pub/douglas-r-hofstadter

even faster – but the understanding of the functioning of the nervous system progresses more or less linearly.[25]

Writing in *Nature*, physicist Paul Davies said that *The Singularity is Near* is fun to read, but should be taken with caution.[26]

For the scientific journalist John Horgan, the view of the Singularity is more of a religious one than a scientific one,[27] while the science fiction writer Ken MacLeod has coined the words – vaguely derogatory – of a "rapture for nerds," referring to the "rapture" expected shortly by many Christian Fundamentalists.

Finally, we add that not everyone is convinced of Kurzweil's prophetic-technological skills – which are often stressed in the media. For example, in 2010, the American journalist John Rennie had fun reviving all of the forecasts that the Transhumanist thinker has made in recent years and putting them up against a reality check, showing how, according to him, the most accurate of Kurzweil's forecasts are also often the most vague and banal ones.[28]

But let us pretend, for a moment, that everything that Kurzweil and associates say is true without a shadow of a doubt. We would thus be dealing – after 2045 – with a world that transcends our capacity for understanding. Despite this, and while sharing these ideas, several Transhumanists have tried to elaborate scenarios related to the world to come, starting with the human types that will live in it.

9.2 The People of Tomorrow

Super-Healthy

With almost absolute control of the human body at the molecular level, it is obvious that our post-human descendants are destined to enjoy a health compared to which ours – even in the best cases – pales. After a while, the last "wild animals" left in our ecosystem will be viruses and bacteria and, as nanotechnologies progress, they too will be "tamed," But it does not end there: by enhancing our natural senses, and giving us new ones, nanomedicine will offer us unprecedented multi-level systemic access to our internal physical and mental states, including the activity and conditions of organs, tissues and cells – even of single neurons, if

---

[25] D. Linden, *The Singularity is Far: A Neuroscientist's View*, «Boing Boing», July 14, 2011. http://boingboing.net/2011/07/14/far.html

[26] P. Davies, *When computers take over*, «Nature», March 23, 2006, n. 437, pp. 421–422. http://www.singularity.com/When_computers_take_over.pdf

[27] J. Horgan, *The Consciousness Conundrum*, "IEEE Spectrum", June 1, 2008. http://spectrum.ieee.org/biomedical/imaging/the-consciousness-conundrum/0

[28] Anyone wishing to judge on their own can take a look at: J. Rennie, *Ray Kurzweil's Slippery Futurism*, 'IEEE Spectrum ', November 29, 2010. http://spectrum.ieee.org/computing/software/ray-kurzweils-slippery-futurism

desired. Several parts of ourselves, to which we have no access today, will end up within the radius of action of our awareness. We will be able to analyze in detail all of the "sub-systems" that make up our mind, our "conflicting desires" and the "contrasting opinions" we have on a certain theme. Reconstructing and "treating" our minds will mean opening a window on the ultrafine structure of our thoughts, thus becoming "naked to ourselves" in a way that we cannot even imagine – in the face of the efforts that today's psychotherapists perform to gain access to the inner lives of their patients. In short, we will end up enjoying a form of free will that is unthinkable today. Nanomedicine will become omni-pervasive, eliminating a series of conditions that, today, we consider "natural," but that could push the post-humans to wonder how we people of the present managed to stay focused on our daily affairs; this includes dependencies of all kinds – caffeine, nicotine, drugs, addictions to food and all of the compulsive behaviors that reduce the quality of our lives, from superstitions to hypochondria, from dependence on work to shopping. And then, all of the allergies, intolerances, nausea, diarrhea, gastrointestinal problems, the annoyances caused by heat and cold, itching, pimples. But also the imperfections of the skin, the superfluous hairs, the dizziness caused by rapid rotation, the asymmetries of the body and face, headaches, ingrown toenails, tinnitus, ear wax, closed nose, flatulence, drowsiness and, obviously, menstruation. Not to mention phobias of all kinds, from the fear of spiders to those of height or closed places, as well as all of our behavioral traits that we do not like – it is obvious that, if you are misanthropes and you like it that way, nobody will force you to change. To conclude: post-humans will not even have an occasional knee or back pain.[29]

Post-human Bodies

In the previous chapters, we have already spoken extensively about all of the possible enhancements and changes to which our biological bodies could be subject in a more or less distant future. Now let's add some more curious stuff. Let's start with Natasha Vita-More, who, as a good artist, has also created a representation of a possible post-human body, which she baptized "Primo Posthuman."[30] There are various adaptations proposed: we go from the possibility of changing sex at will – with its ability to procreate – to a meta-brain equipped with enhanced cognitive self-correction systems and a nanotechnological memory file, to a very efficient "system of waste recycling" – what we have now is "messy," Vita-More underlines – with enhanced senses and the nanotech skin of which we have already spoken.

[29] R. A. Freitas Jr., *Nanomedicine. Volume I: Basic Capabilities*, Landes Bioscience, Austin 1999. www.nanomedicine.com

[30] N. Vita-More, *First Posthuman – The New Human Genre*. http://www.natasha.cc/primo.htm

In an article entitled *Human Engineering and Climate Change*,[31] the Transhumanist thinkers S. Matthew Liao, Anders Sandberg and Rebecca Roache argue that human beings must genetically modify themselves voluntarily so as to reduce the ecological impact of our species; among their proposals is that of inducing in ourselves an intolerance towards meat (whose production does, in fact, have a remarkable ecological impact), acquiring the ocular characteristics of cats (in order to reduce our lighting needs), and reducing our body mass.

The possible changes that can be applied to the human body are practically endless: we can imagine acquiring the physical abilities of this or that animal, creating collective minds, adapting our bodies to outer space, excessively increasing our size, giving our skin the strangest colors, changing the number of our chromosomes, abolishing sleep, making our skeletons flexible or our fingers tentacular, developing selective hearing – to hear only what we wish – or making wings grow; in the excellent novel *Accelerando*, the science fiction writer Charles Stross has fun imagining a hybrid human-corporation.[32] In short: ask and it will be given to you. This is what Transhumanists call "morphological freedom," the freedom to choose whatever form they want, and which, for these thinkers, must have the status of any other human right.[33] Of course, it is a thing we can easily talk about, but with some caution; let's say you find yourself – just to give an example – with a colleague at your office who decided to grow a third, disturbing eye in the middle of their forehead, a thing that could make your working day a bit more animated and, perhaps, not very comfortable.

The Second Uncanny Valley

In psychology and robotics, the "Uncanny Valley"[34] is a term used to indicate the discomfort that we feel when we deal with something that looks very human, but, at the same time, is "not quite there." So, for example, some people find very realistic dolls or human-like robots "disturbing."

---

[31] S. M. Liao, A. Sandberg and R. Roache, *Human Engineering and Climate Change*, 2 February 2012, http://www.smatthewliao.com/wp-content/uploads/2012/02/HEandClimateChange.htm

[32] C. Stross, *Accelerando*, Ace Books, New York 2005.

[33] A. Sanberg, *Morphological Freedom – Why we do not just want it, but need it.* http://www.aleph.se/Nada/Texts/MorphologicalFreedom.htm

[34] https://en.wikipedia.org/wiki/Uncanny_valley

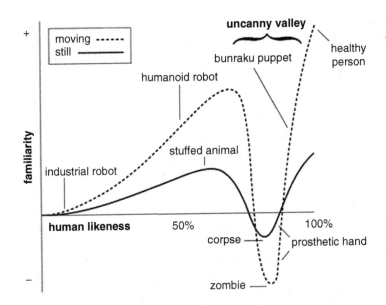

The reason for this is not known, but there are a few hypotheses around – for example, the idea that things that elicit this kind of sensation are tied to our natural avoidance of pathologies and individuals suffering from contagious diseases.

So, what if – asks Jamais Cascio[35] – there is a Second Uncanny Valley waiting for us after the present one, an Uncanny Valley related to our post-human descendants?

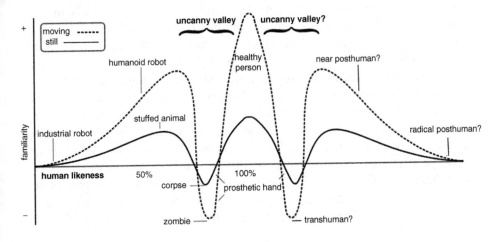

---

[35] J. Cascio, The *Second Uncanny Valley*, http://www.openthefuture.com/2007/10/the_second_uncanny_valley.html

What are the possible implications for the above-mentioned morphological freedom? Would we find slightly different post-humans more unsettling than completely different, more radical post-humans? Oh, well; the only thing we can do to solve this conundrum is to wait and see.

No Sexes, Please, We're Post-human

George Dvorsky and James Hughes take note – from their point of view, at least – of the progressive cultural erosion of sexual genders, as well as of the philosophical elaborations of feminism and its parallel movements. As elaborated in an article,[36] they therefore foresee the concrete disappearance of the sexual genders – seen by them as an arbitrary limitation of human potentials – both from an anatomical and a neurological point of view, through the usual technologies. Nanotechnologies, genetic engineering, in vitro reproduction, artificial wombs, sex change techniques, virtual bodies simulated by computers: the next venture will bring us, according to the two, to the transcendence of not only our humanity, but also our sexual gender. In return, we will get greater psychological fluidity and androgyny that we can alter at will; in essence, postgenderism does not demand the cancellation of sexual and gender differences, but rather acknowledgment of the fact that they are the result of choice and not of genetic and cultural imposition.

In Praise of Neuro-Diversity

Again, Dvorsky[37] introduces us to the interesting topic of neuro-diversity, arguing that, in the future, we will have tailor-made cognitive processing methods; in practice, we will be able to alter, in a specific and highly individualized way, the way in which we perceive the world, from an exquisitely neurological point of view. That is: we will be able to modify our emotional responses, our modalities of personal interaction, our social involvement, our aesthetic tastes and our priorities. In short, the day will come when we will be able to decide the exact way in which we want to relate to the world. The Transhumanist thinker here introduces two terms coined and originally spread widely by the Autism Rights Movement, an international social movement that includes associations and networks that deal with autism, i.e., those of "neuro-typicality" and "neuro-diversity." And autism is a good example of this last concept; the idea behind it is that, far from being interested in social interactions – like most "neuro-typical" human beings, "autistic people are or would be more than happy to devote themselves to their own thoughts

---

[36] G. Dvorsky and J. Hughes, *Postgenderism: Beyond the Gender Binary*, IEET-03, March 2008. http://www.sentientdevelopments.com/2008/03/postgenderism-beyond-gender-binary.html

[37] G. Dvorsky, *Designer Psychologies: Moving beyond neurotypicality*, «Sentient Developments», May 29, 2011. http://ieet.org/index.php/IEET/more/dvorsky20110528

and specific priorities, without any need to be "cared for." Here, Dvorsky is certainly not saying that we should all become autistic, but rather invites us to consider the concept of neuro-diversity, and to see it as a possible enrichment of society; in essence, it is an invitation – so Transhumanist – to develop neuroscience in order to gain the possibility of modifying our personal psychologies at will. The whole is therefore inscribed in a notion dear to the Transhumanists, i.e., that of cognitive freedom, which implies our right to change our minds as we see fit. Among the areas of our psyche we might work on is our aesthetic sense – which we could modify in order to aesthetically appreciate things we would not normally think of. We could also decide to mix different sensory channels, producing a relatively uncommon neurological phenomenon in us, that of synaesthesia. Another rather obvious field of intervention is the manipulation of emotional responses; on the other hand, we could decide to identify and voluntarily eliminate our cognitive biases, obtaining a clearer and more rational capacity for thought. The tools to be used are always the same: tailor-made drugs, genetic manipulations, neural implants, and so on.

About Human Nature

Here is the usual question: what will become of human nature after all of these changes and restructurings? We can handle it in various ways; it can be said that human nature will disappear, and that it is better this way; or that it is in our nature to continually overcome ourselves, so becoming post-human is the most human thing we can do; or even that human nature does not exist, that it is only the product of culture, for which Transhumanism does not risk anything at all. On this theme, an interesting point of view is adopted by Larry Arnhart, a scholar at Northern Illinois University and creator of "Darwinian conservatism," an approach that entrenches the conservative mentality in evolutionary thinking. Do not worry, Arnhart says: human nature is here to stay. Our bodies, our brains and our desires have been shaped by evolution in order to resist genetic manipulation. In particular, our desires are the result of thousands of years of evolution and adaptation to the environment, forming a balanced system that allows us to live, and which is difficult to alter. According to the philosopher, we have about twenty natural desires, such as the desire to possess a sexual identity, to have wellbeing and health, to practice the arts, to gain intellectual understanding – universal desires, although subject to some cultural variation. If ever, one day, it was possible to develop empowering technologies, it would be unlikely to be used *against* these basic desires; it is more probable that humans would use them to better satisfy the desires they already have. In short, nobody – except possibly a madman – would choose to destroy her/his body to take shelter in a virtual world.[38]

---

[38] L. Arnhart, *Human Nature is Here to Stay*, The New Atlantis, n. 2, summer 2003, pp. 65–78. http://www.thenewatlantis.com/publications/human-nature-is-here-to-stay

Enhanced Animals

As our enhancing technologies develop, we will approach the possibility of "intellectually raising" the most "promising" species of our planet, such as chimpanzees and dolphins – but also elephants, whales, and so on. Maybe you find unsettling the idea of going out one day and finding yourself in the middle of a demonstration of enhanced cows protesting against "centuries of exploitation," and the like. The fact remains, at least in theory, that if we accept the possibility of human enhancement, then we must admit that it is also possible for animals. We are talking about the so-called *uplift*, a concept that comes from science fiction – and, in particular, from a homonymous cycle of novels by the American science fiction writer David Brin – one that was promptly adopted by the Transhumanists, especially by George Dvorsky.[39] All of this should be accomplished through the usual familiar nanotechnology, biotechnology, robotics – to provide mechanical limbs for animals that, like dolphins and whales, lack the ability to manipulate – and artificial intelligence. If you are rubbing your hands while thinking about the prospect of enhanced sub-human servants doing all of the jobs "that young people do not want to do anymore," then you are going too fast: Dvorsky is keen to point out that the animals chosen as candidates for uplift must be upgraded to our level and that we must guarantee all of the "human" rights that we have to them. Strange as it may seem, there is also someone – in the scientific community *tout court* – that is attempting, even if only for exploratory purposes, such an experiment. In particular, in 2005, Sue Savage-Rumbaugh and her colleagues from Iowa's Great Ape Trust in Des Moines placed eight bonobo chimpanzees in a facility equipped with different tools that allowed the animals to get food, to decide who entered and who exited, and so on. The monkeys also had access to musical instruments, TV, and much more. The ultimate goal was to verify whether members of this species are able to learn and communicate notions that are more complex than those that they usually do, thus demonstrating that these skills are not a prerogative of *Homo sapiens*. A very rough and primitive form of uplift, in short.

Disembodied Minds

Then, there are the artificial post-human intelligences, which can be of different degrees; J. Storrs Hall has proceeded to classify them.[40] The scholar divides them into: Hypohuman AI, not exactly as intelligent as us, and subject to us; Diahuman AI, comparable to human beings and, like us, able to learn; Parahuman AI, friendly

---

[39] G. Dvorsky, *All Together Now: Developmental and ethical considerations for biologically uplifting nonhuman animals*, « Journal of Evolution and Technology», Vol. 18, n. 1, May 2008, pp. 129–142 http://jetpress.org/v18/dvorsky.htm

[40] J. Storrs Hall, *Kinds of Minds*. http://lifeboat.com/ex/kinds.of.minds

intelligences developed to be a part of us, with which we can enter into symbiosis or even merge; Allohuman AI, minds intellectually at our level, but endowed with modes of perception and motivations that do not overlap with ours, in practice, almost alien; Epihuman AI, artificial intelligences superior to us, but not by that much, and that therefore maintain a relationship of continuity with us. And, finally, we have the Hyperhuman AI, an artificial intelligence capable of overcoming the entire human scientific community in any task, able to understand scientific knowledge in its entirety, as a unitary whole, and whose intellectual productivity can rival that of all of humanity.

Enter the Infomorph

And post-human artificial intelligences don't even represent the last word in terms of evolution; after them, there could be an even more alien kind of entity, the "infomorph." The Russian Transhumanist Alexander Chislenko, in a 1996 article, *Networking in the Mind Age*, theorized it.[41] Technically, it would be a purely "informational" creature, a "distributed info-being"; essentially, it is a "virtual body of information" that can have emergent traits, such as a real personality. It is a "software agent" with distributed intelligence and an independent existence; it represents the logical outlet of our growing dependence on technologies and systems composed of elements characterized by functional proximity instead of physical proximity; one example is the same internet network that we currently use, and, in particular, the prospect of doing future work on our terminal by temporarily assembling data and functions from locations – that is, servers and databases – physically distant from each other. This "ontological precariousness" will deepen the process – already in progress – of the liberation of functional structures from the material substratum, and will lead to the birth of "advanced information entities"; informorphs, in fact, entities without body and with a capacity to handle information almost perfectly. To simplify a decidedly difficult subject: we are used to thinking of ourselves as ontologically "solid," unitary entities with well-defined boundaries, and we tend to apply the same reasoning to the supposed artificial intelligences we will create. Infomorphs will instead be made up of volatile and changing sets of information, software, databases, and servers, which can change at any moment, and, despite this – that is, despite their "I" being fluid and absolutely temporary – they will be capable of intelligent behavior. All of this is actually the result of a process that started a long time ago: interacting with nature, human beings have begun to "expel" an increasing number of structural elements from their organism. And so, if, initially – like all other animals – humans have accumulated energy in their bodies in the form of fat, through cultural progress,

---

[41] A. Chislenko, *Networking in the Mind Age*, 1996. http://www.lucifer.com/~sasha/mindage.html

they began to store energy outside of themselves, creating fire, accumulating wood, procuring sources of energy of all kinds, and so on. In short, energy has become something "exosomatic" – that is, external to the body – distributed and shared. Chislenko jokes about it, saying that, nowadays, people tend to attribute greater value to their bank deposits than to their own fat deposits. And this discourse applies to many other aspects of human life, which have become slowly exosomatic; for the thinker, even medicine can be read in this way, that is, as an exosomatic version of our immune system. In practice, human beings tend to think in an "automorphic" way, which is to identify functional unity with object/material unity, but this will change and, just like energy resources, so could the psychological/mental/cognitive aspects of our species become exosomatic, thus leading to infomorphs. These are entities that could also have some advantages over us: for example, they will not have to go to school, since, if they need to learn something, they will have to do nothing but copy it from their fellows.

Other Minds

In 1984, the philosopher and AI scholar Aaron Sloman promoted, in "The Structure of the Space of Possible Minds,"[42] the idea that there might be more than one kind of mind and tried to describe the possible structures hosted by this conceptual space. In 2015, the Latvian-born computer scientist Roman V. Yampolskiy reprised and extended Sloman's work.[43] Yampolskiy's definition of mind is "an instantiated intelligence with a knowledge base about its environment." So, let's examine what the main traits of the mind(s) are. First of all, infinitude of possible minds: "If we accept that knowledge of a single unique fact distinguishes one mind from another, we can prove that the space of minds is infinite. Suppose we have a mind M, and it has a favorite number N. A new mind could be created by copying M and replacing its favorite number with a new favorite number N + 1. This process could be repeated infinitely, giving us an infinite set of unique minds. (…) Alternatively, instead of relying on an infinitude of knowledge bases to prove the infinitude of minds, we can rely on the infinitude of designs or embodiments. The infinitude of designs can be proven via inclusion of a time delay after every computational step. First, the mind would have a delay of 1 nanosecond, then a delay of 2 nanoseconds, and so on to infinity. This would result in an infinite set of different mind designs. Some will be very slow, others superfast, even if the underlying problem-solving abilities are comparable." From a theoretical viewpoint, the topic is rather

[42] http://www.cs.bham.ac.uk/research/projects/cogaff/sloman-space-of-minds-84.pdf
[43] R. V. Yampolskiy, *The Space of Mind Designs and the Human Mental Model*, http://hplus-magazine.com/2015/09/02/the-space-of-mind-designs-and-the-human-mental-model/. The article is an excerpt from Yampolskiy's book *Artificial Superintelligence: A Futuristic Approach*, Chapman and Hall/CRC, Boca Raton 2015.

complex, so, for the details, I would refer you to the original article; let's say that, according to Yampolskiy, minds can be classified by size, complexity and properties. About the design: "[S]ome minds will be rated as 'elegant' (i.e., having a compressed representation much shorter than the original string); others will be 'efficient,' representing the most efficient representation of that particular mind." There are a few philosophically non-trivial questions that we can ask: "Could two minds be added together? In other words, is it possible to combine two uploads or two artificially intelligent programs into a single, unified mind design? Could this process be reversed? Could a single mind be separated into multiple non-identical entities, each in itself a mind? In addition, could one mind design be changed into another via a gradual process without destroying it? For example, could a computer virus (or even a real virus loaded with the DNA of another person) be a sufficient cause to alter a mind into a predictable type of other mind? Could specific properties be introduced into a mind given this virus-based approach? For example, could friendliness be added post-factum to an existing mind design?"

Studying possible minds is not an easy task: right now – we have more or less 7 billion human minds available, and they are pretty homogeneous, both from the viewpoint of the hardware – the human body/brain – and software – psychological design and knowledge:

"[T]he small differences between human minds are trivial in the context of the full infinite spectrum of possible mind designs. Human minds represent only a small constant-size subset of the great mind landscape. The same could be said about the sets of other earthly minds, such as dog minds, bug minds, male minds, or in general the set of all animal minds. (…) Given my definition of mind, we can classify minds with respect to their design, knowledge base, or embodiment. First, the designs could be classified with respect to their origins: copied from an existing mind like an upload, evolved via artificial or natural evolution, or explicitly designed with a set of particular desirable properties. (…) Last, a possibility remains that some minds are physically or informationally recursively nested within other minds. With respect to the physical nesting, we can consider a type of mind suggested by Kelly (2007b),[44] who talks about 'a very slow invisible mind over large physical distances.' It is possible that the physical universe as a whole or a significant part of it comprises such a megamind." If this last case is true, the other minds should be seen as "nested" inside of this megamind. Another typology of mind architecture consists of self-improving minds, minds able to change and adjust their own architecture; we might even argue that such minds, in doing so, change themselves in other minds, becoming ontologically different – but this is basically the issue of the preservation of personal identity, which is far from having been solved: "Taken to the extreme, this idea implies that a simple act of

---

[44] K. Kelly, *A Taxonomy of Minds*, http://kk.org/thetechnium/archives/2007/02/a_taxonomy_of_m.php

learning new information transforms you into a different mind, raising millennia-old questions about the nature of personal identity."

With respect to their knowledge bases, minds could be separated into those without an initial knowledge base, which are expected to acquire their knowledge from the environment; minds that are given a large set of universal knowledge from their inception; and those minds given specialized knowledge only in one or more domains.

Other classifications are possible: by goal – we can imagine minds that have no-long term or terminal goals; by the absence of goals – minds that change regularly or set goals randomly; by free will – if there is such a thing – and consciousness – whatever this word means. So, will post-humans differ from us by so much that they have not just one, but several different types of mind?

9.3 End of Life: Never

Let us now resume the theme of immortality, which, in our opinion, is the most controversial issue of Transhumanism. The objections raised against the desire to live forever are numerous; here, we will examine only a few. Remember, however, that, for every objection to scientific immortality, it is possible to develop a counter-objection: this is the job of the Transhumanists, and, for a complete overview of these debates, we refer you to the many sites that we mentioned in the chapter on immortality. Generally, the first objection that is raised is the one related to overpopulation: if nobody dies anymore, where will we put everyone? Here, the possible answers go from a moratorium on reproduction – not really a good idea, in our opinion, but we can discuss it – to expansion into space – thanks to the usual nano-machines. Then, there is the "gerontocratic question," that is to say, that a society made of immortals would be destined to stagnate – in reality, here, we forget that, from the perspective of Transhumanism, we do not want to rejuvenate only the body, but also the mind. Moreover, immortality would certainly not prevent the birth of new generations, which could just be born and, if desired, seek their fortunes around the cosmos. Another problem: odious dictators and political leaders would become immortal – but, for that matter, so would the great geniuses of humanity, who could continue to create and discover new things. In more general terms, we can say that, for Transhumanists, human creativity is very fruitful, our species has a great capacity for adaptation and, for every problem, we can find a solution; moreover, if our lives were longer, we would adapt to them, organizing things differently, perhaps making larger plans, and so on.[45]

---

[45] For an analysis of objections to scientific immortality and counter-objections, we recommend: B. Bova, *Immortality. How Science is Extending Your Lifespan and Changing the World*, Avon Books, New York, 1998.

But now, we have to tackle the thorniest issue, namely, that of our relationship with endless time. Just imagine, for a moment, what it means to live *forever*, that is, eternally: this implies that, after billions and billions of years, even after the stars have gone out, the galaxies are dead and the universe has diluted into the void, you would have just begun to touch the time without end that there remains to be lived. As Woody Allen said, "Eternity is a very long time, especially towards the end." The problem – raised by all of the adversaries of Immortalism and Transhumanism – is that of boredom, into which all of the immortals would sooner or later be plunged. On this theme, the journalist and science writer Ed Regis, who is not a Transhumanist, makes some interesting reflections, humorous but not too much so, and asks us to consider the following points. First: everyday life is sometimes boring; and so what? Second: Eternity could be as boring or as exciting as we make it. Third: being dead is more exciting? Four: if you find eternal life boring, you can end it at any time.[46]

Then, there is the problem of memory: after thousands, millions or billions of years, what will we do with the loads of memories that we have accumulated in the meantime? Wouldn't we be at risk of being crushed by them or, contrastingly, of not being able to keep them in our limited cerebral space, transforming us, for all intents and purposes, into an interminable series of Serial Selves, forgetful of each other? In reality – the Transhumanists would answer – it would not be so difficult to imagine a future technology capable of accumulating and adequately managing the memories accumulated over longer and longer periods.

Other answers to the dilemmas mentioned above are also possible: one could imagine an endless evolution – this would be interesting, moving continuously from a state of lesser complexity to a state of greater complexity – or a radical change in our relationship with time – think about the speculations by Bernal that we dealt with in the first chapter – and so on.

There is a more profound question, however, regarding immortality, addressed – although, in our opinion, not fully grasped by critics – by the American philosopher Bernard Williams in his famous article *The Makropulos Case: Reflections on the Tedium of Immortality*.[47] The article starts, as is known, from the famous play by the Czech playwright Karel Capek.[48] It tells the story of a woman, Elina Makropulos, to whom – at the age of 42– is offered an elixir of immortality that "blocks" her at that age, stopping the aging process. Now, the much-desired

---

[46] E. Regis, *Great Mambo Chicken and the Transhuman Condition*, Penguin Books, New York 1990, p. 97.

[47] B. Williams, *The Makropulos Case: Reflections on the Tedium of Immortality*, in: JM Fisher (ed.), *The Metaphysics of Death*, Stanford University Press, Stanford 1993, pp. 71–92.

[48] If you are passionate about science fiction, you will certainly know that it was Capek – in the work *RUR* – who first used the Czech word *robot*, which literally means "forced laborer," in the sense that we all know of.

eternal life turns out to be a curse for poor Elina, who – after 300 years – ends up sinking into a state of coldness, indifference and apathy unimaginable for us ordinary mortals. According to Williams, immortality is just like that, and, for him, such a condition would not be minimally desirable. Let's try to take a closer look at the philosopher's reasoning. An eternal life would be – as the name says – *eternal*, and so it would last forever or, at the very least, for a very, very long time. Now, all of us human beings have a character, a personality, which includes a whole series of regular behaviors, habits, and physical and character traits that make us what we are, that is, individuals; we also have goals, projects, and values. If we succeed in obtaining immortality, we – our character, our personality – should measure ourselves according to that fact; we would accept eternal life only if we could dedicate it to what we want, to pursue our projects, to embody our values – perhaps taking our time, doing things with the necessary calm. Well, if we remain substantially equal to ourselves, even with the necessary adjustments and wisdom produced by age, the situations in which we would find ourselves immersed, and the reactions – relatively predictable – that we would have towards them, would always remain the same. Everything would become enormously repetitive. In essence, we would live the same situations over and over again. For example, if we devote ourselves to a certain sport or anything else, we would end up always witnessing the same dynamics and we could not help but become cold and detached. At this point, you might think: people actually change, they develop new interests, new values, and they are, in the end, entities in progress. And this is the crux of the problem: how many changes will an immortal have experienced after billions and billions of years? It will become everything and the opposite of everything, he will have pursued his own objectives and exactly the opposite ones, she could even undergo radical changes in her identity – first becoming a man, then a woman, then something else, in a progressive psychological fluidization that would deprive it of every essential trait, psychological and existential that is its own. It would no longer be a real person, says Williams, but a simple natural phenomenon. What can Transhumanists say in answer to what, in effect, seems an incontrovertible argument? The only possible answer that we found, we did not find in Transhumanist literature, but in a science fiction novel by Greg Egan, an author very close to Transhumanist themes. In *Diaspora*, Egan describes a post-human society very similar to that desired by Transhumanism, and whose members are facing a problem similar to that posed by Williams. The problem is solved through some special programs that these post-humans insert into their personality, software that allows users to combine flexibility with stability, obtaining, substantially, a second-order control over their own mental processes, so as not to risk "Dissolv[ing] into an entropic confusion."[49] However, this is a sort of stop-gap solution, and the problem raised by Williams remains, in our opinion, still open.

---

[49] G. Egan, *Diaspora*, Orion, London 1997.

On the theme of immortality and boredom, the Australian-Canadian philosopher Mark Walker then makes an interesting and decidedly Transhumanist proposal: instead of wasting time in pointless discussions on this issue, why don't we carry out an "experimental ethics" test, and try to concretely obtain life extension, so as to discover empirically whether immortality is boring or not?[50]

Finally, on the theme of immortality, the Transhumanist Charles Tandy makes a couple of rather interesting observations. First of all – intellectual empowerment aside – immortality would allow humans to work for long periods on the philosophical problems that have always obsessed our thought, and perhaps future post-human philosophers could even come up with a solution, or at least get closer to it than we ever have. Secondly, post-humans could finally manage to do something that has not been possible since the times of the Renaissance, i.e., accumulating al human knowledge within a single individual. In the tradition of the "Renaissance man," we would then find ourselves with the "neo-Renaissance super-man." Thirdly, for Tandy, immortality would overturn history – or, more precisely, our relationship with it. If, in fact, until now, the individual has always been in some way temporally and ontologically dependent on the civilization or the historical community in which he was inserted, an immortal would be in the exact opposite condition; she would last longer than the nation or society to which she belonged. In other words, the unique and undisputed protagonists of history would no longer be historical epochs, civilizations and nations, but individuals – obviously immortal.[51]

9.4   Welcome to Weirdtopia

How do you imagine the far future? Immortality, flying cars, invulnerability, plenty of leisure time, robots taking care of everything else? Maybe. Or maybe evolution will bring about something very different from ourselves; maybe technology will produce post-humans with whom we will not be able to relate at all, given the ontological abyss that will separate us from them. This is the thesis proposed by a British post-humanist philosopher, David Roden, in his interesting – and recommended – book *Posthuman Life: Philosophy at the Edge of the Human*.[52]

---

[50] M. Walker, *Boredom, Experimental Ethics, and Superlongevity*, in: C. Tandy (ed.), *Death and Anti-Death, Vol. 4: Twenty Years After De Beauvoir, Thirty Years After Heidegger*, Ria University Press, Palo Alto 2006, pp. 389–416.

[51] C. Tandy, *Extraterrestrial Liberty and The Great Transmutation*, in: ibid. (edited by), *Death and Anti-Death, Vol. 4: Twenty Years After De Beauvoir, Thirty Years After Heidegger*, Ria University Press, Palo Alto 2006, pp. 351–368.

[52] D. Roden, *Posthuman Life: Philosophy at the Edge of the Human*, Routledge, Abingdon-on-Thames 2014.

The central thesis of the book is that our musings about Humans 2.0 are mistaken; we tend to imagine our post-human descendants simply as better, improved versions of ourselves, smarter, more beautiful, and longer-lived. But, in doing so, we forget about what Roden calls "technological disconnection," the fact that enhancement comes with a separation between the old and the new. For instance, being intellectually enhanced would push you onto a new plateau, from which things would look completely different from before, in a way that would be totally confounding to a non-enhanced human being. As Roden says, a post-human future will not necessarily be informed by our kind of subjectivity or morality. What Roden says is, of course, open to debate, and endorsing it is, in my modest opinion, an act of faith, as much as endorsing the opposite opinion is also an act of faith. The core of the issue is – no surprise – the nature of evolution. If, as Conrad Lorenz proposes, cognitive evolution is a process of integration of lower levels into higher levels – integration, not substitution or disconnection – then, and only then, is there a chance for us, the humans, to at least try to understand or relate to the post-humans.[53] Or, in other words, to see something human in even the weirdest post-human we might meet in the far future. If Lorenz is wrong, then Roden is right, and the post-humans will be as close to us as a Lovecraftian entity. No matter who is right, the future is likely to be very different from our present, so different that it will look "weird."

9.5   After Language

Ben Goertzel likes to speculate somewhat about language and what it will become in the future, when the mind is heavily transformed by the new transhumanist technologies, when our normal way of communicating will be transfigured and telepathy becomes a common phenomenon, as we have seen in the case of techlepathy.

Language, as we know, is what most distinguishes us from animals. To tell the truth, plenty of other species have a more or less sophisticated form of language, even if, in technical terms, rather than a language, it is a "call system," a series of specific signals that mainly concern the here and now, and they do not possess – as we know – the ability to generalize and refer to other times and places. Not only that, but they do not even have the ability to refer to themselves, to talk about talking, and so on.

It is certainly possible, says Goertzel, that dolphins and whales possess a language as sophisticated as ours, but structurally different, perhaps not composed of

---

[53] K. Lorenz, *Behind the Mirror. A Search for a Natural History of Human Knowledge*, Harvest/HBJ, New York 1973.

discrete entities – words – but of "waves of meaning," more apt to express emotions than concepts. The truth is that we do not know much about it for now. In any case, our language, composed of discrete parts, seems to reflect our cognitive propensity to break reality into more parts, to decompose it and reassemble it in our minds. Indeed, there seems to be a parallel between our ability to construct complex sentences and the manual ability that allows us to assemble tools of various kinds. When our ancestors began to use language – who knows when, along the evolutionary scale – they could hardly have foreseen all of the consequences that it would bring, from literature to spam, from gossip to philosophical reflections on logic.

So, the question that Goertzel rightly poses is: what will come after language?[54] Certainly, let us add, something will come, just as language came after the call system of our pre-human ancestors. And his proposal is as simple as it is radical: in the future, the wall that separates the mind from language will collapse; linguistic expression will simply be a mind with a specific configuration and a specific perspective on reality. At this point, we feel the need to underline an essential fact: the mind – this is not Goerzel saying this, we are saying it – should be considered distinct from the Self. It would basically be a kind of equipment that our own self possesses. And so, the Self has memory, thought, imagination, and so on. Let's go back to Goertzel: how did this idea come to him? While working on a specific AI system, OpenCog.[55] This is an open source software that he is working on, together with other people, with the aim of creating a future form of AGI (Artificial General Intelligence). The idea would be to use OpenCog to control video game characters, and robots as well. And what do two Open-Cog systems do when they have to communicate? They exchange reciprocal pieces – pieces of mind, says Goertzel. Now, this is not really an easy thing to do between human beings. The idea of exchanging thoughts, emotions, and mental images directly cannot be done with the same amount of ease. Everyone has their own equipment, and a certain degree of mutual adaptation – that is, translation – is necessary. If it is our destiny to exchange the contents of our minds directly, some form of "shared language" is necessary. In practice, a "standard mind" is needed that serves as common ground for the minds of all communicants. This sort of "mental language" – or, rather, shared mental structure, as the term 'language' would be too simplistic – has been baptized, by Goertzel, "Psy-nese". In practice, a standardized thought structure, suitable for transmitting "subjective pieces of mind." In other words, to communicate, we will synthesize – in ways that are still to be imagined – copies of the parts of our mind that we want to share, and transmit them, using a kind of common mental structure, much more universal than the human languages used now. It is

---

[54] B. Goertzel, *What Will Come After Language?* http://hplusmagazine.com/2012/12/27/what-will-come-after-language/

[55] https://opencog.org/

difficult to imagine such a situation: we can only think of minds that touch momentarily and that exchange parts of their content.

9.6   Colonizing the Universe in Eight Easy Steps

We have become immortal, practically invulnerable, we control our psyche, we have built a technological paradise. So, now what do we do? It's rather obvious: we move on to the conquest of the universe. What else?

And, of course, even here, the Transhumanists have a nice plan ready, which resumes the project of colonization of the galaxy developed at the end of the '80s by the American futurist Marshall T. Savage and illustrated in a super-detailed way in a specific book called, like the above-mentioned plan, *The Millennial Project*.[56] Savage's proposal led to the birth of – guess what? – a special organization, the First Millennial Foundation, which attracted thousands of people from around the world. Savage appeared on several television programs dedicated to the future and, after having achieved some media success, decided to retire to private life and abandon the foundation – for reasons never clarified. In 2006, the futurist and American architect Eric Hunting – member of the aforementioned organization, meanwhile renamed the Living Universe Foundation[57] – decided to update the Savage project, combining it with the ideas of Transhumanism and launching *The Millennial Project 2.0*.[58] Here are the various phases of the project.

Foundation

The objective of this first step is to create an organized global community with a common financial structure for the development of projects related to the Millennial Project; in particular, these ideas would be carefully promoted in the media and within the international community.

Aquarius

As the name implies, this step aims to build a global infrastructure for renewable energy, starting with an arcology – that is, a stable and self-sufficient

---

[56] M. T. Savage, *The Millennial Project: Colonizing the Galaxy in Eight Easy Steps*, Little Brown & Co, New York 1994. The book was originally published in 1992; the second edition was published in 1994, with an introduction by Arthur C. Clarke.

[57] http://www.luf.org/

[58] http://tmp2.wikia.com/wiki/Main_Page

community – in the Ocean, which will serve as a center of experimentation for technologies and non-terrestrial lifestyles that we will then export to space.

Bifrost

This phase involves the large-scale development of an energy supply system based on renewable energy sources, as well as the construction of a space elevator that starts from marine arcologies and leads passengers into appropriate habitats within the circumterrestrial space.

Asgard

At this stage, we are already taking flight, and, in particular, we are witnessing the development of space stations and orbital industrial structures.

Avalon

Finally, we land on other planets, putting special emphasis on attempts to build pressurized habitats on the Moon and Mars, as well as realizing underground structures along the lines of the arcologies already realized on Earth.

Elysium

Nothing to do with the Neill Blomkamp movie of the same name; this phase concerns, in particular, Mars, and aims at its terraformation, i.e., the use of nanotechnologies, biotechnologies and anything else to transform the environment, so as to make it – if possible – habitable for human beings.

Solaria

In this phase, we will move on to the realization of a Solar Civilization, consisting of post-humans who will reside on the planets of our system, as well as in numerous spatial habitats of various shapes and characteristics.

Galactica

This is – for now – the final phase of the project, which aims to develop an interstellar colonization program based on nanotech rockets driven by artificial intelligence and powered by anti-matter; launched in every direction, these machines will take charge of exploring the galaxy, selecting all suitable environments and preparing them for the arrival of the human or post-human beings who will inhabit them.

Hunting emphasizes the merits of merging the Millennial Project with Transhumanist thinking, and, in particular, the possibility – by adapting our bodies and using substrates other than the biological one – to live in environments completely unsuitable for normal human beings, including empty space, which, at this point, could become a place of residence for us like any other. The expansion into space and the physical adaptability of post-humans could eventually lead to a considerable degree of speciation, i.e., humanity could be subdivided into numerous species, artificial or otherwise, even ones very different from each other.

9.7 Great Maneuvers at the Border of the Cosmos

We have already clarified several times that the post-Singularity world would exceed our capacity for understanding; the reality that awaits us will be unimaginable, and, now, relying on the miserable cognitive tools that we have, we can only try to glimpse the surreal light that will illuminate the lives of our descendants of the distant future. Nonetheless, Transhumanists – some of whom, as you already know, are planning to stay in circulation for a very long time – have tried to come up with some real-life scenarios for their next material paradise. Before leaving for our grand trip, there are two or three things we must keep in mind. The first is a general consideration, namely, that if a given thing is not explicitly prohibited by the laws of physics, then it must be possible: it is just an engineering problem. The second consideration is much more trivial, and that is that, if you are immortal, time is certainly not lacking, indeed, you have plenty of it. Which implies that you have all of the time in the universe, which you can use to enjoy it, but also to do things that challenge the wildest imagination, provided that such enterprises do not violate any fundamental physical principle. The third consideration is that of sources: Transhumanists are certainly not the first to try to imagine what the distant future will be like. In fact, for more than a century, science fiction writers from all over the world have made a living doing just that, with mixed results; some forecasts seem very naive, others, instead, seem decidedly far-sighted.[59] And then there are the predictions of this or that renowned scientist, who, between one "serious" research and another, likes to speculate about the world to come. All of these speculations are obviously united by the limits mentioned above, namely, the fact that we cannot absolutely imagine what the future will be like, given that,

---

[59] We will not even try to draw up a list of works and authors of science fiction who have ventured into this enterprise. Let us just mention the British philosopher and writer Olaf Stapledon, author of *Last and First Men* and *Star Maker*. We do not even know whether his works are classifiable as "novels"; rather, they are real visions of tomorrow. Let's recall then another essential author, Arthur C. Clarke, whose *The City and the Stars* took us to a distant future – more than one billion years –, for an imaginative journey on Earth.

in all likelihood, it will be *much stranger* than that described by science fiction scientists and writers, who, nevertheless, have elaborated a shared imagination from which Transhumanists draw heavily; it follows that many of the ideas we will encounter are not, in fact, the prerogatives of Transhumanism – that is, they have not been elaborated by its followers – but have been incorporated into it without any effort.

Conquest of the galaxy aside, the Transhumanists have, in fact, inserted themselves quite well into the scientific community, which allows for a certain "osmosis of ideas": while Transhumanist ideas are gradually penetrating *mainstream* scientific culture, several particularly original ideas have come from the latter and have been welcomed with open arms by Transhumanism. This is the case of the famous "scale of Kardashev" – named after its creator, the Soviet astronomer Nikolai Kardashev.[60] This is a method for measuring the degree of technological progress of the alleged alien civilizations that we will perhaps encounter in the cosmos, a system based on the amount of energy that a given civilization is able to use. And so, a Type I civilization is able to use all of the energy resources available on its planet, a Type II civilization those of its own star system, and a Type III civilization those of its own galaxy. Later, various changes were proposed to this scale; for example, Carl Sagan proposed introducing Type 0, as well as intermediate degrees – so that the Earth would rank at grade 0.7. A Type II civilization would be able to build a Dyson sphere or a Dyson swarm – we'll soon get into what those are – as well as carrying out the so-called *star lift*, the controlled removal of sizeable portions of the material that makes up its own star, to be used for other purposes. A Type III civilization should be able to do the same, but in a much wider context – galactic or intergalactic.

Zoltan Galantai, a scholar at Budapest Polytechnic, proposed the introduction of Type IV to indicate the civilizations that control the energy resources of their own universe[61]; according to the Transhumanist Milan Ćirković, this category should refer instead to the civilizations that control the energy from their local super-cluster.[62] There are those who have come to propose Type V – civilizations that have multiple universes. This is, as you can imagine, simple speculation. Among additional proposals, we have that of the American aerospace engineer Robert Zubrin, based on the use of a different measurement system, i.e., the "mastery" rather than the energy – that is, not the use of energy resources, but the

---

[60] N. Kardashev, *Transmission of Information by Extraterrestrial Civilizations*, « Soviet Astronomy», n. 8, p. 217, 1964. http://adsabs.harvard.edu/full/1964SvA.....8..217K

[61] Z. Galantai, *Long Futures and Type IV Civilizations*, September 7, 2003. http://mono.eik. bme.hu/~galantai/longfuture/long_futures_article1.pdf

[62] M. Ćirković, *Forecast for the Next Eon: Applied Cosmology and the Long-Term Fate of Intelligent Beings*," Foundations of Physics", Vol. 34, n. 2, pp. 239–261, February 2004. http:// arxiv.org/ftp/astro-ph/papers/0211/0211414.pdf

degree of control of your surroundings.[63] Sagan then proposed using a scale based on the amount of information available. Its classification starts at level A – a civilization that has 10^6 different bits of information, that is, a quantity inferior to that of any known human civilization – and ends at level Z – a civilization with 10^{31} bits, a level that, according to him, no civilization in the cosmos has yet reached, as the universe would be too young for it.[64]

In *Parallel Worlds*, Michio Kaku argues that a Type IV civilization should be able to handle "extra-galactic" energies, such as dark energy[65]; for his part, John D. Barrow proposes an inverse scale, based on how "small" a certain civilization manages to work: the members of a Type I civilization would manage to manipulate objects of their own scale, for example, they could build buildings, dig mines, and so on; those of Type II could manipulate genes, replace organs, and so on; those of Type III would be able to manipulate molecules and molecular bonds, creating new materials; those of Type IV could instead manipulate single atoms, creating nanotechnologies and producing complex artificial life forms; the Type V civilizations would be able to manipulate the atomic nucleus and engineer the nucleons that compose it; those of Type VI could manipulate elementary particles, such as quarks and leptons, to create articulated structures of atomic and sub-atomic dimensions; finally, the Omega- type civilizations could manipulate the structure of space and time at the most fundamental levels. For Barrow, our civilization would straddle Types III and IV.[66]

It must be said that several authors – including Zubrin – have criticized Kardashev's scale, for the fact that it assumes that we can know or understand the choices and, in general, the behavior of civilizations much more advanced than ours, reducing it to a classification elaborated in our terms. What interests us is that, according to the Singularity theorists, our civilization is about to make the "great leap," and become a Type I civilization. This would give us ample room for maneuvering, and would allow us to dedicate ourselves to what would seem to be one of the favorite pastimes of post-humans: astronomical engineering.

This is a purely speculative discipline, which concerns the possibility – through unknown technical means – of manipulating stars and planets, restructuring entire star systems and attributing forms normally not present in nature to them. Various scholars have enjoyed imagining different types of stellar-scale mega-structure, to be used as habitats, for computation or as propulsion systems, e.g., the "Globus

[63] R. Zubrin, *Entering Space: Creating a Spacefaring Civilization*, Tarcher, New York 1999.

[64] C. Sagan, *Cosmic Connection: An Extraterrestrial Perspective*, Cambridge University Press, Cambridge 2000.

[65] M. Kaku, *Parallel Worlds: The Science of Alternative Universes and Our Future in the Cosmos*, Doubleday, New York 2005, p. 317.

[66] J. Barrow, *Impossibility: Limits of Science and the Science of Limits*, Oxford University Press, Oxford 1998, p. 133.

Cassus," an artistic project by the Swiss architect Christian Waldvogel that represents the restructuring of the Earth and its transformation into a hollow artificial world much larger than the original. It is an imagination game, but a very detailed one, in which the author outlines the procedures for the progressive dismantling of our planet and the construction of the new world, which, once completed, will reach more or less the size of Saturn. The project also envisages the construction of a new ecosystem and its repopulation with human beings.[67]

Shkadov Propulsion, proposed by the Russian scholar Leonid Mikhailovich Shkadov,[68] is a reflecting surface of enormous proportions, capable of accelerating the movement of a star through space by reflecting or absorbing the light from one side of it.

The Dyson sphere, theorized for the first time, as a simple mental experiment, by the British physicist and mathematician Freeman Dyson, is an artificial shell-shaped mega-structure – constructed using the material derived from the dismantling of the planets of a certain system stellar – that would be wrapped around a star. Dyson, who was openly inspired by Stapledon's *Star Maker*, considers it a logical choice by a technologically more advanced civilization, and therefore one more in need of energy than ours.[69] In fact, by rebuilding a star system in that way, one could capture *all* of the energy emitted by a star, and not just a very small part – like the classic planets. Later, science fiction writers reworked this idea in various ways. In reality, the idea of a real compact and closed sphere is the result of a literal interpretation of the original Dyson article; the scholar thought that such a shell would not have been stable, and he mostly considered the idea of a homogeneous distribution of matter around a given star, in practice, a swarm of objects – satellites, spatial habitats, and so on – uniformly and densely distributed but independent of each other. The clarification of the scholar promptly yielded a further definition, namely, "a Dyson swarm."

The "Alderson disk," invented by NASA researcher Dan Alderson, is a gigantic disk several thousand kilometers thick; the Sun is located in a hole in the center of the disk, whose outer radius would correspond more or less to the orbit of Mars or Jupiter. By equipping this structure with adequate life support systems, it could be made habitable on both sides, thus obtaining a very large amount of space, though it would be immersed in perpetual twilight.

But if you want a *specifically* Transhumanist idea of cosmic engineering, you must address the Matrioshka Brain. This is the brainchild of the American

---

[67] C. Waldvogel, B. Groys, C. Lichtenstein, M. Stauffer. *Globus Cassus*, Lars Müller Publishers, Baden 2005.

[68] L. M. Shkadov, *Possibility of controlling solar system motion in the galaxy*, Thirty-eighth Congress of the International Astronautical Federation, October 10–17, 1987, Brighton.

[69] F. J. Dyson, *Search for Artificial Stellar Sources of Infra-Red Radiation*, « Science», Vol. 131, n. 3414, pp. 1667–1668, 1969. http://www.islandone.org/LEOBiblio/SETI1.HTM

computer scientist and Transhumanist Robert Bradbury, author of an essay that has quickly become very popular: *Under Construction: Redesigning the Solar System.*[70] For Bradbury, our post-human descendants will one day roll up their sleeves and rebuild the universe; after having gotten rid of death, they will be able to engage in *very* ambitious projects. We will begin with the dismantling – via nano-machines – of the asteroids, which will be transformed into solar panels floating in space, suitable for handling the entire energy production of the Sun – achieving, among other things, our arrival at the status of a Type 2 civilization. We will dismantle the planets, transform them into computronium and place them as concentric spheres around the Sun – a bit like the homonymous Russian dolls, hence the name – thus succeeding in optimizing all of the energy of our star. *Et voila*, here is our Matrioshka Brain: a super-mega computer with the dimensions of the solar system, in which we can simulate all of the hyper-realistic virtual universes that we want to build, and in which our descendants can live as simulations.

If the post-humans do expand throughout the universe, gradually transforming the other star systems into as many Matrioshka Brains, the sky will end up darkening. The Matrioshka Brains will be new transcendent entities, with a life expectancy of billions of years; they will also be able to interact with each other, creating true interstellar – or intergalactic – civilizations of beings like them. It is hard to imagine what a Matrioshka Brain would do all day: for example, will they be able to play with the structure of space-time, testing its malleability? It is likely that the "thoughts" that will dwell in them will be far beyond ours.

Not that we like them too much, these Matrioshka Brains; they remind us too much of the "vast and cold intellects" that, according to H.G. Wells, populated the Mars of *War of the Worlds*. They look less-than-human, not more-than-human; but, on the other hand, imagining something that goes beyond the human is fundamentally impossible to us, and it may be that, if these super-brains of computronium ever exist, they will be accompanied by something much more bizarre, but also more human.

We human beings have what Ernst Cassirer called "categories of the Spirit," discrete pieces of our humanity, in short, like art, philosophy, science, and so on. Will these categories of the Spirit survive post-human evolution? Will they change? And, if so, how? Will they be complemented by brand new categories, currently unthinkable for us? Or will our own way of thinking in terms of categories of the Spirit evaporate, replaced by something else? That of the Transhumanists is often a reductionist and patchy futurology, by their own admission; it is just an intellectual game. Or, at least, we like to take it as such.

---

[70] R. Bradbury, *Under Construction: Redesigning the Solar System*, in: D. Broderick (ed), *Year Million. Science at the Far Edge of Knowledge*, Atlas & Co., New York 2008, pp. 144–167.

Let us then continue with our speculative game, and see what Transhumanists expect to find in the distant future. Even Steven B. Harris – American physician, geriatrician and cryonicist – has decided to play with computronium.[71] This is – we remind you – computational matter, which we can grow and develop at will, using it to build our bodies. Bodies that, thanks to the nature of the computronium, will no longer need a brain, because they will be all brain. Thanks to the malleability of this "universal material," which can "compute," that is, assume all kinds of configuration and material characteristics, becoming anything, the body of the distant future will be reshaped as we please. The only limit is our imaginations, which, it seems, will be strengthened together with all of our other faculties. The distinction between living matter and inanimate matter will disappear, and the flesh of tomorrow will be replaced by something much more resistant and reliable. Indeed, it is possible that the main centers of our thinking will not even be located in our mobile unit – that is, the body – but will be kept safe somewhere else, with systems available for regular back up as well, to avoid any possibility of complete destruction. However you put it, our post-human descendants will continue to need two things: matter and energy. And since they plan to stay in circulation for a long time, they are likely to want to optimize the use of the matter and energy that they already have available. This implies two things: the exploitation of the energy of the stars – through the removal and use of their outer layers through technologies that we obviously cannot even imagine – and the aforementioned dismantling of the planets, in order to obtain construction material. Indeed, eager to obtain everything that they need to be able to dedicate themselves in peace to what they will do better – that is, to think – our distant descendants could come to create a real "wall" of probes in every direction, constantly expanding in space, looking for new stars and new planets to "optimize" – read: dismantle and reconstruct according to schemes difficult for us to imagine. Our natural romanticism does not make us very appreciative of the idea of tearing stars and planets to pieces, and it is certainly possible that these feelings will also be cherished by the post-humans who will displace us; the latter could then decide to conserve – just for sentimental reasons – this or that planet. For example, Saturn, with its rings – although one should not bet too much on that, since, after all, the "prince" of the solar system is full of hyper-compressed building materials – or mother Earth, which could rise to the Fedorovian role of a museum of the origins of humanity.

For the American Transhumanist Amara D. Angelica, in the distant future, the Internet will evolve into something larger, which will extend throughout the galaxy and begin to surpass its boundaries, thus creating the Universenet.[72] Billions

---

[71] S. B. Harris, *A Million Years of Evolution*, in: D. Broderick (ed.), *Year Million. Science at the Far Edge of Knowledge*, Atlas & Co., New York 2008, pp. 42–84.

[72] A. D. Angelica, *Communicating with the Universe*, in: D. Broderick (ed.), *Year Million. Science at the Far Edge of Knowledge*, Atlas & Co., New York 2008, pp. 212–227.

of nanotechnological probes will spread throughout known space, thus connecting all of the distant descendants of the current IPI. Given the extremely dilated timing – interstellar e-mails will not go faster than light – the post-humans will be able to choose to use astronomical engineering, for example, by building *wormholes*, i.e., space-time tunnels able to circumvent the limits in question. The comparison with the internet is obviously rather metaphorical: what we are trying to say here is that, in the distant future, it is possible that the post-humans will decide to build an interstellar exchange system that can be visualized through a comparison of this kind.

9.8   Hacking the World

And here we are at the end of the chapter. We have done pretty much everything. We have become immortal, we have colonized the universe, and then, without paying for it, we have reconstructed it from top to bottom. Does there remain anything else to do? Of course. We still have to hack reality. This idea, much appreciated by the Transhumanists, was not invented by them; let us now become acquainted with "digital philosophy." First of all, let us ask ourselves: where did the world come from? What are time, space and thought? And what is "at the bottom" of reality? These are not, as you can well imagine, new problems, since they are as old as humanity is; new answers are, however, always possible, and the so-called digital philosophy is precisely one of these. This is an approach developed over several decades by a small group of mathematicians and information theorists, namely Konrad Zuse,[73] Edward Fredkin[74] and Stephen Wolfram.[75] The basic idea is that everything revolves around information, that space, time, matter, energy and thought are nothing but the fruit of a process of calculation, that is, of our dear, old computation. In short, creation would be – metaphorically, but only slightly so – the product of a sort of cosmic super-computer of unimaginable size and power. It is no coincidence that the theory of computation has tried to take hold in cosmology; in recent years, attempts to frame the most varied natural phenomena, such as DNA, in computational terms have multiplied, and digital philosophy, with its pan-computationalist vision, is simply its logical consequence. Already, quantum mechanics has forced us to think about the very real possibility that, at the sub-atomic level, matter has a "spectral" nature, that is, it is free of many of the "solid" characteristics that we attribute to it in our everyday life. And, descending from level to level, we first arrived at the atoms, then at the particles, then, perhaps, at the super-strings. And "at the bottom"? Is there something further

---

[73] K. Zuse, *Calculating Space*, 1969. ftp://ftp.idsia.ch/pub/juergen/zuserechnenderraum.pdf
[74] http://www.digitalphilosophy.org/
[75] S. Wolfram, *A New Kind of Science*, Wolfram Media, Champaign 2002.

down, at the bottom of reality? Underneath it all, according to digital philosophers, there would only be a series of "ones" and "zeros," of bits, very simple and irreducible entities. Even physicist John Wheeler hypothesized, in the '80s, that all physical phenomena owe their existence to binary-type choices: reality would, in a sense, be the fruit of questions that presuppose answers of the "yes and no" variety. For example, when some atoms hook themselves to each other, it is as if they were calculating, in a very precise way, the distance and the angle at which to engage, the properties that must arise from this match, and so on. And if you are thinking about computer simulations of this or that physical phenomenon, you got it right: for the digital philosophers, the world is just like that. Between computer simulations and the real world, there is no difference in nature, but only in degree. The fundamental ideas of digital philosophers – also shared by many Transhumanists – concern the fact that computation can be used to describe anything – physical and biological phenomena, social and historical realities, artistic and literary works. If, for Zuse, the whole universe "runs" on a computational substratum, for Fredkin, the most concrete thing in the world is information. Let's give an example. If you have a fairly advanced computer, you can use it to simulate, computationally, an older computer – say, the Commodore 64 or the Vic 20 with which you played as children. The most advanced computer can then be simulated by an even more advanced machine – in practice, recreated within the latter, on a formal level. The latter can be simulated, in turn, by an even more advanced computer, up to the largest computational processes, that is humanity, life and the entire cosmos, which hosts our computers. And the universe, who computes that? What "computer" is it that generates the entirety of reality? Fredkin answers by assuming that the computation of physical reality takes place in an unspecified "Other," which could be another universe, another dimension, a meta-universe superior to ours, or a "something" that we have yet to imagine. In more recent times, the American mathematician Stephen Wolfram has taken this discourse into his own hands, deepening it – in all senses, seeing that his reflection reaches the very bottom of reality, which is at the level of Planck's scale, a term with which we define an order of magnitude much smaller than a proton. Below this magnitude, we do not know how things stand, but it seems that, in the sub-Planckian realm, our familiar notions of space and time are no longer valid; they would, in fact, be simple approximations of more fundamental concepts waiting to be discovered. And, for Wolfram, right at that scale, there would reside the computational substratum that allows for the computation of space, time and energy. And some astrophysicists, far from seeing an "invasion of the field by outsiders" in the work of digital philosophers, have chosen to adopt these conceptions.[76]

---

[76] See C. Seife, *Decoding the Universe: How the New Science of Information Is Explaining Everything in the Cosmos, from Our Brains to Black Holes*, Viking, New York 2007; S. Lloyd, *Programming the Universe: A Quantum Computer Scientist Takes On the Cosmos*, Knopf, New York 2006.

Why are Transhumanists so interested in these things? Well, the answer is obvious: to "hack the universe." Or, rather: not so they themselves will be able to do it, but so that the post-human intelligences of the distant future can "awaken" the cosmos. The ability to read and modify the "source code" of creation will allow them to manipulate the laws of physics in a more or less subtle way, i.e., the "software" of reality; to produce, *pardon me*, compute another baby universe; or, failing that, to "open a connection" with another parallel reality.

Well, at this point, why not say it loud and clear? Some Transhumanists would like to replace God, or at least join him in the process of creating the Universe(s).

10

The God-Builders

10.1 The Transhumanist End-Game

Certainly, simply living forever and escaping the entropic process that will slowly extinguish our universe cannot be the final goal of Transhumanists, right? After all, what are you going to do with all of that time to fill? Evolution might be the answer, or one of the possible answers. Another one, which will appear to be full of hubris to casual readers – and not just to them, I have to say – would be to become, or to build, God Himself.[1] Not a bad goal, as a way to fill the eternal life that you have just obtained. But, before jumping to conclusions and labeling the entire Transhumanist movement as just plain crazy, bear with me for a moment. In truth, I have at least to point out that Transhumanists are not alone in this dream – that is, the desire to become just like God. They are in good company, actually, as the concept of "theosis" – of being elevated to the level of God – can be found in a number of different Christian denominations, and not just Christian ones at that.

But let's start from the beginning: the God-Builders. The idea of erasing the classical religions and imposing, or creating, new ones is far from new, and this has become even truer in the modern age. Let's just think, for example, of the "cult of reason" instituted by Robespierre, or the "new religion of the blood" dreamed of by the national-socialist ideologist Alfred Rosenberg. And let's not forget the "religion of humanity" proposed by Ludwig Feuerbach, in which God would be replaced by Humanity as an object of religious worship. Not, of course, this or that individual; rather, humanity as a whole, as a general entity, as a super-individual

---

[1] As an alternative, you might opt to turn yourself into a whole evolving universe, that is, to yourself become a hyper-realistic simulated evolving reality, a process that we can call "universification": http://estropico.blogspot.dk/2011/10/universification-cosa-fa-un-essere.html#axzz53sS0n03X

© Springer Nature Switzerland AG 2019
R. Manzocco, *Transhumanism - Engineering the Human Condition*,
Springer Praxis Books, https://doi.org/10.1007/978-3-030-04958-4_10

reality somehow embodied in every single individual, with all of its potential and all of its achievements so far, an idea that Transhumanists should both appreciate and get to work on. But, besides Robespierre, Rosenberg and Feuerbach, we should really mention the Russian God-Builders, whose name I unashamedly stole for this chapter. After all, it sounds pretty cool, doesn't it?

"God-Building" was an idea proposed by some prominent early Marxists, and especially Anatoly Lunacharsky, who was associated with a central character of the Bolshevik Revolution, Alexander Bogdanov. Instead of being completely abolished, religion should be seen for what it is: a psychological and social source of morality to be harnessed for the greater good of the Communist utopia, which meant leaving behind the coldness of classical Marxism, embracing the religious sentiment concealed in the human mind and instituting a meta-religious frame composed of new symbols and new rituals, to be superimposed onto the old, religious ones. Lunacharsky outlined his theory in a two-volume book, *Religion and Socialism* (1908–1911).[2] Unlike *tovarisch* Lenin, who was atheist, the God-Builders were agnostic. Lenin, of course, strongly disagreed with the God-Building movement, which he believed obfuscated the role held by religion in the exploitation of the masses; his victory in the October Revolution brought about the dismissal of Lunacharsky's school of thought.

While, for Lunacharsky, the process of God-Building was a purely symbolic one, for some Transhumanists, it looks like a real, actual project, although purely speculative – for now – and set in a very far future.

From this viewpoint, Isaac Asimov is another writer – and, yes, thinker – worth mentioning. Let us not forget how much of a debt Transhumanism owes to science fiction, and let us mention the interesting essay authored by Asimov on the topic of immortality and becoming God-like, *The Magic Society*.[3] That is, a future, utopian, technologically-generated society composed of individuals who have left Death behind. And not just that: it would be a society that excludes any kind of constraint – you don't have to live forever, if you don't want to –, one that is boundlessly wealthy and freed from pain, suffering, disease, and strain. One in which you don't have to work to make a living – the robots take care of that – and you can devote all of your time to any kind of pleasurable activity you want. Of course, like others before him, Asimov warns us against the main danger lurking in the shadows of this apparently perfect techno-utopia: boredom. And, being that this society is deprived of any constraint, the immortals inhabiting it are free to choose the obvious way out: assisted suicide. Comparing these people of the future with the leisure classes of the present and the past, we can roughly categorize two kinds of people: those who find pleasure on the outside – pleasurable

---

[2] A. V. Lunacharsky, *Religiia i sotsializm*, Shipovnik, Moscow 1908–1911.

[3] I. Asimov, *The Magic Society*, in: *Science Past – Science Future*, Ace Books, New York 1977, pp. 369–380.

activities of any kind, even dangerous ones – and those who find meaning and purpose in the pleasures of the mind. And, the pleasures from the outside being way more short-lived and ephemeral than those coming from our inside – from philosophy, mathematics, science, art, and so on – it is very likely that the first category of people of the future will choose to end their lives sooner than the second one. That is, the Magic Society will witness a progressive shift from a demographic composed mostly of outward-bound people to one composed of inward-bound people. Still, even the latter will be defeated by boredom, according to Asimov. And that's why he does *not* recommend the pursuit of an immortal life – although we can skip the details of why life would end in boredom here, as we have already heard these arguments multiple times. Our story is not over yet, though. Talking about God, in this same essay, Asimov tackles a classic topic of discussion among theologians in olden times: being eternal, that is, having existed since, well, forever, what was God doing before the Creation? Asimov doesn't know the answers, of course, but he tries to propose a solution: being infinitely complex, God will always find something interesting to contemplate in Himself. And, of course, we would like to add, not only will He always find some trait to contemplate, but also some action to undertake. Besides the Creation of this "Vale of Tears," of course. This statement, about the infinite complexity of God, is one that Asimov does not seem very interested in. But it might be embraced by some eccentric thinkers whom we all know at this point. *Now* this looks like the end-game of the Transhumanist endeavor: not simply to escape entropy, but to pursue an infinite level of complexity.

10.2   The Order of the Cosmic Engineers

Of course, when Transhumanists speak of "God-building," they really mean it. The art and science of creating and becoming Gods must be understood in a literary sense, a goal embodied by a now defunct organization, the "Order of the Cosmic Engineers" – quite a name, if you ask me.[4][5] The starting point of this – vaguely spiritual – group is the idea that the Universe is a cold, indifferent place, devoid of any "magic," and this is the reason why we, or our descendants, will have to do our best in order to introduce it. To make the Universe a sacred place again, basically, using the tools of a technology akin to magic.

Our story is actually relatively old, and it begins with a paper, published more than 30 years ago by an American sociologist of religion and Transhumanist

---

[4] http://web.archive.org/web/20110722024245/http://cosmeng.org/index.php/Main_Page

[5] The Order was formally founded during a conference on religion in Second Life on June 4 and 5, 2008, and launched in World of Warcraft on June 14, 2008.

thinker, William Sims Bainbridge, and re-edited more recently.[67] In *Religions for a Galactic Civilization* and *Religion for a Galactic Civilization 2.0*, Bainbridge's analysis recognizes that the human path toward the stars has basically halted, that our current technological level is insufficient and cannot help us in colonizing our planetary neighbors – much less other planetary systems – and that the main reason for this stalemate is sociological. That is, we humans lack a strong, quasi-religious fervor toward space-colonization and spacefaring – a necessary ingredient, if we are to invest the enormous amount of energy, resources and effort required for such a titanic enterprise. What we need, according to Bainbridge, is exactly that: a religious mentality, able to energize the peoples of the Earth and to push investment and, even more important, innovation. What the sociologist is suggesting here can be summed up by his – quite bold, I must say – statement: "The heavens are a sacred realm, that we should enter in order to transcend death." In Bainbridge's view, religiosity is a universal human trait, and so, if we want to get things done and reach the stars, we have to hijack religion and use it in order to promote spacefaring-friendly values. After all, the first wave of human space-faring, which took us to the Moon, was the brainchild of a small and very motivated group of dreamers and researchers, who managed to exploit the Cold War and the contraposition between West and East for their own purposes. There was a lot of randomness and luck in our first steps in space, and, Bainbridge warns us, the window of opportunity might close very soon. That is, we might settle for a stable and eco-sustainable society very soon, and this could curb or halt any human ambition toward spacefaring. And, of course, the only possible answer to this attractive and dangerous stability is a Transhumanism-based religion, which Bainbridge has christened The Cosmic Order. Of course, if you, the prospective believer in this Cosmic Order, want to join and enjoy the benefits of this religion, there are a few things you should do; basically, you should undergo a massive series of psychological tests – and maybe gather all of the autobiographical records that you can. This process will allow the AI of the future to reconstruct your personality, and this new incarnation of you – although, is it really you? – will join the post-humanity of the far future in its plans for Galactic colonization. Besides, in doing so, you might even get to know yourself better, which can be useful in your present life. The link between Transhumanist immortality and spacefaring depends on the notion that, if you pursue the first project, you must pursue the second as well: in fact, physical immortality – in our original body or in a robotic surrogate – requires plenty of room, of the kind that you can find only in the empty

---

[6] W. S. Bainbridge, *Religions for a Galactic Civilization*, in: E. M. Emme (ed.), *Science Fiction and Space Futures*, American Astronautical Society, San Diego 1982, pages 187–201. war-iscrime.com/new/religions-for-a-galactic-civilization/

[7] W. S. Bainbridge, *Religion for a Galactic Civilization 2.0*, Aug 20, 2009. https://ieet.org/index.php/IEET2/more/bainbridge20090820/

vastness of the cosmos. So, the message is clear: do you want immortality – physical or electronic? You better hurry up and explore space, and, in order to do so, you need a social movement able to tap into the natural religiosity and existential needs of regular human beings. You need The Cosmic Order. However, Bainbridge isn't self-delusional: "My speculations may have seemed outlandish and absurd. But, in the literal meaning of the term, the universe itself is outlandish. The human condition is one of extreme absurdity unless fixed in a cosmic context to provide meaning. Human societies need faith, and if they lose traditional faiths they will struggle to discover new faiths, lest they collapse. (…) Thus it is wrong to feel that irrational religion must always be a hindrance to progress. I have suggested that only a transcendent, impractical, radical religion can take us to the stars. The alternative is one or another form of ugly death. A successful outcome depends on a kind of lucky insanity, and it is quite unlikely. But for our species, at least it is still possible".

Back to the Cosmic Engineers now. Their Prospectus was signed by Bainbridge himself and by many other prominent Transhumanists – or the "College of Architects," as they call themselves in this context: Howard Bloom, Riccardo Campa, Stephen Euin Cobb, Ben Goertzel, Max More, David Pearce, Giulio Prisco, Martine Rothblatt, Philippe Van Nedervelde, and Natasha Vita-More. The goals set in this Prospectus are, of course, very ambitious – I wouldn't expect anything less from a group of Transhumanists. In their Prospectus, our Cosmic Engineers aim to "permeate our universe with benign intelligence, building and spreading it from inner space to outer space and beyond. The milestone grails of our joyful cosmic quest sequentially comprise (…) a roadmap for the exponential ingression of intelligence into inanimate matter; make inner space come alive with intelligence by engineering intelligence computation at the molecular level; engineer, spur and guide the responsible geometric intelligence expansion from inner space to human scale to outer space scale; intimately join, cross-pollinate and cross-leverage our mental resources into a meta-mind society; deeply optimize our material universe for cosmos-wide intelligence computation; answer the ultimate questions of the origin, nature, purpose and destiny of reality; tune universe-creation parameters so as to further improve the maximization of the computational ability of a universe; engineer and engender one or more new baby universes with controlled physics parameters; (…) Engineering 'Magic' Into A Universe Presently Devoid Of God(s)." Quite a plan. "As scientists, we face reality and do not believe in metaphysical 'magic', supernatural 'miracles', direct interventions by or various forms of contact with putative praeter-natural deities, etc. At the same time, we are persuaded reality might well be much stranger and much more complex than even our most radical thinkers dare imagine. In the immortal words of Shakespeare: 'There are more things in heaven and earth, Horatio, than are dreamt of in your philosophy.' There Is No 'God'… Yet." The underlying vision looks quite grim, though: "At present, and at least for the time being, we, sentient human beings, are here on our own in our tiny backwater corner of the cosmos. As a sentient species, and while we believe there are valid reasons for enthusiastic and active *hope*, we presently do find and understand ourselves as left to our own devices in an apparently coldly indifferent, seemingly cruel, and hostile universe. We furthermore share the conviction that there actually never was and also never will be a 'supernatural' god, at least not in the sense understood by theist religions." But there is room for hope: "[I]n the (arguably) very far future one or more *natural* entities -i.e. entities existing *within* our present universe- are highly likely to come into being – plausibly resulting from the agency of our and other species – which will, to all intents and purposes, be very much akin to 'god' conceptions held by theist religions. We refer to conceptions of personal, omnipotent, omniscient and omnipresent super-beings, 'deities' or 'gods'." Clearly, such ambitious projects of cosmic engineering call for a very long lifespan and healthspan, as the required amount of time vastly exceeds the eighty-or-so years allotted to us. The inherent optimism of the Cosmic Engineers caused them to adopt the super-classic and often quoted "Third Law" of Arthur C. Clarke: "Any sufficiently advanced technology is indistinguishable from magic."[8] And this is exactly the main goal of the Cosmic Engineers: "As engineers, we aim to build what cannot be readily found. Adopting an engineering approach and attitude, the *OCE* aims to turn this universe into a 'magical' realm in the sense of *Clarke's Third Law*: a realm where sufficiently advanced technology turns daily reality into what would be considered by most today as a seemingly supernatural 'magical' realm." Doesn't that sound like religion? Well, not really, according to the Engineers themselves: "As much as the *OCE* enthusiastically espouses universe-scale cosmic visions and worldviews, including spiritual sensibilities attendant to such, the *OCE* however is – emphatically – NOT a religion. The *OCE* is not a religion, not a faith, not a belief, not a church, not a sect, not a cult. We are not a faith-based organization. We are a convictions-based organization. We do not worship anyone or anything. Any kind or form of worship is really anathema to us."

This is quite an interesting point, and a critical one: as we are about to see, the relationship between Transhumanism and religion is not that straightforward.

---

[8] Clarke's three laws appeared in Clarke's essay *Hazards of Prophecy: The Failure of Imagination*, published in 1962 in *Profiles of the Future: An Enquiry into the Limits of the Possible* (Gateway, London 2013). The three laws are: 1. When a distinguished but elderly scientist states that something is possible, he is almost certainly right. When he states that something is impossible, he is very probably wrong. 2. The only way to discover the limits of the possible is to venture a little way past them into the impossible. 3. Any sufficiently advanced technology is indistinguishable from magic.

10.3   From Transhumanism to Religion, and Back

Far from being a single, air-tight ideology, the Transhumanist movement has very fuzzy, blurry borders, and accepts among its members anyone who is willing to endorse the main Transhumanist goals, like life-extension, mind-upload, and so on. Which means that you can be Transhumanist and Communist, Transhumanist and Libertarian, Transhumanist and Fascist. And, of course, Transhumanist and Religious. So, the relationship between Transhumanism and religion is, well, quite complicated. On average, Transhumanists are atheists or agnostics, or, at least, this is the official narrative. So, for example, sociologist René Milan, among others, defends a rationalist and non-religious view of Transhumanism[9]; Transhumanist and Mormon, Lincoln Cannon begs to disagree, stressing the fact that Transhumanism and Atheism are not exactly the same thing, and that the former actually has a clear religious component.[10] In fact, a recent poll shows that only half of those identifying as Transhumanist consider themselves either agnostic or atheist; Cannon stresses that many Transhumanists are actually tired of the anti-religious rhetoric of some members of the movement, and, even more importantly, for several members of this movement, Transhumanism functions exactly as a religion, with sacraments (the nutritional supplements religiously taken on a daily basis), rituals (for example, cryonic suspension), the apocalyptic faith in the advent of the Technological Singularity, the prophecies about life extension and the quasi-religious beliefs in the substrate-independent survival of the human mind: "Transhumanism, for many Transhumanists, is clearly [a] postsecular religion, even if [it is] misrecognized."

Besides some interesting attempts to merge Transhumanism with traditional religion, we can even find Transhumanists willing to create new forms of purely Transhumanist religion; one example is Dirk Bruere, who developed his own approach, called The Praxis.[11] "So, who are the Transhumanists?" asks Bruere. "Well, nobody really knows how many people define themselves as such. The best guess is probably less than one hundred thousand, mostly engineers and scientists and not, as one might expect, science fiction fans. No doubt a much greater number agree with at least one or more H+ ambitions but who do not buy the whole package. (…) The one thing almost all Transhumanists agree upon is the desirability of not dying of old age, and remaining healthy indefinitely, or at least until the even more exotic technologies hopefully begin to make an appearance. It is the increasingly high profile of this area of medical technology that is partially

---

[9] R. Milan, *The question of religion and transhumanism*, http://transhumanity.net/the-question-of-religion-and-transhumanism-opinion/

[10] L. Cannon, *Transhumanism is not Atheism and is often misrecognized religion*, https://lincoln.metacannon.net/2014/09/transhumanism-is-not-atheism-and-is.html

[11] http://www.neopax.com/praxis/

responsible for the dissemination of H+ agendas in the media as it hitches a ride on this and another area of increasing public concern – Artificial Intelligence (AI). (…) By now we are way out on the fringes of fringe beliefs, but there is still one more massive step to take, and one which leads us a core doctrine of ancient Gnostic Christianity. Let's go back to digital heaven and the resurrection of the dead. So there we have it – a new religion for a new millennium. Indeed, there are a few small organizations that are explicitly of a religious outlook based around these ideas, most notably Terasem and latterly The Praxis. They try and provide an answer to the question why – why should anyone (or anything) bother to revive those in a cryonic preservation state, or those even more seriously dead and mostly forgotten? Where to go from here? Well, Transhumanism is in one strange theo-logical class of its own, namely, that if it is not true, its adherent[s] believe it can be made so. However, unlike most other religions they will not be knocking on your door trying to convert you, nor will they be asking you for money."[12]

And let's not forget the already-mentioned attempt to merge Transhumanism and Buddhism – see, for example, Michael LaTorra's article *What is Buddhist Transhumanism*.[13] And this is hardly the only attempt to merge a mainstream reli-gion and the fringiest of Transhumanist ideas. In *The Maitreya and the Cyborg: Connecting East and West for Enriching Transhumanist Philosophy*,[14] Miriam Leis attempts an interesting crossover between the Transhumanist concept of "post-human" and the Maitreya, a Buddhist eschatological figure present in many Buddhist philosophical schools. Of course, not everybody in the Transhumanist community agrees with this parallelism between Transhumanism and Buddhism.[15] Let's also mention the Christian Transhumanist Association,[16] originated from a small online community of bloggers and internet activists at the beginning of 2013 and turned into a 501c3 nonprofit organization in 2015. Also known as C+, or Christianity+, this association was founded by Micah Redding, Lincoln Cannon and Christopher Benek. Of course, Christianity+ has been vocally critical of the highly atheistic positions of Transhumanists like Zoltan Istvan. We already men-tioned Lincoln Cannon, and now let's also point out that he is the founder of

---

[12] D. Bruere, *Transhumanism – The Final Religion?*, https://ieet.org/index.php/IEET2/more/bruere20151207

[13] https://www.tandfonline.com/doi/full/10.1080/14746700.2015.1023993

[14] M. Leis, *The Maitreya and the Cyborg: Connecting East and West for Enriching Transhumanist Philosophy*   http://indiafuturesociety.org/the-maitreya-and-the-cyborg-connecting-east-and-west-for-enriching-transhumanist-philosophy/

[15] W. Evans, *If You See a Cyborg in the Road, Kill the Buddha: Against Transcendental Transhumanism*, «Journal of Evolution and Technology» – Vol. 24, Issue 2 – September 2014 – pp. 92–97, https://jetpress.org/v24/evans.htm

[16] https://www.christiantranshumanism.org/

the – very active – Mormon Transhumanist Association,[17] which is what we could call "a match made in Heaven": in fact, the Transhumanist philosophy of self-directed evolution fits nicely into the mainstream Mormon theological view of theosis, the future "exaltation" of Man to the level of God.

Merging Transhumanism and religion is not the only option; other approaches are possible. For example, in *Trans-Spirit: Religion, Spirituality and Transhumanism*,[18] Michael LaTorra tries to develop a specific Transhumanist research program that aims to explore the functional and evolutive origin of religion and spiritual life – that is, religion is seen as a mechanism with adaptive value – in order to develop technologies able to induce spiritual and mystical experiences at will (or, I'd like to add, maybe even prevent them). The basic assumptions of LaTorra's Trans-Spirit program are rooted in a biological and neuroscientific interpretation of religious phenomena, in an emerging field called "neurotheology" derived from the cross-pollination of these two disciplines.

Make no mistake, though; the real purpose of religious Transhumanism is exactly this: not to simply love God or worship Him, but to be "exalted," to reach, one way or another, His level of power, knowledge and wisdom. So, one of the main proponents of this approach, Giulio Prisco, quotes Arthur C. Clarke, saying, "It may be that our role on this planet is not to worship God, but to create him."[19] "I am persuaded that we will go to the stars and find Gods, build Gods, become Gods, and resurrect the dead from the past with advanced science, space-time engineering and 'time magic,'" says Prisco. "I see God emerging from the community of advanced forms of life and civilizations in the universe, and able to influence space-time events anywhere, anytime, perhaps even here and now. I also expect God to elevate love and compassion to the status of fundamental forces, key drivers for the evolution of the universe."

Another proponent of this idea is Ted Chu, who, in his *In Human Purpose and Transhuman Potential: A Cosmic Vision for Our Future Evolution*, promotes what he calls the "Cosmic View," which entails the creation of our successor, CoBe (Cosmic Beings), post-human Gods able to permeate the whole universe with hyper-intelligent lifeforms.[20]

Jamais Cascio quotes Stewart Brand, author of the *Whole Earth Catalog* (1968): "We are as Gods and might as well get good at it" – but with some caution – "We

---

[17] https://transfigurism.org/

[18] M. LaTorra, *Trans-Spirit: Religion, Spirituality and Transhumanism*, in: «Journal of Evolution and Technology» – Vol. 14, Issue 1 – August 2005, pp. 39–53, http://jetpress.org/volume14/latorra.html

[19] G. Prisco, *Religion Fiction Inspires Real Religion*, http://turingchurch.com/2015/01/10/religion-fiction-inspires-real-religion/

[20] T. Chu, *Human Purpose and Transhuman Potential: A Cosmic Vision for Our Future Evolution*, Origin Press, San Rafael 2014, http://transhumanpotential.com/htptwp/

are Gods, but we're the gods of an earlier age. Powerful, yes, but petulant; wise yet warlike; arrogant and utterly capricious… and also able to create sublime beauty. (…) We are as Gods, but we have gotten pretty good at it—as long as we remember that this means we are as likely to be Loki as Athena."[21]

Another prominent Transhumanist, B. J. Murphy, likes to compare our posthuman descendants to Gods, angels and ghosts,[22] and makes an interesting suggestion: what if, using nanotechnology and other Transhumanist devices and artifacts, we could actually spread consciousness and self-awareness everywhere, basically making every single object or piece of matter self-conscious and, in doing so, making panpsychism – the philosophical idea that everything is alive and conscious – real?[23] Technologist Ramez Naam, in speaking of the Transhumanist revolution, also uses a quasi-religious metaphor: "We are, if we choose to be, the seed from which wondrous new kinds of life can grow. We are the prospective parents of new and unimaginable creatures. We are the tiny metazoan from which a new Cambrian can spring. I can think of no more beautiful destiny for any species, no more privileged place in history, than to be the initiators of this new genesis."[24] George Dvorsky also predicts a God-like future for our species: "Future civilizations may eventually figure out how to re-engineer the Universe itself (such as re-working the constants)."[25]

And, in his sci-fi novel *The Transhumanist Wager*, Zoltan Istvan issues his own three Transhumanist laws:

1. A Transhumanist must safeguard one's own existence above all else.
2. A Transhumanist must strive to achieve omnipotence as expediently as possible – as long as one's actions do not conflict with the First Law.
3. A Transhumanist must safeguard value in the Universe – as long as one's actions do not conflict with the First and the Second Laws.[26]

But Giulio Prisco says it best: our purpose is engineering transcendence.

---

[21] J. Cascio, *Pantheon*, https://ieet.org/index.php/IEET2/more/cascio20111128

[22] B. J. Murphy, *A Transhumanist's Journey to Becoming Gods, Angels, and Ghosts*, https://ieet.org/index.php/IEET2/more/murphy20130617

[23] B.J. Murphy, *Engineering Panpsychism: A possibility?*, https://proactiontranshuman.wordpress.com/2014/09/16/engineering-panpsychism-a-possibility/

[24] R. Naam, *More Than Human. Embracing the Promise of Biological Enhancement*, Random House, New York 2005, pp. 233–234.

[25] G. Dvorsky, *How Will Our Universe Die?*, https://ieet.org/index.php/IEET2/more/dvorsky20070525

[26] Z. Istvan, *The Transhumanist Wager*, Futurity Imagine Media LLC, Reno 2013, p. 4.

10.4   Engineering Transcendence and the Cosmist "Third Way"

Giulio Prisco – Italian-born futurist, Transhumanist and IT/virtual reality consultant – differs quite markedly from the classic Transhumanist profile. In fact, unlike characters like Zoltan Istvan, he is a believer. That is, he does believe in God, or, at least, in some form of technologically-mediated Transhumanist spirituality. Of course, his faith doesn't have much to do with the classic, mainstream religions, even though Prisco likes to pay respect to them, and to the intellectual efforts of their theologians. Prisco believes that our species does have a "manifest destiny," which consists in "colonizing the universe and developing spacetime engineering and scientific 'future magic' much beyond our current understanding and imagination. Gods will exist in the future, and they may be able to affect their past — our present — by means of spacetime engineering. (…) Future Gods will be able to resurrect the dead by 'copying them to the future.'"[27] Speaking of resurrection: together with Ben Goertzel, Prisco resurrected the good old Russian Cosmist philosophy, and published the Ten Cosmist Convictions, which we lay out here:

1. Humans will merge with technology, to a rapidly increasing extent. This is a new phase of the evolution of our species, just picking up speed about now. The divide between natural and artificial will blur, then disappear. Some of us will continue to be humans, but with a radically expanded and always growing range of available options, and radically increased diversity and complexity. Others will grow into new forms of intelligence far beyond the human domain.

2. We will develop sentient AI and mind-uploading technology. Mind-uploading technology will permit an indefinite lifespan to those who choose to leave biology behind and upload. Some uploaded humans will choose to merge with each other and with AIs. This will require reformulations of current notions of self, but we will be able to cope.

3. We will spread to the stars and roam the universe. We will meet and merge with other species out there. We may roam to other dimensions of existence as well, beyond the ones of which we're currently aware.

4. We will develop interoperable synthetic realities (virtual worlds) able to support sentience. Some uploads will choose to live in virtual worlds. The divide between physical and synthetic realities will blur, then disappear.

5. We will develop spacetime engineering and scientific 'future magic' much beyond our current understanding and imagination.

6. Spacetime engineering and future magic will permit achieving, by scientific means, most of the promises of religions — and many amazing things that no human religion ever dreamed. Eventually we will be able to resurrect the dead by 'copying them to the future.'

---

[27] G. Prisco, *Yes, I am a believer*, http://turingchurch.com/2012/05/21/yes-i-am-a-believer/

7. Intelligent life will become the main factor in the evolution of the cosmos, and steer it toward an intended path.
8. Radical technological advances will reduce material scarcity drastically, so that abundances of wealth, growth and experience will be available to all minds who so desire. New systems of self-regulation will emerge to mitigate the possibility of mind-creation running amok and exhausting the ample resources of the cosmos.
9. New ethical systems will emerge, based on principles including the spread of joy, growth and freedom through the universe, as well as new principles we cannot yet imagine.
10. All these changes will fundamentally improve the subjective and social experience of humans and our creations and successors, leading to states of individual and shared awareness possessing depth, breadth and wonder far beyond that accessible to 'legacy humans.'[28]

The Ten Cosmist Convictions can also be found in Goertzel's book *A Cosmist Manifesto*[29] – an attempt to develop a more radical view of Transhumanism, and, of course, a recommended reading.

To Prisco, human existence, with all of its drama, its existential emptiness, its absurdity, can be fixed using an engineering approach – a kind of engineering way beyond our present imagination. In *Engineering Transcendence*,[30] Prisco sets his plan in motion, establishing its steps: engineering resurrection, engineering God, engineering hope and happiness. The first step entails somehow finding a way to recover all of the information about, well, ourselves from the past, and copying it into the future; that is, finding a way to reproduce every human being who ever existed in the future, so that he/she can live again, possibly happily ever after. Pure science fiction, for now, but you never know. One possible scenario is Frank Tipler's Omega Point theory (more on this later), but Prisco's – and ours as well – favorite engineering solution is the one described by Arthur C. Clarke and Stephen Baxter in their fascinating novel *The Light of Other Days*.[31] The theoretical background of the novel is highly speculative: micro-wormholes with huge density are embedded in the very same fabric of spacetime, and this allows every point of it to be connected with every other point, meaning that every spacetime "pixel" is connected to every other spacetime "pixel," in the present, past and future. The world described in the novel is deeply shattered by a new, surprising invention, the

[28] G. Prisco, *Ten Cosmist Convictions*, http://cosmistmanifesto.blogspot.com/2009/01/ten-cosmist-convictions-mostly-by.html

[29] B. Goertzel, *A Cosmist Manifesto: Practical Philosophy for the Posthuman Age*, Humanity+, Los Angeles 2010, https://humanityplus.org/projects/press/

[30] G. Prisco, *Engineering Transcendence*, http://giulioprisco.blogspot.com/2006/12/engineeringtranscendence.html

[31] A. C. Clarke; S. Baxter, *The Light of Other Days*, Tor Books, New York 2000.

"wormcam," a device that can use these micro-wormholes to observe the past and recover any kind of information; as a consequence, it becomes possible to resurrect long-dead people, recovering *all* of the information that constituted them – not just their DNA, but every single memory, with an incredibly high level of fidelity. To indicate this process, Prisco coined a suggestive term:

> Time-scanning – someday it will be possible to acquire very detailed information from the past. Once time-scanning is available, we will be able to resurrect people from the past by 'copying them to the future' via mind-uploading. Note: time-scanning is not time travel, and it is free from the "paradoxes" of time travel. Time-scanning is just a form of archaeology --- uncovering the past by means of available evidence and records. Of course the very high definition form of time-scanning proposed here is orders of magnitude more powerful and sophisticated than archeology as we know it, but the concept is the same.[32]

This approach – similar to the proposal of another Transhumanist thinker, Mike Perry[33] – represents a contemporary version of the old Fedorov plan of resurrecting the dead through the gathering of all of their "particles" that are dispersed throughout the cosmic space – a quite naïve plan, seen through today's eyes, but still interesting and anticipatory.

If technological resurrection looks ambitious, wait until you see the second step: engineering none other than God – or Gods. This one is not really a plan; it is just speculation about what could happen in a very far future. Future cosmic civilizations could evolve to such a level of power and knowledge that they could become able to manipulate the whole universe and the very structure of spacetime. Prisco mentions Freeman Dyson, who, in *Infinite in All Directions*, equates the mind to God, or considers God the collection of all minds, present and future, and Socinius, a Renaissance philosopher who considered God to be inherent in the universe. So, basically, to requote Prisco, "we will go to the stars and find Gods, build Gods, become Gods, and resurrect the dead from the past with advanced science, spacetime engineering and 'time magic.'"[34] God, or Gods, will be entities emerging from the evolution of a truly cosmic civilization, or from a hierarchy of more and more advanced civilizations, human and non-human. To put it in another way:

> Science and our transhumanist convictions, taken to their logical conclusion, say that intelligent life can evolve to X-like status, where X means extremely advanced and able to perform "magic" (Clarke's Third Law). The universe is

---

[32] G. Prisco, *Transcendent Engineering*, in «Terasem Journal of Personal Cyberconsciousness», Vol. 6, Issue 2, December 2011, http://www.terasemjournals.com/PCJournal/PC0602/prisco. html

[33] M. Perry, *Forever for All: Moral Philosophy, Cryonics, and the Scientific Prospects for Immortality*, Universal Publishers, Irvine 2000.

[34] G. Prisco, *A minimalist, open, extensible Cosmic Religion*, http://turingchurch. com/2014/08/25/a-minimalist-open-extensible-cosmic-religion/

probably full of Xs, and we can become Xs ourselves. Xs can take control of the dynamics of space-time, create new universes, and resurrect the dead. In the simple language that we use everyday (and we should use whenever possible to keep things clear and immediate) X is called "God" and this is called "Religion."[35]

Which reminds us of similar musings by the Brazilian/American physicist Marcelo Gleiser:

Imagine, then, that in some corner of the galaxy, other intelligent creatures also discovered some version of science. But they did so, say, a million years before us, which, in cosmic time, is not much. These creatures would now be machine-hybrids, completely different from what they once were. (...) Perhaps "they" are only information, free-floating in coded energy fields spread across space. Perhaps they have, much beyond anything we can presently contemplate, the power to create life, choosing its properties at will. They could, for example, have created us, or some of our ancestors, as part of an experiment in their version of evolutionary genetics, or as a test bed in a study of the relation between intelligence and morality. They could, perhaps, be observing us, as we observe animals in a zoo or a laboratory. These entities, immaterial but living as self-sustaining bundles of information, could have been our creators. Would they be gods, even if not supernatural?[36]

As a side effect of all of this, this specific brand of Transhumanism – what do we call it? Transcendent Transhumanism? Spiritual Transhumanism? – could also produce hope and happiness in the present time, as these specific Transhumanist memes can offer everybody the hope that, maybe, someday, a super-advanced God-like civilization will use a small amount of its endless resources to, well, resurrect everybody into a material Paradise, into a Transhumanist Heaven. A memetic niche already filled by an *ad-hoc* Transhumanist association, the Society for Universal Immortalism.[37] What Prisco is offering us here is a kind of mythology, but one in line with what is considered possible, or at least not impossible, by our contemporary scientific standards. This mythology has been christened by its founder with an appropriate name: The Turing Church. The religious nature of some aspects of Transhumanist thinking has been noted by American professor of Religious Studies Robert Geraci as well, in his book *Apocalyptic AI*.[38]

Now it's time to dig a little deeper into this idea of technological resurrection, starting with the concept of Quantum Archaeology.

---

[35] G. Prisco, *Cosmic Religion discussion, and plans*, http://turingchurch.com/2014/09/01/cosmic-religion-discussion-and-plans/

[36] M. Gleiser, *Astrotheology: Do Gods Need To Be Supernatural?* https://www.npr.org/sections/13.7/2012/11/28/165993001/astrotheology-do-gods-need-to-be-supernatural?t=1532420287358

[37] http://universalimmortalism.org/

[38] R. M. Geraci, *Apocalyptic AI: Visions of Heaven in Robotics, Artificial Intelligence, and Virtual Reality*, Oxford University Press, Oxford 2012.

10.5   Do Not Go Gentle into That Good Night

That is, Transhumanism "Plan C." The concept of QA (Quantum Archaeology) originally derives from the work of a few anonymous contributors to Ray Kurzweil's blog, Kurzweilai.net, popularized, with mixed results, by some guy named "Eldras."[39] No matter the origin, the idea of Quantum Archaeology represents probably one of the most extreme ideas to be imagined by Transhumanist thinking. Let's put it simply: this controversial project consists, basically, in finding a way to resurrect the dead. And not just the recently deceased, but every human being who ever existed since the dawn of humankind, bodies, minds and, of course, memories. This – let's face it – crazy idea was inspired by a fictional discipline developed by the science fiction writer Isaac Asimov in his novel *Foundation* (1942–1951), called "psychohistory," an approach that combines sociology, history and mathematics in order to predict the future of the Galactic Empire. In the case of Quantum Archaeology, such an approach has a different purpose, which is to reconstruct the past in all of its minute details, using simulations, physical laws and a deterministic interpretation of the world. The concept used by Eldras is that of "retrodiction," which means "the explanation or interpretation of past actions or events inferred from the laws that are assumed to have governed them."[40] Quantum Mechanics is a very complicated topic, and it lends itself to many different interpretations. Some of them are probabilistic – to put it simply, they accept a degree of objective randomness – some others are not; that is, you can find deterministic interpretations of Quantum Mechanics, and Eldras bets exactly on these. In other words, Eldras's worlds are super-deterministic, memories are nothing mystical (they are just states of the brain) and, if you have computers powerful enough (to date, we don't), you can literally develop a super-precise simulation of the present (environment, individuals, memories, and so on) and, starting from them, retrodict the physical states of the past very precisely. Using super-computers, specific algorithms, physical laws and the metaphysical principle of causality, you can develop what Eldras calls a "Quantum Grid," a reproduction/simulation of the environment and everything that it contains. In turn, this would allow you to engage in super-precise guesswork so as to reconstruct the memories of every member of our species since its birth. To tell the truth, we already have simulation grids that we use for this or that field of investigation, and it helps to imagine Quantum Archaeology as a sort of super-forensic science based on computational power that does not yet exist. The final goal would be to create a Grid of Grids, able to accommodate all possible data, recovered using super-recursive algorithms – special algorithms still in their infancy, but that promise to overcome the limits of traditional computing. And, of course, using the

---

[39] Eldras, *Can Science Resurrect the Dead?* http://transhumanity.net/can-science-resurrect-the-dead/

[40] https://en.oxforddictionaries.com/definition/retrodiction

Transhumanist technologies that we have already covered, these super-precise simulations could be brought back to life, maybe through 3D bioprinting.[41] And there you have it: the technological resurrection of the dead, a process that, unavoidably, has a religious taste. And how many people are we talking about? According to Eldras, over 106 billion dead people from 50,000 B.C.E. onward would be brought into the modern world. That is, if we don't consider the humans who lived before that age, as, apparently, our species – according to the most recent estimates – appeared around 200,000 years ago.

Any way you look at it, this Quantum Archaeology doesn't appear very "grounded," that is, it looks like a titanic job, plus there is absolutely no guarantee that we would be able to calculate the memories of our ancestors in exact fashion. Maybe it's better to look for some other solution; again, better ask Prisco, who likes wild speculation – which is to be expected, from a Transhumanist. Clearly, Prisco's theological Transhumanism represents an explicit attempt to cope with his mortality and with the death of his loved ones: "I cope with the grief from the death of loved ones by contemplating the Cosmist possibility, described by many thinkers including Nikolai Fedorov, Hans Moravec and Frank Tipler, that future generations (or alien civilizations, or whatever) may develop technologies to resurrect the dead. A related idea is that our reality may be a "simulation" computed by entities in a higher-level reality, who may choose to copy those who die in our reality to another reality."[42] From a theoretical viewpoint, the psychological underpinnings of this approach don't matter much: what is important is the conceptual consistency of these theories, which we will now peruse.

We are plunging into deep theoretical physics here, so I will try to keep it as simple as possible. Prisco plays with two different – but maybe related, at least according to his musings – concepts, that is, "quantum entanglement" and "wormholes." According to Wikipedia, quantum entanglement is "a physical phenomenon which occurs when pairs or groups of particles are generated, interact, or share spatial proximity in ways such that the quantum state of each particle cannot be described independently of the state of the other(s), even when the particles are separated by a large distance—instead, a quantum state must be described for the system as a whole."[43] It is a "weird" phenomenon, the "spooky action at a distance" that perplexed Einstein so much – as it entails the possibility of transmitting information (between the two entangled particles) at a faster-than-light speed. Even though the two particles cannot be "pre-set" – that is, you cannot set the kind of information you want to transmit in advance – it is always random and out of

---

[41] Z. Istvan, *Quantum Archaeology: The Quest to 3D-Bioprint Every Dead Person Back to Life*, «Newsweek» 3/9/18, https://www.newsweek.com/quantum-archaeology-quest-3d-bioprint-every-dead-person-back-life-837967

[42] G. Prisco, *How to Cope with Death: The Cosmist 'Third Way,'* http://turingchurch.com/2012/08/04/how-to-cope-with-death-the-cosmist-third-way/

[43] https://en.wikipedia.org/wiki/Quantum_entanglement

your control, and so you cannot use those entangled particles to transmit faster-than-light messages. Wormholes are holes in the fabric of spacetime that might connect two very distant points in space. What if – asks Julian Sonner, an MIT researcher specializing in String Theory – the entangled particles were connected by, well, wormholes?[44] And what if, asks Prisco, this theoretical approach were to allow us to reach into the past? Through entangled particles/wormholes, we might be able – if the physics of the future allow it – to connect the present with the past, and – without generating paradoxes – to recover any amount of detail of any kind of information we want.[45] You can see where we are going with this: we might literally be able to copy long-dead and long-forgotten people into the present – or the future. For now, this is just speculation, and we don't really know whether there is a connection between quantum entanglement and wormholes, and if this "weird" composite entity could really reach into the past – it doesn't seem likely right now, but time will tell.

Prisco's ideas rest on a couple of concepts – not very kosher, from a strict scientific viewpoint – that we are going to look at right now: quantum mysticism and the "Akashic records." If you are interested in New Age thought and/or fringe physics, you probably have heard of the term "quantum mysticism," a set of ideas that attempts, strongly and directly, to associate human consciousness with quantum mechanics, for the purpose of proving the existence of paranormal phenomena and, more generally, tying the human mind to the rest of the universe, thus offering hope for its survival after the death of the physical body. Normally considered pseudoscience, quantum mysticism finds its roots in the early years of quantum mechanics, when physicists like Erwin Schrödinger promoted the idea that consciousness plays a role in quantum theory, an idea objected to by other physicists, like Albert Einstein and Max Planck. It is a long and complicated story, which saw, in 1961, Eugene Wigner suggesting – in a paper entitled *Remarks on the mind–body question* – that the observer did actually play a role in quantum phenomena. To cut it short: in the '70s, the New Age movement included these ideas in its own musings, using them to justify paranormal phenomena, the idea of a general universal interconnectedness, and so on. Again, if you are familiar with these fringy interpretations, you have probably heard of the Fundamental Fysiks Group, a bunch of physicists who welcomed quantum mysticism with open arms, mixing it with meditation, parapsychology, New Age-ism and, of course, Eastern philosophies – and maybe with the help of some LSD too.[46] Among the members

---

[44] J. Chu, *You can't get entangled without a wormhole. MIT physicist finds the creation of entanglement simultaneously gives rise to a wormhole*, December 5, 2013, http://news.mit.edu/2013/you-cant-get-entangled-without-a-wormhole-1205

[45] G. Prisco, *Quantum Entanglement and Wormholes*, http://turingchurch.com/tag/wormholes/

[46] D. Kaiser, *How the Hippies saved Physics. Science, Counterculture, and the Quantum Revival*, W. W. Norton & Company, New York 2011.

of this group, we find Fritjof Capra, author of *The Tao of Physics*,[47] and Gary Zukav, who wrote *The Dancing Wu Li Masters*.[48][49]

About the "Akashic records." Theosophy is a body of occult doctrines introduced in the US during the nineteenth century by the Russian émigré Helena Petrovna Blavatsky (1831–1891); anthroposophy is the esoteric doctrine developed by the Austrian thinker Rudolf Steiner (1861–1925). Both doctrines incorporated the concept of "Akashic records," a non-physical, etheric plane of existence that contains the memory of all human and non-human events, thoughts, and emotions belonging to the past, the present and the future. Of course, this concept was not invented by Blavatsky or Steiner, but originally belonged to the Hindu tradition – *ākāśa* in Sanskrit means "atmosphere," while, in Hindi, *akash* means "heaven" or "sky." It was Alfred Percy Sinnett who first introduced the concept of the "Akashic record" to the West, in his book *Esoteric Buddhism*[50] (London, Chapman and Hall, 1885). According to a follower of Madame Blavatsky, Alice A. Bailey, the Akashic records are like an immense photographic film, registering every human and animal experience, both real and imagined. This idea was also endorsed by Nikola Tesla. A scientific maverick worshipped by many believers in the paranormal, Tesla can also be considered a proto-Transhumanist. In *Man's greatest achievement* – a quite visionary article published in the *Milwaukee Sentinel* on July 13, 1930[51] –, Tesla asks, "What has the future in store for this strange being, born of a breath, of perishable tissue, yet immortal, with his powers fearful and divine? What magic will be wrought by him in the end? What is to be his greatest deed, his crowning achievement?"

In an attempt to merge the Western and Eastern traditions, the scientist stresses that "all perceptible matter comes from a primary substance, of a tenuity beyond conception and filling all space – the Akasha or luminiferous ether – which is acted upon by the life-giving Prana or creative force, calling into existence, in never ending cycles, all things and phenomena. Can Man control this grandest, most awe-inspiring of all processes in nature? Can he harness her inexhaustible energies to perform all their functions at his bidding, more still – can he so refine his means of control as to put them in operation simply by the force of his will?

"If he could do this he would have powers almost unlimited and supernatural. At his command, with but a slight effort on his part, old worlds would disappear and new ones of his planning would spring into being. He could fix, solidify and

---

[47] F. Capra, *The Tao of Physics*, Shambhala Publications, Boulder 1975.

[48] G. Zukav, *The Dancing Wu Li Masters*, William Morrow and Company, New York 1979.

[49] Also, do not miss the 2004 film *What the Bleep Do We Know!?*, which deals with these topics.

[50] http://www.worldcat.org/title/esoteric-buddhism/oclc/894150821

[51] https://news.google.com/newspapers?nid=1368&dat=19300713&id=1l5QAAAAIBAJ&sjid=0Q4EAAAAIBAJ&pg=4431,1664754&hl=en

preserve the ethereal shapes of his imagining, the fleeting visions of his dreams. He could express all the creations of his mind, on any scale, in forms concrete and imperishable. (…) He could make planets collide and produce his suns and stars, his heat and light. He could originate and develop life in all its infinite forms."

Prisco tries to modernize these ideas and to insert them into his Transhumanist *Weltanschauung*, quoting, for example, the very heterodox work of Ervin Laszlo and his *The Akashic Experience: Science and the Cosmic Memory Field*,[52] in which the philosopher promotes the idea of a universal natural memory field in the Universe that contains and saves every possible piece of information about, well, everything. Our Transhumanist theologian also covers a book by a mathematician, Ralph Abraham, and a physicist, Sisir Roy: *Demystifying the Akasha: Consciousness and the Quantum Vacuum*,[53] in which the two propose a mathematical model for the so-called Akashic field. The goal here is straightforward: if all information is saved somewhere, nobody is really dead, but rather people are simply dormant somewhere in the deep structure of reality, waiting to be "fished out" by the far-future Akashic engineering that our post-human descendants will develop.

The book *Technological Resurrection: A Thought Experiment* by Jonathan Jones[54] tries to tackle an issue that we have already mentioned, the "Theseus Ship" paradox. Copying minds and/or bodies into the future is not enough, because these are just copies, and not the originals. Jones proposes so-called "Temporal Resurrection," which consists of a not-well defined "retrieval of the consciousness," using wormholes. Maybe what the author means – although I am not sure – is that, in order to make sure that the consciousness that we saved is actually the original one, what we need to do is actually create a spacetime "bridge," that is, a form of metaphysical continuity between the old Self and the new one.

10.6   Everything that Rises Must Converge

For spiritual Transhumanists, the "worst case scenario" – that is, the inability to resurrect the dead through Quantum Archaeology, Time Scanning or similar highly speculative technologies – consists in having to wait a few billion years, until the advent of the Omega Point. This belief – that we already encountered while talking about Pierre Teilhard de Chardin – has recently been revived by an American mathematical physicist and cosmologist, Frank J. Tipler. Tipler's ideas represent an attempt to prove the basic tenets of Christianity scientifically, that is,

---

[52] E. Laszlo, *The Akashic Experience: Science and the Cosmic Memory Field*, Inner Traditions, Rochester 2009.

[53] R. Abraham; S. Roy, *Demystifying the Akasha: Consciousness and the Quantum Vacuum*, Epigraph Publishing, Rhinebeck 2010.

[54] J. Jones, *Technological Resurrection: A Thought Experiment*, Amazon Digital Services LLC, 2017.

the existence of God, the immortality of the soul, the resurrection of the dead and the Final Judgement. Received with strong skepticism by the scientific community, Tipler's theories imply that either the universe will collapse in on itself in the distant future or – in the case in which its destiny consists in perpetual expansion, or even acceleration – it will be forced to do so by the super-human intelligence that will have permeated it by then. In other words, near the end of the collapse of the universe, the post-human intelligence of the future will bring about God Himself, which will arrange and utilize all of the resources of the Cosmos to compute, to infinitely slow down the subjective perception of the time flow, to simulate – aka resurrect – the dead and to live with them happily forever after.[55] To Tipler, the Omega Point is basically a cosmological state of the very distant future in which the living Cosmos – God – will simulate the Heavens and fulfill the promises of the Bible and the Gospels.[56]

Aside from a few cases, the majority of Transhumanists are skeptical about Tipler's theories, too, but this doesn't mean that they are dismissive toward the concept of simulation *per se*; in fact, the idea that the whole of reality could be a simulation is quite popular in that community. Let's take a look at this theory. If you have watched the movie *The Matrix*, you already know what we are talking about: the whole Universe, with all of its contents, the stars, the planets, and every living entity, might be a simulation generated by computers extremely more powerful than the ones we puny humans have. After all – or so the argument goes – we are developing more and more realistic technologies for virtual reality, and so our world might be, in turn, just the product of a higher level of reality, and this one of an even higher level of reality, and so on and on.

If we talk about the simulation argument, the first name that we find in the Transhumanist arena is that of Nick Bostrom, one of the most respected Transhumanist scholars I am aware of, a philosopher at Oxford and director of the Oxford Future of Humanity Institute.[57] Bostrom's proposal takes the shape of a trilemma, called "the simulation argument" and published for the first time in 2003.[58] Instead of directly supporting the simulation hypothesis, the philosopher argues that, among the following three statements, one must be or is almost certainly true:

1. "The fraction of human-level civilizations that reach a posthuman stage (that is, one capable of running high-fidelity ancestor simulations) is very close to zero." Or, in other words, our species is very likely to go extinct before turning post-human.

---

[55] F. Tipler, *The Physics of Immortality: Modern Cosmology, God and the Resurrection of the Dead*, Doubleday, New York 1994.

[56] F. Tipler, *The Physics of Christianity*, Doubleday, New York 2007.

[57] https://www.fhi.ox.ac.uk/

[58] N. Bostrom, *Are you living in a computer simulation?*, 2003, http://www.simulation-argument.com/simulation.html

2. "The fraction of posthuman civilizations that are interested in running ancestor-simulations is very close to zero." That is, any post-human civilization is extremely unlikely to run a significant number of simulations of their evolutionary history (or variations thereof);
3. "The fraction of all people with our kind of experiences that are living in a simulation is very close to one." In other words, we are almost certainly living in a computer simulation.

The assumption here is that a post-human civilization would have enough computational power to generate a complete, hyper-realistic simulation of the universe, and that, at the end of the day, the number of simulated civilizations – or "Sims" – would greatly exceed the "basement" civilizations. And a basement civilization might not even exist, as every civilization/universe might be a simulation of a higher level civilization, and so on, *ad infinitum*.

And the basic idea is that a civilization either goes extinct or it develops to a point at which it is able to run a simulation of its reality. If this is true, basically almost every civilization must be able reach such a level, and so, in the (multi) verse, the simulated realities would proliferate and be more numerous than the basement realities. In turn, this implies that our reality is more likely than not to be a Sim.

Bostrom himself refuses to choose, as, to him, there is no strong argument in favor of one statement or the other.

I am not going to cover all of the criticisms raised against the simulation hypothesis; I will just add that, for now, there is no proof that it is actually possible to run a high-fidelity simulation of the "real" world and that a high-fidelity simulation wouldn't be a simulation anymore, but an actual God-like "act of creation," which means that the simulated reality would have the same ontological status of a basement reality, and so it should not be considered just a simulation, but rather a parallel reality.

Another "side effect" of the simulation hypothesis that we mentioned is the idea that the world might have a hierarchical nature, maybe an infinite one – and this idea doesn't necessarily have to be connected with the simulation hypothesis itself; it might actually belong to mainstream physics. Prisco likes to play with an onion-like interpretation of the nature of reality: "[O]ur scientific understanding of the universe may grow without bonds, but always find new fractal depths of unexplained phenomena, to be explored by future scientists. Richard Feynman said: 'If it turns out there is a simple ultimate law that explains everything so be it. That would be very nice discovery. If it turns out it's like an onion with millions of layers and we are just sick and tired of looking at the layers then that's the way it is!' And perhaps the onion with millions of layers is really an onion with an infinite number of layers, and we will always find new things to explore and understand. (...) The outer layer of the onion is the world of Newtonian mechanics, where stones thrown in the air move on understandable and predictable paths. (…) To better understand physics, we must move to deeper layers: the worlds of Bohr and Einstein, where quantum and relativity play an important role. But it is difficult to put together relativity and quantum physics (that's why we don't have a theory of quantum gravity yet), so it seems that we need to go down to deeper layers, not explored yet. Probably the "mystery" of quantum entanglement, and the "mystery" of consciousness (here "mystery" just means something that we don't understand yet) play important roles in deeper layers. (…) So perhaps the onion really has infinite layers (visualize a fractal onion with each layer half as wide as the outer layer), in which case the adventure of science will never end but, at any given moment, there will be an infinite ocean, infinitely larger than the known lands, that science hasn't explored yet. Besides finding an infinitely complex universe intellectually convincing, I also find it aesthetically and emotionally appealing — an entirely known universe would be a very boring place."[59] But there is a catch: if you want to actually comprehend – or keep comprehending – the infinite complexity of the world, you need to become more and more complex yourself, which means infinite evolution – unless, of course, you can find a way to "mystically" achieve the infinite complexity of the Universe in one single leap. In the meantime, Prisco's transformation – from a quasi-agnostic member of the Order of the Cosmic Engineers to a believer in God – is now complete. In his Transhumanist theological vision – also based on Fred Hoyle's work –, God might even be talking to us from the far future: "God can tweak complex processes in space-time with messages hidden in random noise to choose the random outcomes of individual quantum events. God is omnipotent indeed, and works 'below' the laws of physics. Fred Hoyle imagined a hierarchy of Gods in the universe, from lower-case local gods all the way to an asymptotic cosmic God that emerges from the physical universe, comes into full being beyond time, controls space and time, seeds the universe with life, and keeps tweaking and fine-tuning the whole of space-time with subtle quantum messages and time loops."[60]

This God or Gods could be us, or, more likely, our hyper-evolved descendants. So, I would like to close this chapter by quoting Kurzweil: "Will the Universe end in a big crunch, or in an infinite expansion of dead stars, or in some other manner? In my view, the primary issue is not the mass of the Universe, or the possible existence of antigravity, or of Einstein's so-called cosmological constant. Rather, the fate of the Universe is a decision yet to be made, one which we will intelligently consider when the time is right."[61]

[59] G. Prisco, *The Big Infinite Fractal Onion Universe*, http://turingchurch.com/2012/12/02/the-big-infinite-fractal-onion-universe/

[60] G. Prisco, *Divine Action and Resurrection: something like that, more or less*, http://turingchurch.com/tag/resurrection/

[61] R. Kurzweil, *The Age of Spiritual Machines*, Penguin, New York 1999, p. 195.

Correction to: Transhumanism - Engineering the Human Condition

Correction to:
R. Manzocco, *Transhumanism - Engineering the Human Condition*,
Springer Praxis Books, https://doi.org/10.1007/978-3-030-04958-4

The below listed late corrections from the author should be considered in following pages of the current version:

1. On page 2, a new footnote has been added, just after the existing one, with the following text:

This background information comes from the seminal research of Natasha Vita-More, as seen in https://www.metanexus.net/introduction-h-transhumanism-answers-its-critics/; http://www.thescavenger.net/media-a-technology-sp-9915/media-a-technology/175-transhumanism-the-way-of-the-future-98432.html; https://books.google.com/books?id=YeFo_20rfz0C&pg=PT39&lpg=PT39&dq=Dante+Transhuman+Eliot+Cocktail&source=bl&ots=C_3Pn1_94S&sig=ACfU3U3h2zmATQIyeeewMP_1L9-7yFXCfQ&hl=en&sa=X&ved=2ahUKEwiei5S9gtjiAhU-FzQIHWg9DPAQ6AEwAXoECAgQAQ#v=onepage&q=Dante%20Transhuman%20Eliot%20Cocktail&f=false

---

The updated online versions of these chapters can be found at
https://doi.org/10.1007/978-3-030-04958-4_1
https://doi.org/10.1007/978-3-030-04958-4_2
https://doi.org/10.1007/978-3-030-04958-4_3
https://doi.org/10.1007/978-3-030-04958-4

2. On page 5, line 27, a new footnote has been added, just after the existing one, with the following text:

Of course ours is a simplification, and Transhumanism genesis and development are more complicated. For instance, it includes a third central philosopher – who developed her own original perspective –, Natasha Vita-More, saluted by The New York Times as "the first female Transhumanist philosopher" (Cf. C. Wilson, Droid Rage, in "The New York Times," October 21, 2007, https://www.nytimes.com/2007/10/21/style/tmagazine/21droid.html.)

3. On page 37, lines 1-2, a new footnote has been added, just after the existing one, with the following text:

"Transhumanism as a philosophy does not refute or affront those who acknowledge or value death. Transhumanism supports diversity and within the spectrum of diversity are varied shades of gray when referring to the meaning of death. For example in this spectrum are those who are terminal and want to die with dignity, those who have religious and spiritual beliefs and within their morality support death as a next step in life, those who want to die regardless of religious or spiritual beliefs but accept death as a natural law, and those who want to continue living as long as possible with ethical technological and evidence-based scientific medical interventions" (Natasha Vita-More, personal communication).

4. On page 37, line 3, the sentence "A part from Ettinger, Transhumanism has another founding father, Fereidoun…" has been changed to "A part from Ettinger, Transhumanism has other founding fathers; among them Fereidoun…"

5. On page 37, line 17, "the artist" has been deleted from the sentence.

6. On page 37, line 17, after "Vita-More", "– scholar, designer and another pivotal personality in the birth of the contemporary Transhumanist movement –" has been added.

7. On page 98, lines 4-5, a footnote has been added with the following text:

About "Mother Nature," see Max More's (1999) seminal essay A Letter to Mother Nature. http://strategicphilosophy.blogspot.com/2009/05/its-about-ten-years-since-i-wrote.html; https://onlinelibrary.wiley.com/doi/abs/10.1002/9781118555927.ch41.

Conclusion: The Nature of the Beast

Or: How I Learned to Stop Worrying and Love Transhumanism

We have swallowed a lot of science fiction and wild speculation in the course of this book; now we would like to make some observations. The first refers to a statement by Peter Medawar in *The Threat and the Glory: Reflections on Science and Scientists*: "It is the great glory as well as the great threat of science that everything which is in principle possible can be done if the intention to do it is sufficiently resolute." And, as we understand it, the bulk of the things that Transhumanists want to do – genetic manipulation, enhancement, life extension, and so on – are not prohibited by particular physical laws. Our doubts are concentrated on cases of "fringe" issues, such as nano-machines, artificial intelligence and mind-uploading; we do not doubt, however, that, if these things should turn out to be impossible, Transhumanists will put their ingenuity to work and they will come up with something else, perhaps something even more surprising.

The second observation concerns the particular position that Transhumanism enjoys – even from its own point of view – in the history of thought. We find ourselves – it has been said many times – in an era characterized by nihilism. The supreme values have been devalued, faith in God has waned – at least in the circles of thought "that matter," but, in reality, also among the general public. If we wanted to be just a bit melodramatic, we could fill our mouths with phrases like "humans face nothingness all alone," and the like. The fact remains, however, that the metaphysical certainties that we used to embrace have evaporated.

What now? It is urgent that we find some substitute; in our age of controlled demolitions, the only thing we have left is precisely the instrument we use every day to carry out these operations of cleaning, i.e., critical and techno-scientific rationality. It's not much, but, in the absence of something better, we will have to

© Springer Nature Switzerland AG 2019
R. Manzocco, *Transhumanism - Engineering the Human Condition*,
Springer Praxis Books, https://doi.org/10.1007/978-3-030-04958-4

make do. So much so that, at a certain point, people who are scientifically prepared, rational and intellectually courageous – perhaps even a little crazy – have decided that they want to live. That they would prefer to send the grim reaper to Hell, perhaps using the tools of techno-science. Along comes Transhumanism, a thought that starts from a disturbing finding – "it is true, we come from nowhere and it is probably there that we are heading" – but prefers to see this situation as something temporary. "It is true, we are in the middle of the ford" – the Transhumanists seem to say – "but we are already beginning to glimpse the other shore. "A shore on which, in their opinion, the omnipotent science of the future awaits us. But do not worry: even though they may sometimes seem unsettling, Transhumanists are practically harmless. What they are engaged in is, in essence, an intellectual game: they are thinkers, in short, who explore new and strange concepts and try, as far as possible, to share them with others, spreading them outside of their restricted circle. Each of their proposals is then weighed over and over, examined from ethical and political points of view.

Our third observation concerns the difference between the general view of Transhumanism and the specific projects proposed by this or that Transhumanist. And so, De Grey's project is a specific one, and may work or not; J. Storrs Hall's Utility Fog is a specific project, and could be crowned with success or failure. All of these projects cannot be dismissed using sociology – for example, by framing them as myths of the age of technology or as pseudo-rational contemporary mythology. A discourse like this can be conducted – coincidentally – regarding the Transhumanist vision in general; the projects in question are instead para-scientific proposals, and, as such, they must *also* be evaluated by the natural sciences, empirically. This observation also seems to be echoed in a recent and interesting reflection by a young Transhumanist philosopher, John Danaher, who supports – and we agree – the essentially utopian nature of Transhumanism, but also the fact that individual projects stand or fall regardless of the underlying nature – utopian or otherwise – of the Transhumanist project.[1]

The Charge of the Amortals

Then, there are other scattered considerations that come to mind. As we have said, beyond some fairly critical concepts – such as mind-uploading – many Transhumanist proposals are probably feasible. The question is not if, but when. For example, radical longevism, by not violating any known physical principle,

---

[1] J. Danaher, *Tranhumanism as Utopianism: A Critical Analysis*, http://philosophicaldisquisitions.blogspot.com/2018/07/tranhumanism-as-utopianism-critical.html

will come about sooner or later; mental enhancement and mind-brain interfaces will come about sooner or later; and so on. Obviously, the Transhumanists hope that these things will arrive in time, that is, soon enough to allow them to benefit from them. Precisely for this reason, many of them prepare themselves in meticulous ways – maniacal, some would say; they participate in sports, eat healthily, take supplements of all kinds, and keep themselves informed about all of the latest medical progress. They want to live, and they hope to make it to that fabled day.

In this, however, they are no different from those people – and there are many – who, in our society, have decided to live as if time never passes. These are the so-called "amortals" – well portrayed by the British journalist Catherine Mayer[2] – men and women, professionals of all kinds, entertainment and business people, but also many ordinary people, all united by the refusal to grow old or even to consider the prospect of falling out of the scene. It is a fast-growing human group that knows no national borders and is willing to invest time and money to provide themselves with health and vigor for as long as possible. Just like the Transhumanists.

And it is no coincidence that, among the criteria that Mayer uses to distinguish an amortal from a more traditional person, there is also a more or less unconscious confidence in the fact that, in some way, science will succeed sooner or later in resolving the problem of death. In other words, it seems that the Transhumanist mentality has come out of the closet in which it was hiding and, like it or not, is gradually, silently impregnating the common sensibility.

Meanwhile, the Transhumanists continue with their work of theoretical speculation. We do not know whether the post-human will arrive, but if it does, it will find a theoretical apparatus ready and able to defend it, support it, and promote it. Transhumanism is no longer laughable. On the contrary: in the ideal sense, the post-human is already here, has become global, and strongly suggests that the future will be a far stranger place than we imagine. Indeed, far stranger than we can imagine.

---

[2] C. Mayer, *Amortality: The Pleasures and Perils of Living Agelessly*, Random House UK, London 2011.

Further Readings

Abraham, R.; Roy, S., *Demystifying the Akasha. Consciousness and the Quantum Vacuum*, Epigraph, New York 2010.

Alexander, B., *Rapture. A raucous tour of cloning, transhumanism, and the new era of immortality*, Basic Books, New York 2003.

Ansell Pearson, K., *Viroid Life: Perspectives on Nietzsche and the Transhuman Condition*, Routledge, New York 1997.

Appleyard, B., *How to live forever or die trying*, Simon & Schuster, London 2007

B. Bova, *Immortality:: How Science Is Extending Your Life Span – and Changing The World*, Avon Books, New York 1998.

Barrow, J. *Impossibility: Limits of Science and the Science of Limits*, Oxford University Press, Oxford 1998, p. 133.

Becker, E., *The Denial of Death*, Simon & Schuster, New York 1973.

Bell, G., Gemmell, J., *Total Recall: How the E-Memory Revolution Will Change Everything*, Dutton, Boston 2009.

Berube, D. M., *Nano-Hype. The truth behind the nanotechnology buzz*, Prometheus Books, Amherst 2006.

Bostrom, N., *Superintelligence. Paths, Dangers, Strategies*, Oxford University Press, Oxford 2014.

Bostrom, N.; Cirkovic, M. (ed.), *Global Catastrophic Risks*, Oxford University Press, Oxford 2011.

Bostrom, N; Savulescu J. (ed.), *Human Enhancement*, Oxford University Press, Oxford 2009.

Broderick D. (ed), *Year Million. Science at the Far Edge of Knowledge*, Atlas & Co., New York 2008

Broderick, D., *The Last Mortal Generation*, New Holland, Sydney 1999.

Broderick, D., *The Spike*, Tom Doherty Associates, New York 2001.

Campa, R., *La Società degli Automi*, D Editore, Rome 2017.

Campa, R., *Trattato di filosofia futurista*, Avanguardia 21 Edizioni, Rome 2012.

Caronia, A., *Il cyborg. Saggio sull'Uomo Artificiale*, Theoria, Rome-Naples 1985.

Cave, S., *Immortality: The Quest to Live Forever and How It Drives Civilization*, Crown Publishers, New York 2012.

Chace, C., *Artificial Intelligence and the Two Singularities*, CRC Press, Boca Raton 2018.

© Springer Nature Switzerland AG 2019
R. Manzocco, *Transhumanism - Engineering the Human Condition*,
Springer Praxis Books, https://doi.org/10.1007/978-3-030-04958-4

Chu, T., *Human Purpose and Transhuman Potential*, Origin Press, San Rafael 2014.

Clarke, A. C., *Profiles of the Future: An Inquiry into the Limits of the Possible*, Harper & Row, New York 1973.

De Grey, A; Rose, M., *Ending Aging. The Rejuvenation Breakthroughs That Could Reverse Human Aging in Our Lifetime*, St. Martin Press, New York 2007.

Delgado, J., *Physical Control of the Mind: Toward a Psychocivilized Society*, Harper and Row, New York 1969.

Dell'Aglio, L., *Progetto Faust*, Mondadori, Milan 1990.

Dery, M., *Escape Velocity: Cyberculture at the End of the Century*, Grove Press, New York 1997.

Dooling, R., *Rapture of the Geeks. When AI outsmarts IQ*, Harmony Books, New York 2008.

Drexler, K. E., *Engines of Creation. The Coming Era of Nanotechnology*, Anchor Books, New York 1986.

Drexler, K. E., *Nanosystems: molecular machinery, manufacturing, and computation*, John Wiley & Sons, New York 1992.

Egan, G., *Diaspora*, Orion, London 1997.

Esfandiary, F. M. *Up-Wingers: A Futurist Manifesto*, John Day Company, New York 1973.

Esfandiary, F. M., *Optimism One. The Emerging Radicalism*, Norton & Company, New York 1970.

Esfandiary, F. M., *Are You a Transhuman? Monitoring and Stimulating Your Personal Rate of Growth at a Rapidly Changing World*, Warner Books, New York 1989.

Farrell J. P.; de Hart, S. D., *Transhumanism: a grimoire of alchemical agendas*, Feral House, Port Townsend 2011.

Ferrando, F., *Il postumanesimo filosofico e le sue alterità*, ETS Edizioni, Pisa 2016.

Ferrando, F., *Philosophical Posthumanism*, Bloomsbury Academic, London, Forthcoming.

Forward, R. L., *Indistinguishable from Magic*, Baen Books, Riverdale 1995.

Freitas Jr., R. A., *Nanomedicine, Vol. IIA: Biocompatibility*, Landes Bioscience, Austin 2003.

Freitas Jr., R. A., *Nanomedicine. Volume I: Basic Capabilities*, Landes Bioscience, Austin 1999.

Fukuyama, F., *Our Posthuman Future: Consequences of the Biotechnology Revolution*, Farrar, Straus & Giroux, New York 2002.

Garreau, J., *Radical Evolution. The Promise and Peril of Enhancing Our Minds, Our Bodies – and What It Means to Be Human*, Doubleday, New York 2004

Geraci, R. M., *Apocalyptic AI: Visions of Heaven in Robotics, Artificial Intelligence, and Virtual Reality*, Oxford University Press, Oxford 2012.

Glover, J., *What Sort of People Should there Be?* Penguin Books, New York 1984.

Goertzel, B. *A Cosmist Manifesto: Practical Philosophy for the Posthuman Age*, Humanity+, Los Angeles 2010

Gruman, G. J., *A History of Ideas About the Prolongation of Life*, Springer Publishing Company, New York 2003.

Habermas, J., *Die Zukunft der menschlichen Natur. Auf dem Weg zu einer liberalen Eugenik?*, Suhrkamp, Frankfurt 2001

Haldane, J. B. S., *Daedalus, or Science and the Future*, E. P. Dutton and Company, Inc., New York 1924.

Hanlon, M., *Eternity. Our Next Billion Years*, Macmillan, New York 2009.

Haraway, D, *Simians, Cyborgs and Women: The Reinvention of Nature*, Routledge, New York 1991.

Harrington, A., *The Immortalist*, Random House, New York 1969.

Hirstein, W., *Mindmelding. Consciousness, Neuroscience and the Mind's Privacy*, Oxford University Press, Oxford 2012.

Hughes, H. C., *Sensory Exotica: A World beyond Human Experience*, MIT Press, Cambridge 1999.

Hughes, J., *Citizen Cyborg: Why Democratic Societies Must Respond to the Redesigned Human of the Future*. Westview Press, Boulder 2004.

Hugues, T., *American Genesis: A Century of Invention and Technological Enthusiasm, 1870–1970*, Penguin, New York 1989

Huxley, J., *Transhumanism*, in: *Religion without revelation*, E. Benn, Londra 1927. Revised edition in: *New Bottles for New Wine*, Chatto & Windus, Londra 1957.

Istvan, Z., *The Transhumanist Wager*, Futurity Imagine Media LLC, Reno 2013.

Jones, J. *Technological Resurrection: A Thought Experiment*, Amazon Digital Services LLC, 2017.

Kaiser, D. *How the Hippies saved Physics. Science, Counterculture, and the Quantum Revival*, W. W. Norton & Company, New York 2011.

Kaku, M. *Parallel Worlds: The Science of Alternative Universes and Our Future in the Cosmos*, Doubleday, New York 2005, p. 317.

Katz, B. F., *Neuroengineering the Future. Virtual Minds and the Creation of Immortality*, Infinity Science Press, Hingham 2008.

Kekich, D., *Life Extension Express*, Max Life Foundation, Charleston 2010.

Kelly, K., *The Inevitable. Understanding the 12 technological forces that will shape our future*, Viking, New York 2016.

Kurzweil, R., *How to Create a Mind: The Secret of Human Thought Revealed*, Viking Press, New York 2012.

Kurzweil, R., *The Age of Spiritual Machines*, Penguin Books, New York 1999.

Kurzweil, R., *The Singularity is Near*, Viking Books, New York 2005

Kurzweil, R.; Grossman, T., *Fantastic Voyage: Live Long Enough to Live Forever*, Rodale Books, Emmaus 2004.

Kurzweil, R.; Grossman, T., *Transcend: Nine Steps to Living Well Forever*, Rodale Books, Emmaus 2004.

Laszlo, E., *The Akashic Experience: Science and the Cosmic Memory Field*, Inner Traditions, Rochester 2009.

Leakey, R.; Lewin, R., *The Sixth Extinction: Patterns of Life and the Future of Humankind*, Anchor, New York 1996.

Lem, S., *Summa Technologiae*, University of Minnesota Press, Minneapolis-London, 2013.

Leonhard, G., *Technology vs. Humanity*, Fast Future Publishing, London 2016.

Lévy, P., *Collective Intelligence. Mankind's Emerging World in Cyberspace*, Plenum Trade, New York 1997.

Lloyd, S., *Programming the Universe: A Quantum Computer Scientist Takes On the Cosmos*, Knopf, New York 2006.

Longo, G. O., *Il simbionte. Prove di Umanità futura*, Meltemi, Rome 2003.

Longo, G. O., *Homo technologicus*, Meltemi, Rome 2005.

Lorenz, K., *Behind the Mirror. A Search for a Natural History of Human Knowledge*, Harvest/HBJ, New York 1973.

Lyotard, J.F., *La Condition Postmoderne: Rapport sur le Savoir*, Les Editions de Minuit, Paris 1979.

Lyotard, J.F., *Postmodern Fables*, University of Minnesota Press, Minneapolis 1997.

Lyotard, J.F., *The Inhuman. Reflections on Time*, Stanford University Press, Redwood City 1991.

MacLean, P. D., *The Triune Brain in Evolution: Role in Paleocerebral Functions*, Springer, New York 1990.

Macrì, T., *Il corpo post-organico*, Costa & Nolan, Milan 1996.

Mann, T., *Doctor Faustus*, Vintage Books, New York 1999.

Marchesini, R., *Etologia Filosofica*, Mimesis, Milan 2016.

Marchesini, R., *Post-Human. Verso Nuovi Modelli di Esistenza*, Bollati Boringhieri, Turin 2002.

Marinetti, F. T., *Fondazione e Manifesto del futurismo*, in Various authors, *I manifesti del futurismo*, Edizioni di «Lacerba», Florence 1914.

Maslow, A. H., *Toward a Psychology of Being*, John Wiley & Sons, New York 1998

Mayer, C., *Amortality: The Pleasures and Perils of Living Agelessly*, Random House UK, London 2011.

McGinn, C., *The Mysterious Flame. Conscious Minds in a Material World*, Basic Books, New York 1999.

McKibben, B., *Enough: Staying Human in an Engineered Age*, St. Martin's Griffin, New York 2003.

McKinney, L. O., *Neurology: Virtual Religion in the 21st Century*, American Institute for Mindfulness (Harvard University), Cambridge 1994.

Merali, Z., *A Big Bang in a Little Room. The quest to create new universes*, Basic Books, New York 2017.

Miah, A., *Genetically Modified Athletes. Biomedical Ethics, Gene Doping and Sport*, Routledge, London 2004.

Midgley, M., *Science as Salvation: A Modern Myth and its Meaning*, Routledge, Abingdon-on-Thames 1994.

Mitchell, S., *Gilgamesh: A New English Version*, Atria Books, New York 2006.

Moore, P., *Enhancing Me. The hope and the hype of human enhancement*, John Wiley & Sons, Chichester 2008

Moravec, H: *Mind Children The Future of Robot and Human Intelligence*, Harvard University Press, Cambridge 1988.

Moravec, H., *Robot: Mere Machine to Transcendent Mind*, Oxford University Press, New York 1998.

More, M.; Vita-More, N. (ed.), *The Transhumanist Reader*, Wiley-Blackwell, Malden 2013.

Naam, R., *More Than Human. Embracing the Promise of Biological Enhancement*, Random House, New York 2005.

Newberg, a. b. *Principles of Neurotheology*, Ashgate Publishing, Farnham 2010.

Persinger, M., *Neuropsychological Bases of God Beliefs*, Praeger, Westport 1987.

Preston, C., *The Synthetic Age*, The MIT Press, Cambridge 2018.

Regis, E., *Great Mambo Chicken and the Transhuman Condition*, Penguin Books, New York 1990.

Rifkin, J., *Algeny, A New Word – A New World*, Viking New York 1983.

Roco, M. C.; Bainbridge, W. S. (ed), *Converging Technologies for Improving Human Performance*. Springer, New York 2004.

Roden, D., *Posthuman Life: Philosophy at the Edge of the Human*, Routledge, Abingdon-on-Thames 2014.

Sagan, C. *The Cosmic Connection: An Extraterrestrial Perspective*, Cambridge University Press, Cambridge 2000.

Sandberg, A; Bostrom, N., *Whole Brain Emulation: A Road map*. Technical Report # 2008 3, Future of Humanity Institute, Oxford University, Oxford 2008. http://www.fhi.ox.ac.uk/brain-emulation-roadmap-report.pdf.

Savage, M. T., *The Millennial Project: Colonizing the Galaxy in Eight Easy Steps*, Little Brown & Co, New York 1994.

Seidensticker, B., *Future hype. The Myths of Technology Change*, Berrett-Koehler Publishers, San Francisco 2006.

Seife, C., *Decoding the Universe: How the New Science of Information Is Explaining Everything in the Cosmos, from Our Brains to Black Holes*, Viking, New York 2007;

Sherrington, C. S., *Man on his nature*, Cambridge University Press, Cambridge 1942, p. 178.

Silver, L., *Remaking Eden*, Harper, New York 1997.

Sterling, B., *Schismatrix*, Ace Books, New York 1986.

Stock, G., *Redesigning Humans: Choosing our genes, changing our future*, Mariner Books, Boston 2003.

Storrs Hall, J., *Beyond AI. Creating the conscience of the machine*, Prometheus Books, Amherst 2007.

Strole, J.; Bernadeane: *Just Getting Started: Fifty Years of Living Forever*, D&L Press, Phoenix 2017.

Stross, C., *Accelerando*, Ace Books, New York 2005.

Tandy, C. (ed.), *Death and Anti-Death, vol. 4: Twenty Years After De Beauvoir, Thirty Years After Heidegger*, Ria University Press, Palo Alto 2006.

Tipler, F., *The Physics of Christianity*, Doubleday, New York 2007.

Tipler, F., *The Physics of Immortality: Modern Cosmology, God and the Resurrection of the Dead*, Doubleday, New York 1994.

Tuncel, Y. (ed.), *Nietzsche and Transhumanism: Precursor or Enemy?*, Cambridge Scholars Publishing, Cambridge 2017.

Waldvogel, C.; Groys, B.; Lichtenstein, C.; Stauffer, M, *Globus Cassus*, Lars Müller Publishers, Baden 2005.

Warwick, K., *I, Cyborg*, University of Illinois Press, Urbana 2004.

Warwick, K., *March of the Machines: The Breakthrough in Artificial Intelligence*. University of Illinois Press, Urbana 2004.

Weiner, J., *Long for this world. The strange science of immortality*, Harper Collins, New York 2010.

West, M. D., *The Immortal Cell: One Scientist's Quest to Solve the Mystery of Human Aging*, Doubleday, New York 2003.

Wolfram, S., *A New Kind of Science*, Wolfram Media, Champaign 2002.

Young, G. M., *Nikolai F. Fedorov: An Introduction*, Nordland Publishing Company, Belmont 1979.

Young, G. M., *The Russian Cosmists. The esoteric futurism of Nikolai Fedorov and his followers*, Oxford University Press, Oxford 2012.

Zubrin, R., *Entering Space: Creating a Spacefaring Civilization*, Tarcher, New York 1999.

Zuse, K. *Calculating Space*, 1969. ftp://ftp.idsia.ch/pub/juergen/zuserechnenderraum.pdf.